POLYMER SCIENCE AND TECHNOLOGY
Volume 6

PERMEABILITY OF PLASTIC FILMS AND COATINGS

TO GASES, VAPORS, AND LIQUIDS

POLYMER SCIENCE AND TECHNOLOGY

Volume 1 • STRUCTURE AND PROPERTIES OF POLYMER FILMS
Edited by Robert W. Lenz and Richard S. Stein • 1972

Volume 2 • WATER-SOLUBLE POLYMERS
Edited by N. M. Bikales • 1973

Volume 3 • POLYMERS AND ECOLOGICAL PROBLEMS
Edited by James Guillet • 1973

Volume 4 • RECENT ADVANCES IN POLYMER BLENDS, GRAFTS, AND BLOCKS
Edited by L. H. Sperling • 1974

Volume 5 • ADVANCES IN POLYMER FRICTION AND WEAR (Parts A and B)
Edited by Lieng-Huang Lee • 1974

Volume 6 • PERMEABILITY OF PLASTIC FILMS AND COATINGS
TO GASES, VAPORS, AND LIQUIDS
Edited by Harold B. Hopfenberg • 1974

A Continuation Order Plan is available for this series. A continuation order will bring delivery of each new volume immediately upon publication. Volumes are billed only upon actual shipment. For further information please contact the publisher.

POLYMER SCIENCE AND TECHNOLOGY
Volume 6

PERMEABILITY OF PLASTIC FILMS AND COATINGS

TO GASES, VAPORS, AND LIQUIDS

Edited by

Harold B. Hopfenberg

Department of Chemical Engineering
School of Engineering
North Carolina State University
Raleigh, North Carolina

PLENUM PRESS • NEW YORK AND LONDON

Library of Congress Cataloging in Publication Data

Borden Award Symposium, Los Angeles, 1974.
  Permeability of plastic films and coatings to gases, vapors, and liquids.

  (Polymer science and technology; v. 6)
  "Proceedings of the Borden Award Symposium of the Division of Organic Coatings
and Plastics Chemistry of the American Chemical Society, honoring Professor
Vivian T. Stannett."
  1. Plastic films—Permeability—Congresses. 2. Plastics—Permeability—Congresses.
I. Hopfenberg, Harold B., ed. II. Stannett, Vivian. III. American Chemical Society.
Division of Organic Coatings and Plastics Chemistry.
TP1183.F5B67 1974   668.4'95   74-23823
ISBN-13: 978-1-4684-2879-7   e-ISBN-13: 978-1-4684-2877-3
DOI: 10.1007/978-1-4684-2877-3

Proceedings of the Borden Award Symposium of the Division of Organic Coatings
and Plastics Chemistry of the American Chemical Society, honoring Professor
Vivian T. Stannett, held at Los Angeles, California, April, 1974.

© 1974 Plenum Press, New York
Softcover reprint of the hardcover 1st edition 1974

A Division of Plenum Publishing Corporation
227 West 17th Street, New York, N.Y. 10011

United Kingdom edition published by Plenum Press, London
A Division of Plenum Publishing Company, Ltd.
4a Lower John Street, London, W1R 3PD, England

Foreword

The subject of this book has a long and venerable history. The first definitive description of the mechanism of permeation of small molecules through polymer films was presented by Sir Thomas Graham in 1866 in a study of gases in rubber membranes. He called the process "colloidal diffusion" and his picture is similar to that which we are still using, known as the solution – diffusion model. The gas dissolves at one interface followed by diffusion through the membrane and evaporation at the other surface, exposed to lower pressure. In this remarkable paper many of the concepts of permeability which we know today are presented for the first time. In particular a correlation was demonstrated between the permeability and the boiling point of the penetrating gas. The effect of temperature on the gas permeability was correctly related to increased "diffusion" due to the "increasing softness" of the rubber with increasing temperature. This overcomes the decrease in solubility due to the decreasing "ease of condensation" at higher temperatures. The nonporous nature of rubber and its similarity to a liquid was pointed out, and the reduced solubility of gases with crosslinking demonstrated. Finally the potential value of rubber membranes for gas separations was pointed out and quantitatively demonstrated. Many of the chapters of this book are concerned with just these approaches.

The subject was placed on a quantitative basis by Wroblewski in 1879 who showed that the solubility of gases in rubber obeyed Henry's Law, i.e. the solubility is proportional to the partial pressure of the gas. He also first applied Ficks Diffusion Law and combined it with Henry's Law and the known dependence of the permeation rate on the area and thickness of the membrane to derive the now familar expression

$$J = D.S.\frac{p_1 - p_2}{\ell}$$

where J is the amount of gas passing through unit area of the membrane per second, p_1 and p_2 the partial pressures of the permeating gas at both sides of the membrane, ℓ the thickness, D the diffusion constant and S the Henry's Law solubility coefficient.

Daynes in 1919 extended Wroblewski's equation to non-steady state conditions and showed that the solubility and diffusion coefficients could both be determined by measuring the increase of pressure with time starting with no gas initially in the film or at the downstream side of the membrane. Daynes used his method to determine the solubilities and diffusivities of a number of gases in natural rubber at various temperature. His values are in excellent agreement with those obtained today using more refined and sophisticated equipment. The "time lag method" of Daynes was further developed and effectively used by Barrer from 1937 onwards in a long series of elegant experiments leading to much new information about the mechanism of gas permeability in high polymers. Substantial further advances have been made since then both in the theory, methods of measurement, and the various factors influencing the permeability of polymer films to gases.

It must be confessed, however, that a complete understanding of the relative importance of these factors does not exist at the present time and much more work is still needed.

When the transport process is extended to water and organic molecules, the mechanisms of solution and diffusion are much more complete. In particular, interactions between the polymer and the penetrant become important. In addition, at least in the case of water, non-ideal mixing or clustering of the penetrant can occur. In spite of the complications most separation procedures based on membranes involve water and organic molecules, rather than gases, as penetrants and substantial advances have been made in this field.

The symposium, so ably arranged by Professor Hopfenberg and the proceedings of which constitute this book clearly reflect the present status of our knowledge in the various areas outlined above. Most of the chapters are written by the acknowledged leaders in the field and it will be evident, on reading this book, that spectacular advances have been made in our knowledge of transport processes. Equally important is the fact that successful industrial and medical developments have been made based on this knowledge. It is my belief that all those interested in this field will learn much from reading this volume which will form the basis for many further successful developments in the field.

To end on a more personal note, I would like to thank Professor Hopfenberg for organizing such a fine symposium, and the contributors, many of whom have been personal friends and colleagues for many years.

V. T. STANNETT
Raleigh, North Carolina
September, 1974

Preface

In 1952 T.A.P.P.I. and the U.S. Army Quartermaster Corps sponsored the research of Professor Vivian Stannett and Professor Michael Szwarc to characterize and explain the permeation properties of polyethylene and other films and barrier coatings. Their papers, describing their early work, became the keystone of the emerging research in the 1950's and 1960's dealing with the explanation and exploitation of small molecule transport in polymers.

More important than Stannett's accumulation of data was his stimulation of developing scientists. Stannett and Szwarc's student, Charles E. Rogers - of Rogers, Stannett and Szwarc fame - is now Professor Charles E. Rogers at Case Western Reserve University. Professor H. L. Frisch was part of the exciting and productive group of faculty at Syracuse University with Professor Stannett during Stannett's early involvement in membrane processes. Professor Alan Rembaum was conducting his graduate work with Professor Szwarc at the same time.

Stannett was the associate director of the Camille Dreyfus Laboratory while Anton Peterlin was the Director. Dr. Joel Williams worked at the Dreyfus Lab and received his Ph.D. under Stannett's direction. Koje Yasuda and Bob Kesting received their Ph.D.'s under Stannett's direction. Dr. Yasuda has continued his research in this area at the Camille Dreyfus Laboratory and Dr. Kesting has done pioneering and creative work in the development of novel reverse osmosis membranes.

The incestuous development of this explosively growing field is rather intriguing and this autocatalytic growth of people and ideas forms the basis not only of the research area but moreover the contents of this book.

While Stannett was on Sabbatical leave in Paris in 1958, Alan Michaels, then a Professor at M.I.T., visited Stannett to discuss Vivian's work. Michael's interest in polymers, an extension of his important earlier work in colloid and surface science, was further stimulated by his Paris discussion with

Stannett regarding transport in polyethylene. Michael's first work
in the area was an extension of Stannett's early work with polyethylene.

     One of Michael's early students was Wolf R. Vieth, now Head of
the Department of Chemical and Biochemical Engineering at Rutgers.
Dr. James Barrie of Imperial College, London, England, who has
worked with Professor R. M. Barrer for many years, spent his
Sabbatical in Michael's lab working with Vieth, among others. Mike
Lysaght worked for Michaels at the Amicon Corporation and Dr. N. S.
Schneider was a Visiting Professor at M.I.T. while Michaels was
developing his large reserach program in membrane phenomena.

     Professor Allan Hoffman was a colleague of Michaels at M.I.T.
and at Amicon Corporation and preceded Vivian in his Sabbatical
position in Paris. Professor Donald Paul studied undergraduate
chemical engineering at North Carolina State University where I
have been teaching and researching and enjoying since 1967.

     For my part, Alan Michaels inspired my graduate research
program and remains a close colleague. Gerald Gordon was the
demanding graduate assistant in charge of my senior chemical
engineering lab at M.I.T. Morris Salame was a stimulating, happy,
and hardworking classmate.

     Vivian Stannett has been my best friend and closest colleague
since I joined him at North Carolina State University in 1967 as a
consequence of recommendations to Vivian and myself by Hoffman and
Schneider that our colleagueship would be both constructive and
pleasant. Hoffman and Schneider were right and I am forever
grateful for their prescience and kindness. Professor Ralph
McGregor, an Englishman who studied under Professor Peters as did
John Petropoulos and Peter Roussis, is also a member of the faculty
of North Carolina State University. Mitch Jacques is part of the
next generation; he is one of a group of talented and imaginative
students that have worked with me at North Carolina State University.

     In any event, for Vivian and myself, science is facilitated
and made enjoyable by cooperative interactions between people and
the people mentioned here have contributed steadily to the broad
research area of membrane science and to this book. Of course
there are many, many more scientists doing equally creditable
research in the field. Although there are seventy-five contributors
to this book, it is clear that these well known scientists are only
a fraction of the large and growing group of international
researchers in the field.

     Professor Stannett was early but more importantly he gave
counsel and support to everyone who sought his help. He remains

first in colleagueship and the ability to provide stimulation and support to his colleagues. The Borden Award which he received, recognizes his pre-eminence, not only as a researcher, but as a stimulating and inspiring leader who helped define this broad area and contributes steadily to its still explosive growth.

HAROLD B. HOPFENBERG
Raleigh, North Carolina
August 1974

Contents

PART II - INDUSTRIAL MEMBRANE AND
BARRIER FILM APPLICATIONS

PART III – MEMBRANE MODERATED
BIOMEDICAL DEVICES

I. Fundamentals

INTERACTION BETWEEN PENETRATION SITES IN DIFFUSION THROUGH THIN MEMBRANES[†][*]

Stephen Prager and H. L. Frisch

Department of Chemistry, University of Minnesota

State University of New York at Albany

Two models of permeation through thin membranes are proposed: in the first, the membrane is represented as a plate with randomly distributed perforations, and the permeation rate is controlled by the rate at which penetrant diffuses through an external medium; in the second, the penetrant is considered to be adsorbed on one face of the membrane, and the rate controlling process is lateral diffusion across that face to a set of randomly distributed penetration sites. These models are treated by a self-consistent field procedure, and it appears that interaction between penetration sites influences the permeability significantly even when the number of penetration sites per unit area is quite low.

[†]Dedicated to Professor V. T. Stannett whose leadership in the field of diffusion and permeation through membranes has been an inspiration to all of us.

[*]Submitted to the J. Chem. Phys.

Theories of membrane transport which model the membrane as a
set of penetration sites or pores distributed over an impermeable
plate usually consider the processes occurring within each pore in
some detail, but then assume that the contributions of different
pores can simply be summed to get the total penetrant flux through
the membrane. If the rate of permeation is controlled by diffusion
to or away from the pore entrances rather than by diffusion within
the pores, this additivity assumption can become risky, because the
perturbation of the penetrant distribution near a pore extends to
quite large distances from the entrance. Significant overlap between the diffusion zones of different pores may therefore be expected to occur even when the pores are quite widely separated.

In this paper we will show that this type of long range interaction can, under certain circumstances, produce deviations from
simple additive behavior even at low values of n, and that the permeability cannot in fact always be expressed as a Taylor series in
n.

Model I

The first model which we use has been employed by several
authors[1,2,3] to represent transpiration of water through the surfaces of plant leaves. In its simplest form, the leaf surface is
considered as an impermeable plane at z = 0, perforated by holes
(stomata) through which water can evaporate into the surrounding
(stagnant) air, the rate of evaporation being controlled by the
rate at which water vapor can diffuse away.

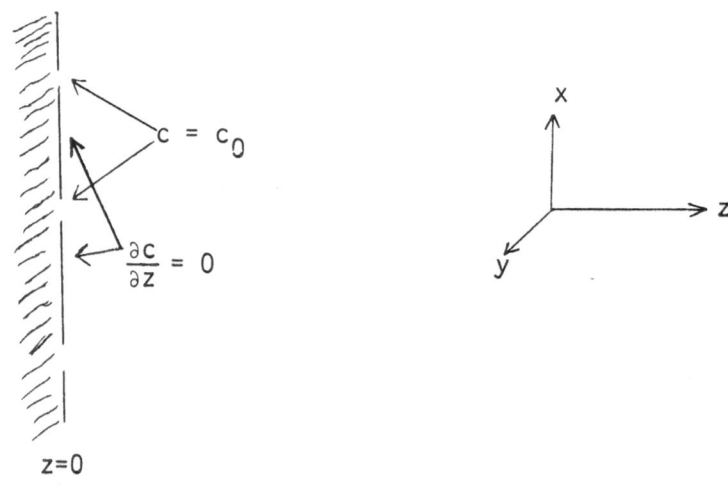

At the leaf surface, the concentration c of water vapor is maintained at the saturation value c_0 at any point falling within a hole, while $\partial c/\partial z$ vanishes at points located in the impermeable area. Far from the leaf surface ($z \to \infty$), $\partial c/\partial z$ approaches a limiting value ($-q/D$), where q is the flux of water vapor, and D its diffusion coefficient in air. In the steady state, the average concentration τ at the leaf surface will be less than c_0, and the effective permeability of the surface is

$$\kappa = q/(c_0 - \tau) \tag{1}$$

When the holes are arranged in a regular two-dimensional lattice, this is a special case of a problem solved by Keller and Stein,[1] and can be treated by their methods. If on the other hand the holes are randomly spaced, an analytic solution is no longer possible. One can, however, obtain approximate results, valid in the limit of low n, by invoking a self-consistent field procedure similar to that used by Frisch and Collins[4] for diffusion-controlled growth of aerosol particles.

The idea is to treat each hole as though it were an isolated perforation in a membrane of permeability κ, to calculate the rate of transpiration Q through a given hole on this basis, and then to require that it be consistent with the assumed value of κ:

$$nQ = \kappa(c_0 - \bar{c}) \tag{2}$$

A source at the origin introducing water vapor into the region $z > 0$ at a rate Q would in the steady state produce a concentration $Q/2\pi Dr$ at a point $\underline{r} = (x,y)$ on the leaf surface if it were impermeable. On a surface with permeability κ this must be modified to include the flux $\kappa(c_0 - c(\underline{r}))$ at points other than the origin; in terms of the concentration fluctuation $c'(\underline{r}) \equiv c(\underline{r}) - \bar{c}$, we have

$$c'(\underline{r}) = \frac{Q}{2\pi Dr} - \frac{\kappa}{2\pi D} \int \frac{c'(\underline{\rho})}{r-\underline{\rho}} d^2\underline{\rho} , \tag{3}$$

the integration extending over the entire xy plane.

The integral equation is readily solved by Fourier transforms:

$$c' = \frac{Q}{2\pi Dr} [1 - \frac{\pi}{2} \frac{\kappa r}{D} (H_0(\frac{\kappa r}{D}) - Y_0(\frac{\kappa r}{D}))] \tag{4}$$

where H_0 and Y_0 are respectively the zeroth order Struve and Neumann functions. Now the average value of c' over the hole should be $(c_0 - \bar{c}) = nQ/\kappa$; for a circular hole of radius a,

$$C_0 - \bar{c} = \frac{2}{a^2} \int_0^a rc'(r)dr = nQ/\kappa \tag{5}$$

(strictly speaking, c' should be $c_0 - \tau$ at every point within the hole, but this refinement would make for a much more difficult calculation without changing the main features of the result). Substitution of (4) into (5) gives an equation to be solved for κ:

$$\pi n a^2 = \frac{\kappa a}{D} - \frac{\pi}{2} \int_0^{\kappa a/D} u(H_0(u) - Y_0(u))du \xrightarrow[n \to 0]{} \frac{\kappa a}{D}(1 + \frac{1}{2} \frac{\kappa a}{D} \ell n \frac{\kappa a}{D} +$$

$$0(\frac{\kappa a}{D})) \tag{6}$$

or

$$\kappa \xrightarrow[n \to 0]{} \frac{D\sigma}{a} (1 - \frac{1}{2} \sigma \ell n \ \sigma + 0(\sigma)) \ , \tag{7}$$

where $\sigma \equiv \pi n a^2$ is the area fraction of open surface. Eqn. (7) may be compared to the result obtained by Keller and Stein for a regular array of holes[3] (in eqns. 10 and 12 of ref. 3, let $\delta \to 0$, $\lambda_1 \to \infty$, $\beta \to \infty$):

$$\kappa \xrightarrow[n \to 0]{} \frac{3\pi}{8} \frac{D\sigma}{a} (1 + \frac{9}{64} \pi^2 \sigma^{\frac{1}{2}} + \cdots) \tag{8}$$

Both (7) and (8) predict that interaction between pores will enhance the permeability, although at low σ the effect is much larger for regular than for random arrays of holes (at $\sigma = .01$ the enhancement is 2% from equation (7) and 15% from equation (8)). Neither (7) nor (8) expresses κ as a simple series in integer powers of σ; at low σ the deviation of κ/σ from its limiting value increases faster than σ, as $(-\sigma \ell n \ \sigma)$ for random, as $\sigma^{1/2}$ for regular arrays.

Model II

Whereas in the first model the permeation rate was controlled by diffusion to pore entrances from an external medium, in our second model the rate controlling step is lateral diffusion across the membrane surface. We consider that penetrant is adsorbed on one face of the membrane at a rate of ν molecules per second per cm^2, and then passes through the membrane by diffusing across the surface to one of a set of randomly distributed penetration sites. These sites are circular regions of radius a, such that a penetrant molecule making contact with the boundary will immediately be transmitted to the opposite face and removed. If the mean concentration of adsorbed penetrant is $\bar{\sigma}$ molecules per cm^2, then the permeability may be defined as

$$k \equiv \nu/\bar{\sigma} \tag{9}$$

To calculate k, we once again use a self-consistent field approach (this is in fact the two-dimensional analog of the problem treated by Frisch and Collins[4]). Let us consider a single penetration site surrounded by a region of uniform permeability κ. If r is the distance from its center, then the steady state diffusion process can be described by the equations

$$D(\frac{d^2s}{dr^2} + \frac{1}{r}\frac{ds}{dr}) - ks + \nu = 0 \tag{10}$$

$$s(a) = 0 \qquad s(\infty) = \nu/k$$

The solution is

$$s = \frac{\nu}{k}(1 - \frac{K_0(\sqrt{\frac{k}{D}}\,r)}{K_0(\sqrt{\frac{k}{D}}\,a)}) \tag{11}$$

where K_0 is a modified Bessel function of order zero. A given penetration site will transmit $2\pi aDs'(a)$ molecules per second, so that, if there are n sites per cm^2,

$$\nu = 2\pi anDs'(a) = -2\pi an\nu \sqrt{\frac{D}{k}}\;\frac{K_0'(\sqrt{\frac{k}{D}}\,a)}{K_0(\sqrt{\frac{k}{D}}\,a)} \tag{12}$$

This is an equation which may be solved for the unknown permeability κ; at low n it becomes, if we again let $\sigma = \pi na^2$,

$$\sigma \xrightarrow[n\to 0]{} - \frac{ka^2}{D}\ell n\,\frac{ka^2}{D} \tag{13}$$

It appears that model II does not give a permeability proportional to σ at any n, no matter how small. Indeed, the initial slope of the κ vs.σ plot is horizontal, reflecting the increasing distance than an average molecule must travel to reach a penetration site as σ goes to zero.

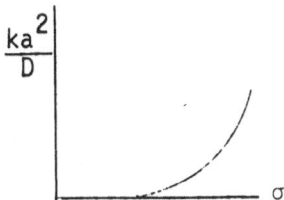

We wish to acknowledge support by the NSF Grants PO 42614X and HO 42516 and the ACS Petroleum Research Fund 3519-C5,6.

References

1. H. Brown and F. Escombe, Phil. Trans. Roy. Soc. London, B193, 233 (1900).

2. R. Lee and D. M. Gates, Am. J. of Botany, 51, 964 (1964).

3. K. H. Keller and T. R. Stein, Mathematical Biosciences, 1, 421 (1967).

4. H. L. Frisch and F. C. Collins, J. Chem. Phys., 20, 1797 (1952)

STEADY STATE TRANSPORT PHENOMENA IN NON-IDEAL

PERMEANT-POLYMER SYSTEMS

A. Peterlin

Camille Dreyfus Laboratory, Research Triangle Institute
Research Triangle Park, North Carolina 27709
Present Address: National Bureau of Standards
Washington, D. C. 20234

Summary

The so generally used Fick's first law of diffusion with the
material current density proportional to concentration gradient of
the penetrant is inapplicable to even the simplest uniform homogeneous polymer membranes which as a rule are not ideal. The non-
ideality shows up in a nonlinear increase of sorbed penetrant with
applied pressure of gas or concentration of liquid. That modifies
the concentration gradient in the membrane. The same effect can
obtain as a consequence of the finite compressibility of swollen
membrane under the applied pressure gradient in the hydraulic experiment. In both cases the membrane becomes nonuniform under the
influence of concentration or pressure gradient in spite of the
fact that it is uniform in a gradient free environment. But the
true driving force is in all cases the gradient of the chemical
potential of the penetrant and not that of concentration. The deviations of sorption, partition and diffusion coefficients from
ideality in highly swollen and in plastically deformed polymer films
can be to some extent described by the change of fractional free
volume as a consequence of the finite amount of sorbate present. A
new effect may occur in the hydraulic experiment. With increasing
swelling the membranes exhibit a substantial viscous flow permeability which may be orders of magnitude larger than the diffusive
permeability. It cannot be described in terms of diffusive transport.

Introduction

The pioneering period of study of diffusive transport of low
molecular weight material through homogeneous, nonporous polymeric

9

membranes was characterized by the indiscriminate use of Fick's first law (FFL) for the current.[1-7] The assumption that the diffusive current density j is the product of diffusion coefficient D and the negative concentration gradient never causes any problem as long as the membrane-permeant system is ideal and the membrane perfectly uniform as far as sorption (S) or partition (k) coefficient are concerned. They must be constant and independent of concentration, pressure and location. Deviations from ideality, however, do not matter if they only involve variations of D.

Experiments have amply demonstrated that neither the diffusion coefficient nor sorption or partition coefficient of real membrane-permeant systems are constants as required for ideality. By good luck the errors caused by the sloppy treatment of transport equations inadvertently just compensate so that with uniform membranes every-thing works well. But the troubles with FFL start as soon as one makes any attempt to treat vectorized membranes or even to treat uniform membranes in a more rigorous manner. One obtains in the former case results which violate the first or second law of thermo-dynamics and in the latter case wrong expressions for the evaluation of transport coefficients.

This is a sure indication that something is wrong with FFL. In the search for a satisfactory formulation of transport phenomena one finds it, among other places, in the excellent book on "Non-Equilibrium Thermodynamics in Biophysics" by Katchalsky and Curran.[8] It is indeed surprising that in spite of so much lip-service to thermody-namics this thermodynamically derived diffusion equation was and is so little used by the most visible workers in the field. Properly handled it avoids not only the errors of Fick's formulation in the case of non-ideal membrane-permeant systems but also wrong conclu-sions in almost ideal systems as for instance highly swollen mem-branes. I would feel that my talk has succeeded beyond expectation if it has convinced at least a fraction of the audience that the use of thermodynamics in transport problems brings not only benefits to the field but also makes life easier for the research worker in the field.

The formulation of transport quantities and formulas is simpli-fied if one uses volume concentrations and fluxes, i.e. the fractional volume of penetrant v instead of concentration in grams per milliliter and measures the flux of density j in flow volume per unit area time (cm.s^{-1}). Since only steady state flow through flat membranes with the normal parallel to the x-axis will be treated one can replace the gradient by total derivatives with respect to x.

Ideal Permeant-Membrane System

The diffusion transport through homogeneous, nonporous polymer membranes was quite generally treated as if the membranes were ideal so that the flow could be described by Fick's laws based on the concentration of the permeant. In such a case the current density j through the membrane is the product of the negative concentration gradient and the diffusion coefficient   (Fick's first law of diffusion)

$$j = -D \; \partial v/\partial x. \tag{1}$$

The material conservation yields Fick's second law

$$\partial j/\partial x = - \; \partial v/\partial t \tag{2}$$

which together with Eq. 1 yields the time (t) and location (x) dependence of concentration and current inside the membrane.

Since in the usual experiment the membrane is fixed on a rigid support it does not move after the steady state of flow is reached. Hence one has no need to consider explicitly the transient flow of membrane material caused by the initial membrane expansion as a consequence of swelling by the permeant if one is interested only in the steady state. One has only to replace the dry state dimensions (x† and d†) by those of the equilibrium swollen membrane (x and d)[9]. In the steady state which will be considered exclusively in that which follows the net flow of the membrane vanishes and only that of the permeant remains. In a plane membrane its density is constant so that the flux per unit area of the membrane reads

$$j = -D \; \Delta v/\Delta x = D(v_1 - v_2)/d \tag{3}$$

if the membrane with a thickness d is uniform and D is independent of v (ideal "permeant/uniform membrane" pair). The subscript 1 relates to the upstream and the subscript 2 to the downstream boundary of the membrane.

In the permeability experiment and in application of membranes one needs the dependence of the flux density on the parameters of the permeant outside the membrane, i.e., on v_1^*, v_2^* or on pressure p_1, p_2. Since v_1 and v_2 are concentrations of the permeant in the membrane at x = 0 and d, respectively, which in turn are functions of the properties of the membrane and the adjacent liquid or gas one has first to formulate this dependence.

In the case of an ideal "gas-membrane" pair one has the simple relationship (Henry's law)

$$v = Sp. \tag{4}$$

The concentration of the gas in the membrane is proportional to the pressure of the gas. Since the sorption coefficient S is constant, independent of p, one obtains from Eq. 3

$$j = -DS \; \Delta p / \Delta x = DS(p_1 - p_2)/d = P \Delta p / \Delta x. \tag{5}$$

The constant permeability coefficient

$$P = DS \tag{6}$$

independent of p, v, and the location in the membrane fully characterizes the transport through the membrane as shown in Eq. 5. The same consideration applies to gas or vapor mixtures with partial pressure $p^{(i)}$ and partial volume fraction $v^{(i)}$ substituting the total pressure p in Eq. 4 and 5 and total volume fraction v in Eq. 4, respectively. Note that in this formulation the fluxes of components are considered completely independent so that all interactions among them can be neglected.

The simplest two cases of diffusive transport of a liquid through a membrane are the transport of a labelled component of the liquid under the concentration gradient of that component (pure diffusive case) and that of a pure liquid under a pressure gradient (hydraulic case). In the first case the liquid on both sides of the membrane contains a different volume fraction v'* of labelled molecules while the pressure is the same. In the second case the liquid has the same composition on both sides of the membrane but the applied pressure differs. The observations can be described by

$$j = P'(v_1'^* - v_2'^*)/d = -P' \; . \Delta v'^* / \Delta x \tag{7}$$

$$j = K(p_1 - p_2)/d = -K . \Delta p / \Delta x \tag{8}$$

Note that the definition and dimension of K in the liquid case are identical with those of P in the gas or vapor case.

The diffusive permeability coefficient P' of the liquid with the dimension $cm^2.s^{-1}$ is equal to the product of diffusion coefficient D and partition coefficient k.

$$P' = k.D \tag{9}$$

where k is the ratio between volume fraction of the tagged component of the liquid inside and outside the membrane at the upstream or downstream boundary

$$v' = k.v'* \qquad\qquad (10)$$

Since the untagged component has exactly the same properties and the same interaction with the membrane one has in complete analogy with Eq. 10 the relations $v'' = kv''*$ and $v = kv* = k$ since $v* = v'* + v''* = 1$.

Equation 9 simply reflects the fact that only the liquid inside the membrane participates in the transport. Note that according to Eq. 7 the current of the untagged component (j'') having the same partition (k) and diffusion (D) coefficient is the same but opposite to that of the tagged component (j'). The algebraic sum of both currents through the membrane just cancels out. Hence there is no change in the volume of total liquid on each side of the membrane, only a change in composition.

In this formulation one completely neglected the interaction between the flows of tagged and untagged component in perfect analogy with the above mentioned diffusion transport of gas and vapor mixtures.

Thermodynamic Formulation of Diffusive Transport

The formulation of the current j as function of partial pressure, concentration or applied pressure of the permeant outside the membrane, Eqs. 5, 7, and 8, involves the membrane permeability, P, P' or K, and thickness d as parameters characterizing the transport properties of the membrane but not causing the transport. The membrane indeed acts as a black-box yielding a current if there is a difference in pressure or concentration between the gas or liquid on both sides separated by the membrane but no current if there is no such difference. This statement derived for ideal "penetrant/ uniform membrane" pair is a straightforward consequence of thermodynamics which precludes the possibility of a perpetuum mobile of the first (flow without a pressure or concentration difference) or second (demixing without any energy input) kind. Hence it automatically applies to more complicated nonideal and nonuniform membranes. The nonideality and nonuniformity modify the parameters P, P' and K which are expected to be functions of c_1*, c_2*, p_1, p_2, and d, but they do not affect the basic Eqs. 5, 7, and 8. Any treatment yielding either directly or indirectly a different result is basically wrong and has to be rejected on the basis of thermodynamic principles.

According to basic mechanics the molecular transport through a homogeneous medium in a potential energy field μ is taking place by motion of the molecules of the permeant, occupying the volume fraction v, with an average velocity w which is the ratio of force F on, and frictional resistance W, of the permeant[1,9-14]

$$j = v.w = v.F/W = -(v/W).d\mu/dx \qquad (11)$$

F, W, and μ relate to a mole of the permeant. The molar resistance W depends on location x in the membrane and the content v of the permeant. In the case of gas and liquid mixtures W depends on the presence of all components of the gas or liquid.

In the case of diffusive transport the proper thermodynamic potential energy function is the chemical potential μ of the permeant. For a pure gas it reads

$$\mu = \mu^\circ + RT\ell np \qquad (12)$$

and for a liquid

$$\mu = \mu^\circ + pV + RT \ell n\ c_m'. \qquad (13)$$

Here μ° is a constant independent of p and c_m', T is absolute temperature, R is gas constant, V is the molar volume and c_m' is the molar concentration of the penetrant (mols of penetrant per mol of liquid) being 1 if the liquid has only one component, i.e. the permeant, and v'/v for a tagged liquid. In the case of a partially tagged liquid, the volume concentration v' of the permeant inside the membrane as appearing in Eq. 11 is the product of total liquid volume fraction v in the volume element and molar concentration $c_m'*$.

In the chemical potential for the gas p is the gas pressure in equilibrium with the local permeant concentration v. In Eq. 13 relating to a liquid it is the driving pressure completely independent of c_m' and v.

It is extremely important to note that the chemical potential of the gas, Eq. 12, and liquid, Eq. 13, does not depend explicitly on the volume fraction v (gas) or v' (liquid). It only depends on pressure in equilibrium with the sorbed gas or on pressure p applied to the liquid and molar concentration of the permeant in the liquid inside and outside the membrane. As a consequence of this fact the diffusion flux is not caused by, or even connected with, the gradient of volume concentration v or v' but only with that of pressure and composition. That is in excellent agreement with the black box concept of diffusion through a fixed membrane which does not predict any flux if the liquid is identical and under the same pressure on

both sides of the membrane even if as a consequence of vectorized
membrane properties the swelling is high on one and small on the
other side. It also tells that any attempt to derive the diffusion
current from an existent or imagined concentration gradient is
rather dubious and must be dropped as soon as it yields results
differing from those derived thermodynamically.

One may ask about the role of the membrane in the expression
for chemical potential of the permeant. According to general rules
of thermodynamics the change of chemical potential of the gas or
liquid caused by mixing with the polymer of the membrane vanishes
as soon as the sorption, i.e. volume fraction v of the gas or
liquid, assumes the equilibrium value. This value v, in turn, is
just determined by the zero of chemical potential of mixing. This
means that Eq. 12 and 13 are not affected by the presence of the
polymer if one is concerned only with the steady state as is the
case in this investigation. The sole term affected could be the
molar concentration if the polymer is considered as an additional
component. But since its molecular weight is extremely high or
even "infinity" in a cross-linked membrane its molar fraction disappears and hence does not influence c_m'.

A direct effect, of course, can and does occur in the volume
fractions sorbed if the components of the gas or liquid mixture have
different Flory-Huggins interaction parameters χ with the polymer or
if some chemical reaction occurs. The latter case can be excluded
from this consideration because we are here not concerned with membranes which may act as sources or sinks of energy or material.
The same applies to the former case since this presentation is
limited to isotopic mixtures so that the components have practically
the same interaction with the polymer, i.e. the same parameter χ.
In such a case the presence of the polymer does not change the composition of the mixture.

By putting μ from Eq. 12 into Eq. 11 and considering that
$D = RT/W$ one obtains for a gas

$$j = - (vD/p) \, dp/dx = - P \, dp/dx \qquad (14)$$

with P being a function of x and p as derived from $vD/p =$
$S(x,p)D(x,p)$. In the case of a tagged liquid Eq. 13 yields

$$j = - (RT/W)(v'/c_m') \, dc_m'/dx = - kD \, dc_m'/dx = -vD \, dc_m'/dx \qquad (15)$$

for a concentration gradient. The flux is proportional to the negative gradient of molar composition c_m' and not to that of the volume
fraction $v' = kc_m' = vc_m'$ of the tagged component. In exactly the same
manner one obtains the hydraulic flux of the pure one-component

liquid ($c_m' = 1$)

$$j = - (vV/W)\, dp/dx = - K\, dp/dx \tag{16}$$

for the case of a pressure gradient.

By integration of Eq. 15 and 16 over the whole thickness ℓ of the membrane one finds

$$v_2'* - v_1'* = c_{m2}'* - c_{m1}'* = -(j/RT) \int_o^\ell (W/v)\, dx = - j\ell/<P'> \tag{15a}$$

$$P_2 - P_1 = -(j/V) \int_o^\ell (W/v)\, dx = - j\ell/<K> \tag{16a}$$

yielding the relationship between the average diffusive and hydraulic permeability

$$<K> = <P'>V/RT \tag{17}$$

which is valid as long as under the influence of applied pressure gradient neither the membrane is compressed, i.e. the volume fraction of the permeant decreased, nor any other type of transport obtains. In an ideal uniform membrane the averages $<K>$ and $<P'>$ are identical with values K and P' of the volume element, respectively.

The main consequence of the thermodynamically correct formulation of the diffusion current is the fact that it is proportional to the local concentration of the permeant and not to its derivative. Since in a real membrane the presence of the permeant acting as a plasticizer increases the polymer chain mobility it reduces the friction coefficient W and hence increases D. In the hydraulic case the applied pressure compresses the membrane. As a consequence of the reduced liquid content the flux density and the diffusion coefficient decrease below the values of the tracer diffusion which goes on without any pressure differential. The gradient of permeant concentration in gas and hydraulic diffusion transforms even an initially uniform membrane into a vectorized one exhibiting a gradient of diffusion coefficient.

No such effects occur in the very unrealistic ideal membrane-permeant systems. In the non-ideal systems the explicit dependence of S, k, and D on p or v has to be considered in the integration of Eq. 14-16.[9-12]

Non-Ideal Uniform Membrane-Permeant Systems

The non-ideality means that the sorption and diffusion coefficients are not more constants but dependent on concentration v or p.

The same applies to permeability $P(p) = D(p).S(p)$.

Starting with FFL, Eq. 1, one obtains in a correct, straight-forward manner[15]

$$j_F = -D \, dv/dx = -D(S.dp/dx + p.dS/dx)$$

$$= -D(S + p.dS/dp)dp/dx$$

$$= -DS(1 + d \ln S/d \ln p)dp/dx \qquad (18)$$

which differs from Eq. 14 as a consequence of the $d \ln S/d \ln p$ term. If, as usually, S increases with p the expression in the parentheses is larger than one thus yielding at a given j and $\Delta p/\Delta x$ a smaller value for $<P>$ than Eq. 14. But the situation is not disastrous as long as the membrane is uniform because any dependence of S and D on p can be always accepted as a measure of penetrant-membrane deviation from ideality. There is indeed no simple way to prove the incorrectness of Eq. 18 and hence that of Eq. 1 on the sole basis of diffusion experiments on uniform membranes.

It is fortunate that gas diffusion in non-ideal systems was practically never based on Eq. 18 but treated in a rather peculiar way which seemed strange to anybody with better mathematical training. After introducing the gas pressure p via Henry's law (Eq. 4) in the expression for the diffusion current (Eq. 1) one simply assumed that S is a constant and so derived Eq. 5. One was not at all worried about the strange fact that independent determination of sorption have demonstrated beyond doubt that in practically all polymer systems S is not a constant so that the correct application of FFL would yield the wrong Eq. 18. By good luck the sloppy treatement indeed produced the thermodynamically correct Eq. 14 identical with Eq. 5 of the ideal system.

In the case of a tagged liquid the volume fraction of the liquid inside the membrane is constant so that the molar and the total concentration of the tagged component are strictly proportional. As a consequence the current calculated from FFL agrees with that derived from the gradient of chemical potential (Eq. 15). This result is independent of the non-ideality of the membrane-penetrant system as long as the diffusive and sorption properties of the tagged component are identical with those of the rest of the liquid.

Under hydraulic conditions FFL is not very helpful because its formulation applies to the concentration and not to the pressure gradient. One obtains from Eq. 1 in a purely formal way

$$j_F = -D \, dv/dx = -D(dv/dp)dp/dx = -K_F dp/dx \qquad (19)$$

The Fickian hydraulic permeability

$$K_F = D.dv/dp = D.dk/dp \qquad \qquad (19a)$$

has a non-vanishing value only in the case that the partition coefficient depends on pressure. It disappears[16] in a completely incompressible membrane with constant, pressure independent k which is a little difficult to believe. As a further consequence of the peculiar proportionality between K and membrane compressibility, Eq. 19a, the relationship, Eq. 17, between K and P' is not more valid. The thermodynamic correlation between K and P' is replaced by one dependent on specific properties of the membrane. The way out of this impasse is the conclusion that FFL is inapplicable to the hydraulic permeability of homogeneous membranes.

Vectorized Membranes[9,11,12]

In a vectorized membrane the transport coefficients depend on location, i.e. on x. One can substantially simplify the comparison between the thermodynamically deduced diffusion current and that derived from FFL if one restricts oneself to the ideal case where all the parameters are independent of concentration and pressure. The modifications caused by deviations from linearity can be easily introduced in all cases where it is of interest.

In the case of gas diffusion through a vectorized membrane S and D are functions of x. On the basis of Fick's first law one derives for the current

$$j_F = - D \, dv/dx = - D(S.dp/dx + p.dS/dx)$$

$$= - P \, dp/dx - Dp.dS/dx \qquad \qquad (20)$$

while the thermodynamically derived value is given in Eq. 14. The important difference resides with the additional term $-Dp.dS/dx$ which yields a current even in the case that there is no pressure gradient. Such a perpetuum mobile of the first kind obviously proves that Eq. 20 is wrong. It is a convincing argument against the applicability of FFL to diffusive transport through vectorized membranes. Together with the results derived from the discrepancy between Eq. 14 and 19 one can conclude that FFL fails in all cases with exception of the ideal homogeneous membrane-permeant system which by itself is hardly a realistic model of polymer membranes.

The unsurmountable difficulties arising from FFL can be best demonstrated if one tries to formulate the transport on the basis of the correctly calculated concentration distribution of the

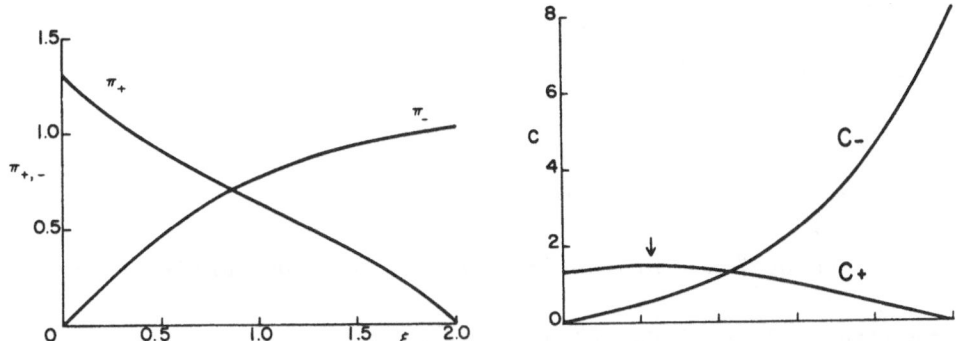

Figure 1. Variation of chemical potential as measured by the driving
pressure $\pi = p-p_2$ (Eq. 14), and concentration c (v in Eq. 15) in a
vectorized membrane under a pressure gradient. The arrow points to
the maximum of concentration where dc/dx = 0. The subscript + (-)
refers to a flow in positive (negative) x direction [12].

permeant. In the case of sufficiently rapid variation of S(x) such
a distribution may exhibit a maximum inside the membrane [12,17] with
dv/dx = 0 and consequently a section where dv/dx has the same direction as the current in complete contradiction with Eq. 1 requesting
the opposite direction of current and gradient (Fig. 1). If in
spite of all that one wishes to express the current by Eq. 1 one
obtains a very peculiar variation of diffusion coefficient from
positive to negative values with the singularity $D_F = \infty$ at the
point with dv/dx = 0. It is obvious that such a controversial
formulation of diffusive transport does not serve any useful purpose
and even contradicts the basic definition of D = RT/W because neither
vanishing nor negative friction resistance W makes any sense.
Moreover, one is almost unable to predict in advance the strange
dependence of such a D_F on location. From Eqs. 14 and 20 it can
be derived as

$$D_F = D/[1 + (d \ln S/dx)/(d \ln p/dx)] \qquad (21)$$

The trouble arises from the second term which changes sign if the
current is reversed. That yields two sets of D_F values for the two
possible current directions. The sign reversal and infinite value
of D_F may occur with negative values of the second term, i.e., if
the gradient of sorption is opposite to the gradient of driving
pressure. One can hardly believe that such a mess offers any

advantage in the study of transport properties as compared with
the straightforward thermodynamic formulation of Eq. 11.

Very much the same argumentation can be applied to the transport of liquids under a composition gradient dc_m'/dx. Equation 1
together with Eq. 10 yields

$$j_F = -D \ (k \ dc_m'/dx + c_m' \ dk/dx)$$

$$= -Dk \ dc_m'/dx - Dc_m' \ dk/dx = -P' \ dc_m'/dx - Dc_m' \ dk/dx \qquad (22)$$

in complete analogy to Eq. 20 with the partition coefficient k play-
ing the role of sorption S. The difference between Eq. 22 derived
from FFL and the thermodynamically correct Eq. 15 resides with the
second term. One sees that FFL leads to a completely wrong result
for the current density and in particular to the term with dk/dx
yielding a current without any driving force applied.

This conclusion is extremely important in connection with some
attempts in the literature[16,18,19] to explain the hydraulic transport
through a uniform membrane not by the applied pressure gradient but
by the concentration gradient which may get established in the
membrane as a consequence of non-uniform swelling being higher upstream than downstream[12]. It was even stated[16] that "if the membrane
is not swollen by the solvent at all, then no concentration gradient
can be induced and as a result there can be no diffusion transport".
Strangely enough the same authors[18,21] believe that the pressure
borne by the membrane, i.e. the compacting pressure p_c, is uniform
through the membrane and hence equal to the total pressure difference
$p_1 - p_2$. The concentration difference they use for explaining the
hydraulic flow is therefore determined not by the difference of up
and down stream pressure on the polymer network of the membrane,
because there is none according to their views, and also not by the
compressibility of the membrane but by the requirement that FFL
yields the thermodynamically expected current. That yields a Δv as
calculated in Eq. 24 from the correct expression for the chemical
potential if one believes that the chemical potential cannot give
rise to a current as described in Eq. 11 but has first to create
a concentration gradient which in turn provokes the current.

The applied pressure in the hydraulic experiment compresses
the polymer membrane at the supported side, i.e., at the lower
pressure side, more than at the high pressure side[21]. The compacting
pressure[9] p_c is indeed the difference between the applied pressure
p_1 and the driving pressure at the location x which enters Eqs. 13
and 15.

$$p_c(x) = p_1 - p(x) \qquad (23)$$

thus being 0 at $x = 0$ and $p_1 - p_2$ at $x = \ell$. As a consequence of
membrane compression k is higher at $x = 0$ than at $x = \ell$. The
originally uniform membrane becomes vectorized with a gradual decrease of liquid fraction with increasing x.

In order to obtain the hydraulic flux, Eq. 16, one needs
according to FFL a concentration difference

$$\Delta v = K\Delta p/P' = (V/RT)\Delta p \qquad (24)$$

so that

$$j = - K\Delta p/\ell = -P'\Delta v/\ell \qquad (25)$$

The factor V/RT correlating the concentration and pressure difference
and derived from the thermodynamic description of material transport
through the membrane has absolutely nothing to do with any physical
or chemical property of the membrane. It is the same for a completely
incompressible and for a highly compressible material. This is by
itself a contradiction arousing most serious doubts about the applicability of FFL which leads to such unrealistic consequences. One
has indeed to do with the strange conclusion that in order to describe
the hydraulic flow through a homogeneous membrane according to FFL
one has to assume that a concentration difference obtains between the
upstream and downstream side of the membrane which is completely
independent of any structural property of the membrane.

The only justification for such a procedure is the fact that
Eq. 1 and 24 together yield the correct thermodynamically derived
hydraulic flow as shown in Eq. 16. It is indeed a vicious circle
not proving anything beyond the proper use of elementary algebra.
Moreover, the whole procedure becomes completely irrational if one
assumes a constant compacting pressure throughout the membrane[18,22]
because in such a case one cannot see any reason for a difference
in swelling, v or k, between the upstream and downstream side of
the membrane.

The extent of actual compression depends on the swelling properties of the membrane. It is high for a highly swollen and hence
very soft membrane and small for a moderately swollen and hence
rigid membrane. In order to simplify the matter, let us consider
the swelling to be dependent on the cross-link density only, i.e.
on the average molecular weight M_c of the chain segment between two

consecutive cross-links. On the basis of Treloar theory of swelling of rubbers under uniaxial pressure[23] one has the connection

$$\chi(1 - v)^2 + (1 - v) + \ln v + (\rho V/M_c)/(1 - v) + p_c V/RT = 0 \quad (26)$$

between compacting pressure and liquid volume fraction. Here ρ is the density of the polymer. This equation permits the calculation of the Flory-Huggins energy interaction parameter χ for the fully relaxed system with a given liquid content v and cross-link density. In a highly swollen membrane ($v \to 1$) one obtains in the first approximation

$$\chi = 1/2 + (1 - v)/3 - (\rho V/M_c)(1 - v)^{-3} + \ldots \quad (27)$$

The omitted terms are higher powers of the small quantity $(1 - v)$. For water ($V = 18$ cm^3/mole) and cellulose membrane ($\rho = 1.5$ g/cm^3) with $M_c = 10^6$ g/mole and $1 - v = 0.1$ the first concentration dependent term in Eq. 27 is 0.03, the second 0.027 and the first omitted term smaller than one tenth of the second term.

One derives from Eq. 26

$$dv/dp_c = -(V/RT)/[2\chi(1 - v) + 1 - (1 - v)^{-1} - (\rho V/M_c)(1 - v)^{-2}]$$

$$= -(V/RT)/[(1 - v)^2/3 + (3\rho V/M_c)(1 - v)^{-2} + \ldots] \quad (28)$$

which for highly swollen membrane reduces to

$$dv/dp_c = -(M_c/3\rho RT)(1 - v)^2 \quad (29)$$

Integration over the thickness of membrane yields in first approximation

$$\Delta v = (1 - v)^2 (M_c/3\rho RT) \Delta p$$

for the change of liquid content between upstream and downstream side of the membrane. The concentration differential Δv is indeed proportional to Δp but with a proportionality factor which depends on cross-link density, i.e. on a material property of the membrane, but has very little or nothing in common with the value in Eq. 24 derived from FFL. In particular there is no change of concentration in a rigid membrane ($M_c = 0$) and a rapid increase with softer membrane (large M_c).

The compaction of the membrane increases the resistance W

and hence makes the membrane less permeable than one would expect on the basis of Eq. 14. In a vectorized membrane the compaction plays quite an important role in the asymmetry of transport properties[9],[11]. But the change of liquid concentration does not constitute a source of flow although it changes the local diffusion coefficient.

That which matters more than anything else is the fact that the artificial explanation of the transport process in membranes by a concentration gradient is in straightforward conflict with thermo-dynamics as already demonstrated in connection with Eq. 21 and 22. No apparent agreement with any experimental data can save the erroneous concept that the diffusive or hydraulic flux is caused by the concentration gradient of the permeant in the membrane as suggested by FFL. The matter is so important that two examples will be presented demonstrating this statement.

Let us support a uniform membrane at a point x_o between 0 and ℓ. Such a support can be made by a rigid grid built into the membrane which is strong enough to hold the membrane in place and has such a small mesh size that the membrane is not pressed through. Under applied pressure p_1 the membrane section between 0 and x_o is non-uniformly compressed while the section between x_o and ℓ exposed to constant pressure $p_2 < p_1$ is completely relaxed. The swelling of this section is identical with that on the upstream side of the membrane. In spite of the equality $v_o = v_\ell$ a finite hydraulic diffusion flux will obtain according to Eq. 16 and will not be zero as demanded by Eq. 25.

The second example is concerned with a vectorized membrane ex-hibiting a swelling gradient i.e. high swelling on one side and low swelling on the other side[24]. That can be achieved by a gradient of the fraction of OH groups which determine the degree of swelling in water or by a gradient of cross-linking. Immersed in the liquid it would swell nonuniformly exhibiting a high liquid concentration on one side and a low one on the other side in spite of the fact that the liquid is the same and under the same pressure on both sides of the membrane. Nobody would expect a steady state flow through such a membrane in agreement with the black-box prediction that there is no flow without a pressure or concentration difference between the free liquids on both sides of the membrane. The flow is definitely not caused although as a rule it is modified by a con-centration gradient in the membrane.

Highly Swollen Membrane

According to Eq. 17 the hydraulic flow through a membrane (Eq.

16) under a pressure gradient and the pure diffusive flow (Eq. 15) under a concentration gradient are equivalently described by the corresponding K, P' pair. This is correct for modestly swollen membranes. At higher swelling, however, K grows faster than P' (Fig. 2). Such an effect is a consequence of incipient viscous flow through the more or less gel-like structure.[9,25] The permeant flows against the friction resistance of the long chain molecules. The hydraulic permeability hence reads

$$K = K_d + K_v \qquad (31)$$

where K_d is the diffusive (eq. 17) and K_v the viscous or flow permeability. For the latter flow one can derive an expression[25]

$$K_v = (M_o/f_o \rho N)(x^{-1} - x_o^{-1}) \qquad (32)$$

$$x = (1-k)/k$$

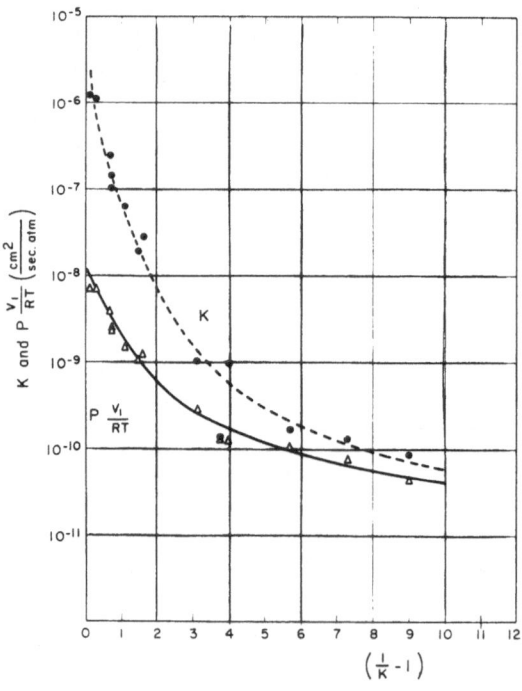

Figure 2. The hydraulic (K) and diffusion (P'V/RT) permeability of a series of hydrophilic membranes plotted versus polymer/water ratio $x = (1-k)/k$.[25]

where M_o is the molecular weight and f_o is the friction coefficient of the building unit of the network, ρ is the density of the dry membrane, N is Avogadro number and k, the partition coefficient, is actually the volume fraction of the permeant in the swollen membrane. The hydration x^{-1} is the volume ratio of liquid and polymer. The limiting value k_o corresponds to a swelling where the fraction of permeant molecules is not sufficient for complete solvation of the membrane molecules and still less for a viscous flow between them. In such a case they can be transported only by the diffusive mechanism, i.e., by jumping through a channel created by the local fluctuation of chain packing. Hence below k_o one expects $K_v = 0$.

According to Eq. 32 a plot of $K_v = K - K_d$ versus $1/x$ yields a straight line with an abscissa intercept at $1/x_o$ and slope $M_o/f_o\rho N$ (Fig. 3). In the case of membranes of widely differing chemical composition (cellulose acetate and various polymethacrylates) the permeabilities K and P' for water turned out to be solely dependent

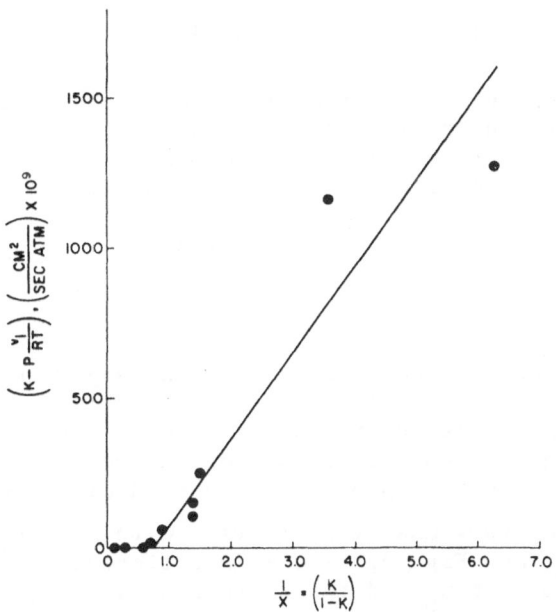

Figure 3. Viscous flow permeability K_v of the membranes of Fig. 2 versus water/polymer ratio $1/x = k/(1-k)$.[25]

on the partition coefficient which, of course, is determined by
chemical composition. The viscous flow permeability[26] K_v indeed
yields the straight line required by Eq. 32 yielding $1/x_o = 0.75$,
i.e., $k_o = 0.43$, and $M_o/f_o\rho N = 2.94 \times 10^{-13}$ cm^3 g^{-1} sec. By
assuming $\rho = 1.5$ one obtains $f_o/M_o = 3.8 \times 10^{-12}$. The molecular
model corresponding best to this value consists not of single chain
sections of the network but of bundles of about five parallel chains
of an average length of 35Å which is equivalent to 14 monomer units
on each chain. The length which is not critical in this evaluation
would roughly correspond to the link length of the network. Such
bundling may occur as a consequence of the hydrophobic nature of
the polymer backbone and the hydrophilic nature of the ester side
chains with the great many OH groups. The bundle of five parallel
chains is a cylinder with hydrophilic surface and hydrophobic core.

There was recently an attempt[19] to explain the difference
between K and P' at high swelling ratio by the use of an equation
for the diffusive flow which was developed for diffusion inside
multicomponent gas or liquid mixtures.[27] By neglecting any correlation among the fluxes and deviations from ideality one obtains
for the flux of the i-th component

$$\vec{j}_i = v_i \Sigma \vec{j}_k - D_i \text{ grad } v_i \qquad (33)$$

The first term describes the convection flow of the volume element
and the second term the diffusion flow according to FFL. In the
case of a two-component system, liquid-membrane, with fixed membrane
($j_2 = 0$) one has for the flux of the diffusing liquid

$$j_i = j = vj - D \, dv/dx$$

$$j = - D(1-v)^{-1} \, dv/dx \qquad (34)$$

which yields an enormous increase of flux and apparent diffusion
coefficient if v is close to 1, i.e. for highly swollen membranes.
Since such an effect only occurs in the case of permeation under
a pressure gradient and not for tracer diffusion (3 component system
with $\Sigma j_k = 0$) it seems to explain the observed differences between
K and P' increasing very nearly as $(1-v)^{-1}$ with increasing swelling
as shown in Eq. 32 and Fig. 3. The main difference between Eq. 32
and 34 is the term with x_o and the different slope $M_1/f_o\rho N$ in the
former and DV/RT in the latter case.

Such a conclusion is based on wrong premises, in this case on
the use of the incorrect diffusion equation. The first term in

Eq. 33 is the flux of component i as a consequence of the translation of the whole volume element. The total flux by diffusion is comprised in the second term. This is perfectly OK and self evident in a liquid mixture. In a fixed membrane case under a pressure gradient the situation is completely different. There is no convection of liquid through the homogeneous membrane although such a convection obtains in the liquid outside the membrane. All the transport has to be performed by diffusion. Equation 33 derived for diffusion in a multicomponent gas or liquid mixture is inapplicable to the diffusion through a fixed membrane.

There are some attempts to save the validity or applicability of Eq. 33 derived for diffusion in a liquid to the case of diffusion through a fixed membrane by claiming that the term $v_i j$ is not caused by convection but is a consequence of coordinate transformation from laboratory to center-of-mass system.[28,29] This is an obvious fallacy as can be easily proven. By definition the frictional resistance W is proportional to, and depends only on, relative velocity of penetrant and membrane which is the same in both frames of reference and hence completely independent of any translation of the whole system at constant velocity. That means that it is not proportional to the velocity of the penetrant or membrane in the center of mass system. Also the chemical potential and membrane thickness are constant in any inertial coordinate system so that Eq. 11 is valid without modification as long as the whole system is either in uniform translation or at rest.

If one insists on the validity of Eq. 33 and claims its derivation from the coordinate transformation to the center of mass system one is faced with a particular density effect. With a large difference of density between the permeant and the polymer the center of mass motion and hence the velocity of permeant in the center of mass system substantially depend on their density ratio. Equation 33 reads in such a case

$$j_1 = [\rho_1 v_1/(\rho_1 v_1 + \rho_2 v_2)]j - D_1 \; dv_1/dx \qquad (35)$$

yielding

$$j_1 = -\frac{\rho_1 v_1 + \rho_2 v_2}{\rho_2 v_2} D_1 \; \frac{dv_1}{dx}$$

$$= -\frac{1 - v_1(1-\rho_1/\rho_2)}{1 - v_1} D_1 \; \frac{dv_1}{dx} = -D_1' \frac{dv_1}{dx} \qquad (36)$$

The effective diffusion coefficient $D_1' = RT/W'$ and the effective frictional resistance

$$W' = \frac{1-v_1}{1-v_1(1-\rho_1/\rho_2)}\, W \tag{37}$$

would depend also on the density ratio ρ_1/ρ_2 and not only on the volume ratio of both components. This is a completely new effect never considered in the analysis of, and also never observed in, Brownian motion. The role of the denominator $1 - v_1$ which according to Eq. 33 is allegedly responsible for a large part of the difference between diffusive and hydraulic permeability at v_1 close to 1 would be drastically reduced by a very high density of membrane material or a low density of the permeant.

Free Volume Concept of Transport Properties

The deviations of transport properties from ideality of polymer membranes reside with the sorption and diffusion coefficients. Both seem to depend on fractional free volume which is a combined result of the polymer and penetrant contribution taking already into account the energy and entropy interactions of both components[24,30,35]. If one assumes a linear relationship one can write

$$\phi = v\phi_1 + (1-v)\phi_2 = \phi_2 + v(\phi_1-\phi_2) = \phi_2 + v\delta\phi \tag{38}$$

where ϕ_1 and ϕ_2 are fractional free volumes of permeant and polymer, respectively. The latter is usually smaller than the former. As a consequence of positive $\delta\phi$ the fractional free volume of the membrane increases with the concentration of the permeant.[32] Since with increasing ϕ the chain mobility is enhanced and the glass transition lowered the permeant acts as a plasticizer.

In first approximation the sorption of a gas or vapor is proportional to the fractional free volume[6]

$$S = A\phi = A(\phi_2 + v\delta\phi) \tag{39}$$

yielding with Eq. 4 the value

$$S = S_o/(1 - Ap\delta\phi) = S_o[1 + Ap\delta\phi + (Ap\delta\phi)^2 + \ldots]$$

$$= S_o/(1-S_o p\delta\phi/\phi_2) \tag{40}$$

increasing more than linearly with the pressure of the gaseous penetrant in satisfactory agreement with many experimental data on simple systems without clustering or other secondary effects.

The diffusion coefficient seems to depend on fractional free volume in a more conspicuous manner[30,34-37]

$$D \sim {}^{-B/\phi} \tag{41}$$

where B is a constant characteristic for the permeant-membrane system increasing rapidly with the size of the permeant. Such a dependence tells that the probability for a diffusive jump decreases with increasing volume of the penetrant and decreasing fractional free volume of the membrane. By introcucing the expression for Eq. 38 one obtains

$$B/\phi = B/\phi_2(1 + \delta\phi/\phi_2) = (B/\phi_2)(1 - \delta\phi/\phi_2 \ldots) \tag{42}$$

which in first approximation yields

$$D = D_o \exp[B\delta\phi/d_2{}^2)v] = D_o e^{\gamma v} \tag{43}$$

$$D_o \sim \exp(-B/\phi_2)$$

$$\gamma = B\delta\phi/\phi_2{}^2 .$$

The extrapolated value D_o drops extremely rapidly with decreasing fractional free volume. The exponential increase of D with concentration is a generally established fact with sorption of vapors into polymers above the glass transition temperature[35-39]. It is well describable by the fractional free volume concept of the membrane/penetrant system. The coefficient γ is directly proportional to the difference of fractional free volume of the components and inversely proportional to the square of that of the polymer.

The best examples of the above-mentioned results can be found in highly swollen membranes[25] and in drawn polymers[8-42] where the majority of the influence on the transport properties is caused by the tighter packing of amorphous chains in the fibrous structure. In a swollen membrane the fractional free volume can be assumed to be in first approximation the sum of contributions of the liquid and polymer as shown in Eq. 38. By introducing this value into Eq. 41 for D one was indeed able to describe the diffusive permeability P' of a great many membrane systems in practically the whole range of v from 0 to 1 (Fig. 2). The fit was not perturbed by the fact that the membranes differed widely in chemical composition (cellulose acetate and various polymethacrylates).

According to Eq. 41 the change of initial sorption of drawn crystalline polymer at v = 0 can be directly interpreted as a reduction of fractional free volume ϕ_2 per unit volume of the amorphous component

$$S_o = A\phi_2 \ (1 - \alpha) \qquad\qquad (44)$$

where α is the volume fraction of the crystalline component. Experiments on drawn linear polyethylene (Fig. 4) show a nearly constant amorphous fraction $(1 - \alpha)$ between 20 and 30 percent, a small initial increase of sorption followed between the draw ratio $\lambda = 8$ and 9 by a rapid drop to a new almost constant value about 15 percent of that before drawing[39]. That can be interpreted as a reduction of fractional free volume of the polymer to about 15 percent of ϕ_2 at draw ratio $\lambda = 1$ as a consequence of transformation of the initial spherulitic to the final fibrous structure. The transformation seems to be completed with $\lambda = 9$. The initial increase is most likely the consequence of the loosening of the spherulitic structure as detected by small-angle X-ray scattering[43] and by a decrease in density[46] which both occur at small draw ratios ($\lambda \sim 2$) when the sample has still the original spherulitic morphology.

The effects are, as expected, much more drastic in the diffusion coefficient[38,42] where the fractional free volume is in the exponential function. Up to a draw ratio 8 the diffusion coefficient drops by a factor 10 while γ is even smaller than in the undrawn sample. Between the draw ratio 8 and 9 the drop of D is again by a factor of 10 with a 20 times increased γ. Further drawing does not change γ perceptibly. The change of γ agrees best with fractional free volume changes derived from sorption data: slight increase at small λ and a drastic drop at λ between 8 and 9. The change of D is a little more complicated because it depends so much on morphological detail.

The gradual formation of fibrous material in almost the whole drawing range up to $\lambda = 9$ lengthens the diffusion path through the rapidly vanishing but significantly more permeable spherulitic material. As a consequence the tortuosity factor τ reduces the observed diffusion coefficient at draw ratio λ

$$D_\lambda = \tau D_1 \qquad\qquad (47)$$

quite drastically below the value D_1 of the undrawn material.[24,42] The true diffusion coefficient of the still remaining spherulitic structure remains practically constant or may even increase as one is tempted to guess from the concentration dependence, i.e., γ, and the slight increase of sorption in this range. With completed transformation into fibrous structure D drops to the extremely small value characteristic for it and derivable from Eq. 45.

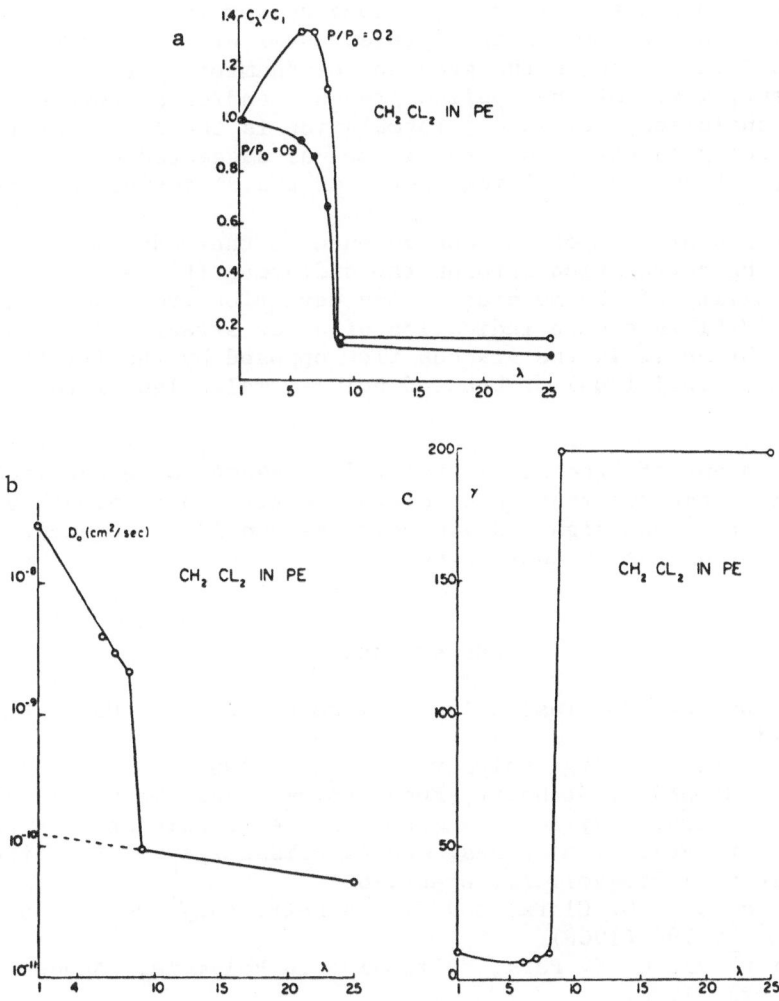

Figure 4. Relative weight gain
$v_\lambda/v_1 = c_\lambda/c_1$ at gas activity
0.2 and 0.9 (a), D at zero sorption
(b), and γ (c) versus draw ratio[36]
for methylene chloride in linear
polyethylene.

Conclusion

The diffusive process through a simple homogeneous membrane, uniform or vectorized, can be described by conventional thermo-dynamics of irreversible processes if the membrane in contact with the permeant does not show complications as for instance clustering, preferential absorption, chain rupture, chemical reaction and so on. The driving force is the gradient of chemical potential of the permeant, i.e., of the applied pressure and/or of composition. The main conclusion from such a formulation is the fact that the concentration gradient is not the cause, as suggested by Fick's first law, but only a modifying factor of the diffusive current.

A particularly important consequence of thermodynamic formu-lation is the correlation between the diffusive (P') and hydraulic (K) permeability of the membrane. Any deviation from the relation-ship K = P'V/RT is a sure indication of a new effect. In highly swollen membrane it is the viscous flow opposed by the friction resistance of individual or bundled polymer molecules of the membrane.

The concept of fractional free volume seems to be particularly successful in the description of diffusive transport through highly swollen membrane and highly drawn polymers and in the concentration dependence of diffusion coefficient.

References

1. R. M. Barrer, "Diffusion in and Through Solids", Univ. Press, Cambridge, 1951.
2. H. Fujita, Adv. High Polymer Res. $\underline{3}$, 1 (1961).
3. H. Yasuda and V. Stannett, Ency. Polymer Sci. Techn. $\underline{2}$, 316 (1965).
4. C. E. Rogers, "Physics and Chem. of the Organic Solid State," Ed. by D. Fox, M. M. Labes, and A. Weissberger, Interscience Publ., New York, 1965, Vol. II, Chapt. 6.
5. H. Yasuda, H. G. Clark, and V. Stannett, Ency. Polymer Sci., Techn. $\underline{9}$, 794 (1968).
6. J. Crank and G. S. Park, "Diffusion in Polymers," Academic Press, New York, 1968.
7. R. Ash and R. M. Barrer, J. Phys. $\underline{D4}$, 888 (1971).
8. A. Katchalsky and P. E. Curran, "Nonequilibrium Thermodynamics in Biophysics," Harvard Univ. Press, Cambridge, Mass., 1965, Chapt. 10.
9. A. Peterlin and J. L. Williams, J. Appl. Polymer Sci. $\underline{15}$, 1493 (1971).
10. J. H. Petropoulos and P. P. Roussis, J. Chem. Phys. $\underline{47}$, 1491, 1496 (1967); $\underline{50}$, 3951 (1969); $\underline{51}$, 1332 (1969).
11. A. Peterlin, J. Appl. Polymer Sci. $\underline{15}$, 3127 (1971); Kolloid-Z and Z. Polymere $\underline{250}$, 553 (1972).

12. A. Peterlin and H. G. Olf, J. Macromol. Sci. B6, 571 (1972).
13. V. Stannett, H. B. Hopfenberg, and J. H. Petropoulos, "Macromol. Sci.," Ed. by C. E. H. Bawn, Univ. Park Press, Baltimore, Maryland 1972, p. 329.
14. H. Yasuda and A. Peterlin, J. Appl. Polymer Sci. 17, 443 (1973).
15. S. Sternberg and C. E. Rogers, J. Appl. Polymer Sci., 12, 1017 (1968).
16. D. R. Paul, Coating and Plastics Preprint 34, 436 (1974).
17. C. E. Rogers, V. Stannett, and M. Scwarc, Ind. Eng. Chem. 49, 1933 (1951).
18. D. R. Paul, J. Appl. Polymer Sci. 16, 771 (1972).
19. D. R. Paul, J. Polymer Sci. Phys. 11 289 (1973).
20. D. R. Paul and O. M. Ebra-Lima, J. Appl. Polymer Sci. 14, 2201 (1970).
21. D. R. Paul and O. M. Ebra-Lima, J. Appl. Polymer Sci. 15, 2199 (1971).
22. S. A. Stern, S. M. Fang, and H. L. Frisch, J. Polymer Sci. A2, 10, 201 (1972).
23. L. R. G. Treloar, "The Physics of Rubber Elasticity, " Clarendon Press, Oxford 1958, p. 165.
24. A. Peterlin and H. Yasuda, J. Appl. Polymer Sci. (in press).
25. H. Yasuda, C. E. Lamaze, and A. Peterlin, J. Polymer Sci. A2, 9, 1117 (1970).
26. A. Peterlin, H. Yasuda, and H. G. Olf, J. Appl. Polymer Sci. 16, 865 (1972).
27. R. B. Bird, W. E. Stewart, and E. N. Lightfoot, "Transport Phenomena," J. Wiley and Sons, New York (1960).
28. H. Frisch, personal communication.
29. D. R. Paul, Lecture at ACS Meeting in Los Angeles, Cal., April 1-5, 1974.
30. M. H. Cohen and D. Turnbull, J. Chem. Phys. 31, 1164 (1959).
31. W. Jost, "Diffusion," Academic Press, New York (1969).
32. H. Fujita and A. Kashimoto, J. Chem. Phys. 34, 393 (1961).
33. J. D. Ferry, "Viscoelastic Properties of Polymers", J. Wiley & Sons, New York (1961).
34. A. K. Doolittle, J. Appl. Phys., 22, 1971 (1951).
35. S. Prager and F. A. Long, J. Amer Chem. Soc. 73, 4072 (1951).
36. C. E. Rogers, V. Stannett, and M. Scwarc, J. Polymer Sci., 45, 61 (1960).
37. H. Fujita, A. Kishimoto, K. Matsumoto, Trans. Far. Soc., 56, 424 (1960).
38. A. Peterlin, J. L. Williams, and V. Stannett, J. Polymer Sci. A-2, 5.
39. J. L. Williams and A. Peterlin, J. Polymer Sci., A-2, 9, 1483 (1971).
40. Y. Takagi and H. Hattori, J. Appl. Polymer Sci., 9, 2167 (1965).
41. A. Peterlin and J. L. Williams, Brit. Polymer J., 4, 271 (1972).
42. A. Peterlin, IUPAC Symp. Macromol., Aberdeen, Scotland, September 10-14, 1973.

43. F. J. Baltá-Calleja and A. Peterlin, J. Macromol. Sci. 4B, 519
     (1970).
44. W. Glenz, N. Morosoff, and A. Peterlin, J. Polymer Sci. B9,
     211 (1971).

DIFFUSIVE TRANSPORT IN SWOLLEN

POLYMER MEMBRANES

D. R. Paul

Department of Chemical Engineering

The University of Texas, Austin, Texas 78712

ABSTRACT

Recent work on transport of solvent and solutes in swollen, hydrocarbon-type network polymers is reviewed, and some generalizations for other systems are made. Network polymers are useful as model membranes since they are well understood thermodynamically and can be free of inhomogeneities such as pores and crystallinity which obscure results in many systems. The absence of pores assures a diffusional transport mechanism. The degree of swelling of the membrane by solvent is a very important parameter which for hydrocarbon networks can be varied over a wide range without altering polymer structure by selecting from a large number of suitable organic solvents. Solvent transport by an applied pressure differential as in reverse osmosis is considered as well as by downstream vacuum as in pervaporation. Emphasis is given to how the outside driving force causes solvent diffusion in the membrane and relationships between the two different processes. The major factors affecting the diffusion coefficient of solutes in such membranes are discussed.

INTRODUCTION

Most membrane processes involve transport of solvent and various solutes. The driving force for solute transport is usually a difference in concentration of this species between the two regions of fluid the membrane separates; however, electrical fields may be employed with ionic solutes. Solvent transport, however, may be induced by an applied hydrostatic pressure, an osmotic pressure difference, or by a downstream vacuum as in pervaporation. Depending on the membrane structure various transport mechanisms may be operative. In all of these, however, the amount of solvent in the membrane, i.e. the degree of swelling, is an important variable.

Recent work (1-9) has shown that swollen networks consisting of suitably crosslinked organic polymers can serve as useful model membranes to elucidate various fundamental phenomena. The purpose here is to review some of the essential features of these studies and to draw attention to the implications for other systems. One advantage of simple network polymers is that their thermodynamics of swelling is well understood. When non-polar, hydrocarbon polymers form the network, the degree of swelling can be varied over a wide range without altering the molecular structure of the polymer simply by chosing from a large number of organic solvents. In this way, one can also systematically vary the molecular characteristics and macroscopic properties of the solvent which may affect transport rate. This allows a whole new dimension of investigation not possible when only water is used as the solvent. However, aqueous based systems hold a place of unique importance for many applications. The swollen networks of interest here are free of inhomogeneities such as crystallinity or macroscopic pores common to many membranes. The lack of pores assures that transport will occur by molecular diffusion which allows establishment of relationships of value in defining the transport mechanisms in more complex systems.

A crosslinked network will imbibe a good solvent until the elastic forces from chain extension, as the network dilates, are balanced by the osmotic forces generated by dilution of polymer segments with solvent. The volume fraction of solvent at equilibrium, v_{10}, depends on the interaction of chain segments with the solvent, the polymer crosslink density, and the solvent molar volume (1). The equilibrium uptake will be less than this value if the activity of the solvent, a_1, is decreased below one by addition of a diluent that does not enter the network or if the solvent is exposed to the network as a vapor at a partial pressure, p_1, less than the equilibrium vapor pressure, p_1^*, since $a_1 = p_1/p_1^*$. Figure 1 shows the equilibrium swelling of a crosslinked natural rubber as a function of a_1 for $CC\ell_4$ (8). Note that pure solvent in this case swells the network so that approximately 84% of the system is solvent.

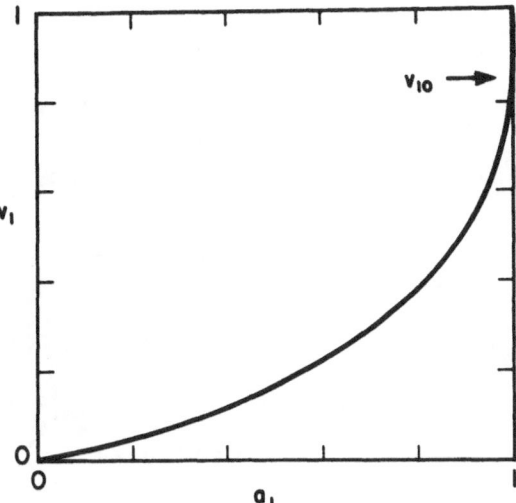

Fig. 1. Swelling versus activity for the CCl_4-crosslinked natural rubber system. The curve was calculated from a form of the Flory-Huggins equation using crosslink density and interaction parameter data given in reference 1. The interaction parameter has been assumed to be constant over the entire concentration range.

HYDRAULIC PERMEATION OF SOLVENT

In reverse osmosis, ultrafiltration, hemodialysis, etc. a hydrostatic pressure difference between the fluid on either side of the membrane causes transport of solvent. Figure 2 shows a common test configuration where the membrane rests on a porous support plate and the upstream and downstream pressures are p_o and p_ℓ. If the fluid on either side is pure solvent, a positive pressure difference will produce a volumetric flux, $n_1 V_1$, of solvent to the low pressure side. If a solute is present, a positive flux may not be realized until a pressure difference related to the osmotic pressure difference is exceeded.

The relation between flux and pressure differences defines a hydraulic permeability as shown in Fig. 2. This relation will be linear initially but may become non-linear later for a host of reasons (3). When solvent transport is by diffusion, this relation must depart from linearity at some pressure as shown by the solid line in Fig. 2   since as will be discussed later a ceiling

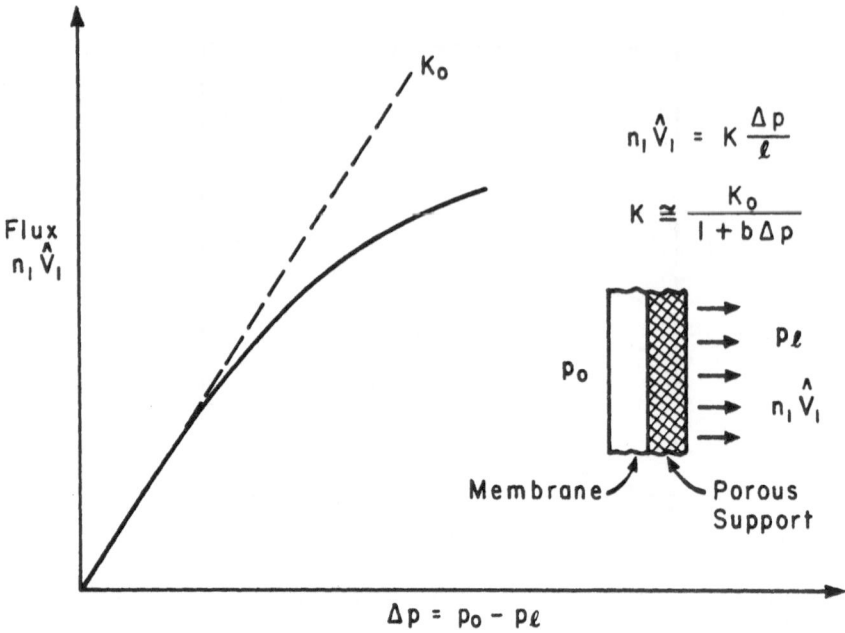

Fig. 2. Schematic of the hydraulic permeation experiment, definition of the hydraulic permeability, and illustration of typical results for a swollen, homogeneous membrane (solid curve). The dotted line is an extension of the initial linear region usually observed. Its slope gives the initial hydraulic permeability K_o.

flux must be approached. The pressure range where this occurs depends greatly on the degree of swelling. For membranes that employ a pore flow mechanism of solvent transport, this limitation does not exist and linearity could in principle continue indefinitely.

When there is no pore structure, solvent transport occurs by a solution-diffusion mechanism since the applied pressure induces a concentration gradient of solvent within the membrane which leads to Fickean diffusion. In previous papers, a detailed thermodynamic analysis of this induced concentration gradient was presented (1-4). It is instructive here to present an alternate

set of simple arguments, backed up by individual experiments,
to show how the applied pressure induces the required gradient.
This is aided by Fig. 3 where four different situations for a
swollen membrane are depicted. In each case a schematic profile of the solvent volume fraction in the membrane, v_1, is
shown (axes are labeled in 3a but omitted elsewhere). The first
three are equilibrium situations involving no transport. In 3a
the membrane rests on a porous support with solvent upstream
and downstream at the same pressure, p_ℓ. Solvent below the
porous support freely communicates with the downstream membrane surface. The membrane contains its equilibrium content

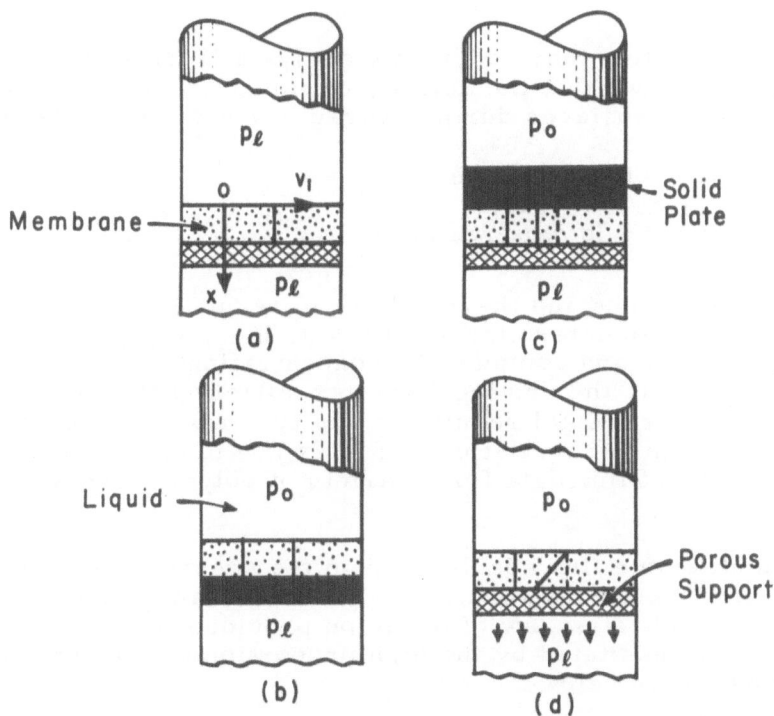

Fig. 3. Illustrations to demonstrate how the solvent concentration gradient is generated in the membrane during hydraulic
permeation. Each figure shows a schematic of the solvent
concentration profile that exists in the situation depicted.

of solvent, v_{10}, throughout. In Fig. 3b the porous plate of 3a is replaced by a solid plate that does not transmit solvent, and the pressure upstream is raised to p_o. Experiments (10) similar to this have shown that the volume fraction of the liquid in the membrane is unchanged from 3a except in certain cases where the molar volume and the partial molar volume of the solvent are different. (This is a secondary effect that may increase or decrease the swelling and is only apparent at very high pressures). This shows that just inside the upstream membrane surface in Fig. 2 that the liquid content is the equilibrium swelling value v_{10}. In Fig. 3c there is a solid plate above the membrane and a porous one below it. A pressure p_o is applied to the upper plate which the downstream liquid is at a pressure p_ℓ. Experiments (11) have shown that this will squeeze liquid from the membrane reducing its liquid content to a value $v_{1\ell}$ throughout at equilibrium. In Fig. 3d the upstream membrane surface is identical to 3b so $v_1 = v_{10}$ there while the downstream surface is identical to 3c so $v_1 = v_{1\ell}$ there. Because of this difference in solvent content, transport occurs and a gradient of v_1 results as illustrated in 3d.

A quantitative thermodynamic analysis (1-4) of the above situation has shown that the activity of solvent, a_1, in the membrane at its two surfaces during hydraulic transport are given by

at x = 0 (upstream) $a_1 = 1$ 1

at x = ℓ (downstream) $a_1 = e^{-(p_o - p_\ell)V_1/RT}$ 2

With these equations and thermodynamic results such as shown in Fig. 1, one can determine v_1 at x = 0, i.e., v_{10}, and at x = ℓ, i.e., $v_{1\ell}$, and compute the concentration difference $v_{10} - v_{1\ell}$ which is the driving force for solvent diffusion. Calculations and experimental results of this type have been used in combination with a formulation of Fick's law to obtain diffusion coefficients from flux data for a variety of polymer and solvent types (1-6).

The diffusive flux resulting from the pressure-induced gradient of solvent concentration in the membrane can be described by Fick's first law of diffusion provided frame of reference terms necessitated by the high proportion of solvent are included as shown below

$$n_1 = \omega_1(n_1 + n_m) - \rho D \frac{d\omega_1}{dx}$$ 3

In this equation the mass flux of solvent, n_1, and the membrane, n_m, are expressed relative to <u>stationary</u> laboratory coordinates.

The solvent content is expressed here as mass fraction, ω_1, and ρ is the density of membrane-solvent system and D is the mutual diffusion coefficient. At steady-state the membrane is stationary, i.e. $n_m = 0$, so we have

$$n_1 = \frac{-\rho D}{(1-\omega_1)} \frac{d\omega_1}{dx} \qquad 4$$

or

$$n_1 \hat{V}_1 = \frac{-D}{(1-v_1)} \frac{dv_1}{dx} \qquad 5$$

when $\hat{V}_1$ is the solvent specific volume. The frame of reference terms $\omega_1(n_1 + n_m)$ in Eq. 3 lead to the terms $(1 - \omega_1)$ and $(1 - v_1)$ in Eqs. 4 and 5 respectively. Very often the frame of reference terms can be omitted with little error since ω_1 or v_1 are very small compared to unity; however, this is far from the case here.

The necessity to include such terms in the first place can be understood by two examples. First, since D is a <u>mutual</u> diffusion coefficient an equation analogous to Eq. 3 should apply to the second component, the membrane in this case, viz.

$$n_m = \omega_m(n_m + n_1) - \rho D \frac{d\omega_m}{dx} \qquad 6$$

If we substitute into Eq. 6 the known facts that $n_m = 0$, $\omega_m = 1-\omega_1$, and $\frac{d\omega_m}{dx} = -\frac{d\omega_1}{dx}$, rearrangement yields Eq. 4 once again as we expect. However, if the frame of reference terms are not included, we do not get this reciprocal behavior for the second component. In fact, Eq. 6 without these terms would say there must be a flux of membrane material since there is a gradient of its concentration. Second, we can see the necessity for these terms by the contrived example of where the experimental apparatus is moving relative to fixed laboratory coordinates at a velocity V in the x direction. The fluxes with the apparatus moving, denoted by primes, are related to those in the stationary case as follows

$$n'_1 = n_1 + \omega_1 \rho V$$
$$n'_m = n_m + \omega_m \rho V \qquad 7$$

where now n'_m is not zero owing to the bulk translation of the system but n_m is zero. Simple rearrangement shows the following terms are equal regardless of the motion of the system

$$n'_1 - \omega_1(n'_1 + n'_m) = n_1 - \omega_1(n_1 + n_m) \qquad\qquad 8$$

and give the flux of 1 relative to the mass average velocity of the system. Only nonsense results when one omits the frame of reference terms in such cases.

For some purposes, it may be more appealing to use a transport rate law based on a gradient of chemical potential, $\frac{d\mu_1}{dx}$, rather than a concentration gradient; however, even when this is done the frame of reference terms should still be retained in the analog of Eq. 3 so that all transformations of coordinates can be executed properly. When there is only a gradient of concentration, such a formulation is equivalent to Fick's law since

$$\frac{\partial\mu_1}{\partial x} = \frac{\partial\mu_1}{\partial v_1}\frac{\partial v_1}{\partial x} = RT\left(\frac{\partial \ln a_1}{\partial v_1}\right)\frac{dv_1}{dx} \qquad\qquad 9$$

At low enough applied pressure differentials, the flux-pressure relation will be linear as shown in Fig. 2. The resulting hydraulic permeability, K_o, in this region can be related to the mutual diffusion coefficient for the solvent in the membrane, D, through the use of Fick's law and the thermodynamics of the situation as described above. This relation is (9)

$$K_o = \frac{DV_1 v_{10}}{RT(1 - v_{10})\left(\frac{\partial \ln a_1}{\partial \ln v_1}\right)} \qquad\qquad 10$$

The term $\left(\frac{\partial \ln a_1}{\partial \ln v_1}\right)$ expresses the change in solvent activity in the membrane with solvent volume fraction at $v_1 = v_{10}$. It can be calculated from data such as that shown in Fig. 1. The term $(1 - v_{10})$ arises from frame of reference considerations in Fick's law and is very important for highly swollen systems since v_{10} is near one; however, it is not so important in slightly swollen membranes where v_{10} is near zero. Fig. 4 illustrates graphically the relation between K_o and D for a crosslinked network membrane with various crosslink densities as the equilibrium swelling is varied. These results were computed assuming that the thermodynamics of the solvent-crosslinked membrane system are described by a Flory-Huggins type equation (see refs. 1, 3, 4, and 9). Each curve is for a fixed product of the solvent molar volume, V_1, and the network crosslink density, ν_e/V_o. The method of presentation does not explicitly show the

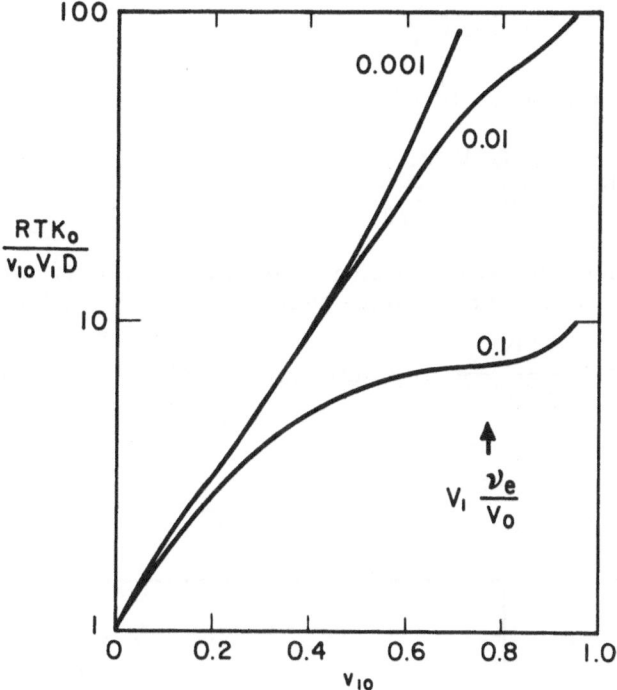

Fig. 4. Reduced hydraulic permeability, $\dfrac{RTK_o}{v_{10}V_1D}$, based on the mutual diffusion coefficient, D, as a function of the equilibrium swelling of the membrane. Each curve is for a fixed product of the solvent molar volume, V_1, and polymer crosslink density, v_e/V_o. For a given curve, the level of equilibrium swelling, v_{10}, reflects the thermodynamic interaction between the solvent and polymer.

thermodynamic interaction parameter, χ_1; however, its effect appears by determining the level of swelling possible, v_{10}, for a given crosslink density.

    In certain situations the tracer diffusion coefficient for solvent in the membrane, D_1, is related to the mutual diffusion coefficient by (9)

$$D = D_1 \left(\frac{\partial \ln a_1}{\partial \ln v_1} \right) \qquad\qquad 11$$

When this is so, Eq. 10 becomes

$$K_o = \frac{D_1 V_1 v_{10}}{RT(1 - v_{10})} \qquad\qquad 12$$

Figure 5 shows the relationship between K_o and D_1 as a function of v_{10}. It is important to note that as $v_{10} \to 0$, both

$\dfrac{RTK_o}{v_{10} V_1 D}$ and $\dfrac{RTK_o}{v_{10} V_1 D_1}$ become unity. This is because both

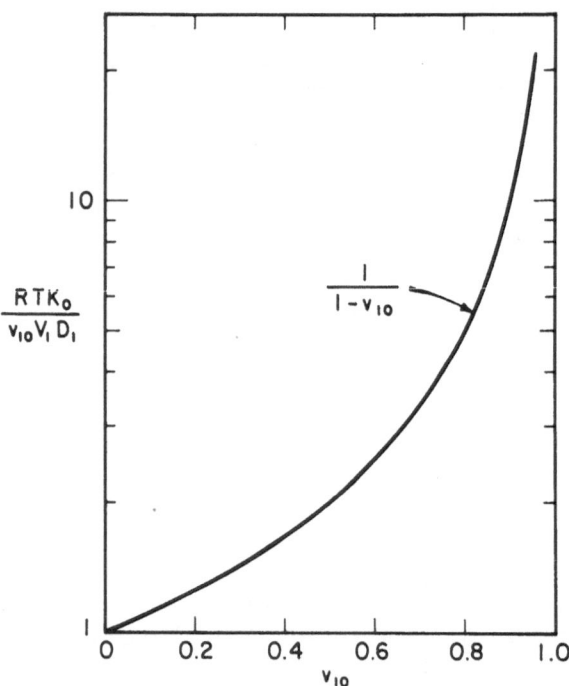

Fig. 5. Reduced hydraulic permeability based on the solvent tracer diffusion coefficient, D_1, measured in a membrane swollen to $v_1 = v_{10}$. The curve shown assumes that the approximate result of Eq. 4 holds.

$(\dfrac{\partial \ell \, na_1}{\partial \ell \, nv_1})$ and $(1-v_{10})$ become one in this limit. In this limit K_o must be zero. For finite levels of swelling, both permeability functions are greater than one as shown in Figs. 4 and 5. A large segment of the literature has insisted that both should be unity for all values of v_{10} which has lead to some erroneous conclusions about the mechanism of hydraulic permeation (4).

These results make very clear the importance of the degree of swelling to diffusive transport by an applied pressure in homogeneous membranes. If the membrane is not swollen by the solvent at all, then no concentration gradient can be induced and as a result there can be no diffusive transport. The larger the amount of swelling at equilibrium then the larger is the possible gradient. This leads to increased rate of solvent transport; however, as will be shown later increased swelling also leads to a greater rate of solute transport through the membrane.

At high enough pressure differentials, the flux-pressure relation must become non-linear for all cases where transport is by diffusion along an induced concentration gradient (for many low swelling systems this non-linearity is often not apparent because of the pressure ranges employed). The reason is that the induced concentration differential is a non-linear function of applied pressure (1) except for the initial region where a linear approximation is adequate. There is a finite maximum concentration differential possible which corresponds to $v_{1\ell}$ approaching zero. According to Eq. 1, this would require an infinite pressure; however, in highly swollen systems this limit can be accomplished very nearly by rather modest pressures (1). Because of this limit, there is a ceiling flux that cannot be exceeded regardless of the pressure. Figure 6 shows schematically what happens to the profile of v_1 at various levels of pressure. At $\Delta p = 0$, $v_{1\ell} = v_{10}$ and there is no flux. At intermediate pressure, $v_{1\ell}$ lies between v_{10} and zero. At very high pressures $v_{1\ell} \simeq 0$ and the flux reaches its maximum possible value asymptotically. The technology of reverse osmosis recognizes maximum practical pressures that can be applied to a membrane which arise from mechanical considerations. The point here, which is not generally recognized in the reverse osmosis field, is that there is a maximum flux that can be obtained theoretically, and as it is approached, increasing pressure will be less effective at increasing flux. Most reverse osmosis systems employ membranes that are only slightly swollen; hence, this occurs at very high pressures. In ultrafiltration where solvent transport occurs by pore flow, no such theoretical limit exists; however, it may occur in practice because of compaction of the pore structure by the applied pressure. This depends solely on the mechanical properties of the membrane.

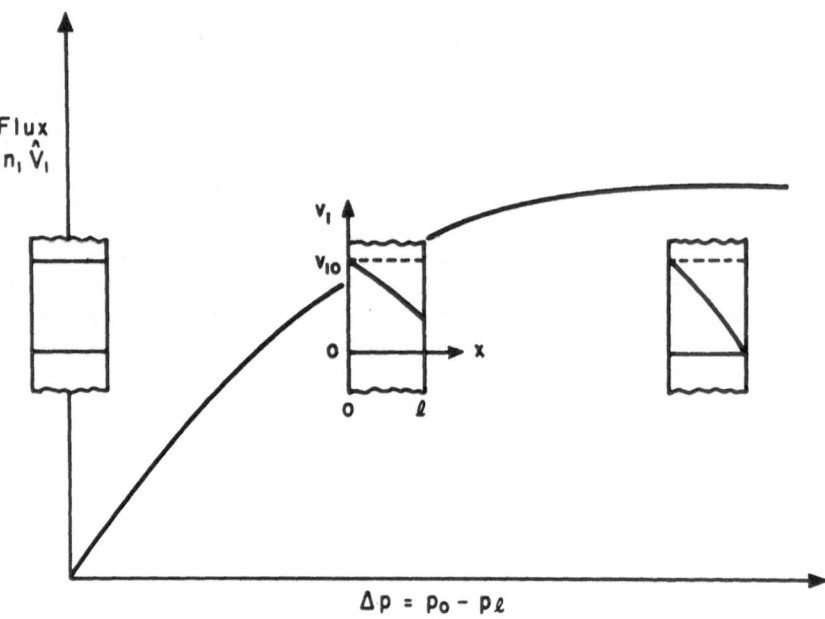

Fig. 6. Illustration of extreme non-linearity in hydraulic permeation and the presence of a ceiling flux. The inserts show the
profile of solvent concentration in the membrane at various stages.

SOLUTE TRANSPORT
IN SWOLLEN NETWORK MEMBRANES

Solutes are generally transported through membranes by
the difference in concentration of this species in the fluid upstream and downstream of the membrane. It is the objective of
reverse osmosis to minimize this transport, whereas in dialysis
the objective is to maximize the rate for some species while
minimizing it for others. Some membrane applications demand
simply a regulated rate of transport. Both the distribution coefficient for solute between the fluid phase and the membrane and the
diffusion coefficient of the solute in the membrane are important
in governing the transport rate. The diffusion coefficient should
depend on the level of swelling of the membrane, the nature of the
liquid swelling the membrane, the solute size, plus its interaction with the environment in the membrane. To date these

factors have not been sorted out fully mainly because most research has employed systems of some particular interest rather than experiments designed to understand each individual factor. Membranes composed of hydrocarbon-type network polymers offer an attractive way to do this since a range of liquids and swelling levels can be selected. Recent work employing dye molecules (7) has made an initial attack on this complex problem. Some data on related systems are also available in the literature (7) for comparison. In lightly crosslinked networks, sieving effects are absent except for very large solutes of macromolecular size. This concept has immense practical interest, but further fundamental discussion of it is not appropriate until the transport of smaller solutes retarded only by the rates of molecular and segmental brownian motion is better understood.

Experimental research to date (7) has revealed that for a given solute the most influential factors affecting the diffusion coefficient of solute, D, in the membrane milieu are the degree of swelling, v_{10}, and the viscosity (hence mobility related factors) of the solvent contained in the membrane. At high levels of swelling, many of the main features of the current data are adequately described by a theoretical equation presented by Meares (12)

$$D = D_o \left(\frac{1 - v_{ro}}{1 + v_{ro}} \right)^2 \qquad\qquad 13$$

to explain solute diffusion in homogeneous ion exchange resins. In this equation D_o is the diffusion coefficient for the solute in pure solvent which can be estimated from various empirical relations (7). The term in parentheses gives the dependence on swelling where v_{ro} is the polymer volume fraction, i.e. $1 - v_{10}$. As v_{10} goes to one, D becomes D_o as expected; however, when v_{10} goes to zero, D becomes zero. The only dependence on solute size comes through D which is approximately inversely proportional to (solute molar volume)$^{0.6}$. At the present time it is not clear that this method of accounting for solute size and shape is adequate even at high swelling; however, from available data no alternate suggestion is possible. The effect of pure solvent viscosity enters directly in D_o as an inverse proportional factor. This seems to give a quite adequate accounting of this parameter at least for high swelling.

At low swelling the Meares' equation breaks down totally because it does not recognize a finite diffusion rate in the undiluted polymer. The effect of the amount of swelling and the nature of the swelling liquid on transport in this region is an interesting and important subject that can be clarified only by further investigation.

REFERENCES

1. D. R. Paul and O. M. Ebra-Lima, J. Appl. Polym. Sci., 14, 2201 (1970).

2. D. R. Paul and O. M. Ebra-Lima, J. Appl. Polym. Sci., 15, 2199 (1971).

3. D. R. Paul, J. Appl. Polym. Sci., 16, 771 (1972).

4. D. R. Paul, J. Polym. Sci. (Phys.), 11, 289 (1973).

5. D. R. Paul and D. H. Carranza, J. Polym. Sci., 41C, 69 (1973).

6. O. M. Ebra-Lima, Ph. D. Dissertation, University of Texas at Austin, 1973.

7. M. Garcin, M. S. Thesis, University of Texas at Austin, 1973.

8. J. D. Paciotti, M. S. Thesis, University of Texas at Austin, 1974.

9. D. R. Paul, J. Polym. Sci. (Phys.), in press.

10. J. S. Ham, M. C. Bolen, and J. K. Hughes, J. Polym. Sci., 57, 23 (1962).

11. S. D. Gehman, Rubber Chem. Tech., 38, 1039 (1965).

12. P. Meares, J. Polym. Sci., 20, 507 (1956).

ACKNOWLEDGMENTS

This work was supported by a grant from the National Science Foundation. Appreciation is extended to J. D. Paciotti for his assistance with various computer calculations.

GENERALIZED DUAL SORPTION THEORY

Wolf R. Vieth and Mary A. Amini

Department of Chemical and Biochemical Engineering

Rutgers University, New Brunswick, New Jersey

ABSTRACT

A theory is formulated to explain certain classes of negative and positive deviations from Henry's law which are frequently observed in sorption plots of penetrants in polymers. Sorption is visualized as a process in which there are dual modes: either the penetrant molecule is normally dissolved and is free to diffuse or it is immobilized, as in a sink or well. It is the second process which gives rise to deviations from normal behavior. To explain negative deviations from Henry's law, the solubility of the penetrant is analyzed as the sum of the contributions from these two modes. A diffusion equation, which is a modification of Fick's second law, is derived for the rate of sorption. Numerical solutions are found for the resulting nonlinear partial differential equation using finite difference techniques. A very good correspondence of the theory and data is observed for several polymer-penetrant pairs. In the case of positive deviations from Henry's law, the polymer network swells to expose more sites, increasing the sorption level synergistically. A statistical analysis of the phenomenon of clustering is used to aid in interpretation of the sorption isotherms. Fick's law is modified to include a rate equation for clustering (assuming a first order reversible reaction). The correlation of the data with theory is quite good; not only in predicting the behavior of sorption transients, but also in yielding the effective steady-state diffusion coefficients.

INTRODUCTION

When the solubility of a penetrant in a polymer is plotted versus the penetrant activity or, conveniently, its partial pressure (equal to the activity times the vapor pressure), Henry's law predicts linearity for normal sorption behavior.

$$C = K_d\, p \tag{1}$$

where C = solubility
       K_d = the Henry's law dissolution
            constant
       p = pressure

Sorption data, however, frequently display nonlinear, negative and positive deviations from Henry's law. A generalized theory is proposed to account for both kinds of non-ideal behavior [1,2].

Sorption is visualized as a process in which there are dual modes: either the penetrant molecule is normally dissolved and is free to diffuse or it is immobilized, as in a sink. It is this second process which gives rise to deviations from normal behavior. Negative deviations can be associated with site binding and low penetrant levels, where some of the penetrant molecules are immobilized on sites or in sinks in the matrix. This would be characteristic of a glassy polymer. Positive deviations can be accounted for by a different mechanism. In this case an opening of the network accompanies site binding and leads to relatively high levels of penetrant in the matrix.

NEGATIVE DEVIATIONS FROM HENRY'S LAW

Sorption

The solubility can be written as the sum of two terms:

$$C = C_d + S \tag{2}$$

where C_d represents ordinary dissolution
      S   represents sorption in the sinks,
          e.g., microcavities or holes (S_h),
          if we restrict attention momen-
          tarily to glassy polymers.

A Henry's law relationship is written for the free species (equation 3). A more complicated model must be used, however, in order to explain the sorption behavior of the trapped species. The sim-

plest nonlinear model that can be used is the Langmuir isotherm
(equation 4).

$$C_d = K_d \, p \tag{3}$$

$$S_h = S_h' \, b \, p \, / \, (\, 1 + b \, p \,) \tag{4}$$

where S_h' = the hole saturation constant
b = the hole affinity constant

At low pressures, where $bp \ll 1$, the sorption isotherm tends
to linearity.

$$C = (\, K_d + b \, S_h' \,) \, p \tag{5}$$

It is assumed that the sinks in the matrix have a finite capacity
for gas. When that capacity is saturated, they will no longer sorb
additional penetrant. In the limit of high pressures, where $bp \gg 1$,
the solubility in the holes reaches the saturation limit, S_h'.
Therefore, at high pressures, the isotherm should once again become linear.

$$C = K_d \, p + S_h' \tag{6}$$

Thus, a plot of C versus p will have two linear regions with a
connecting nonlinear region. The isotherms for methane in polystyrene (Figure 1) do indeed correspond to this sort of picture.
They are nonlinear and display a negative deviation from Henry's
law which is temperature dependent. In accord with equation 6, K_d
can be estimated from the slope of the curve at high pressures.

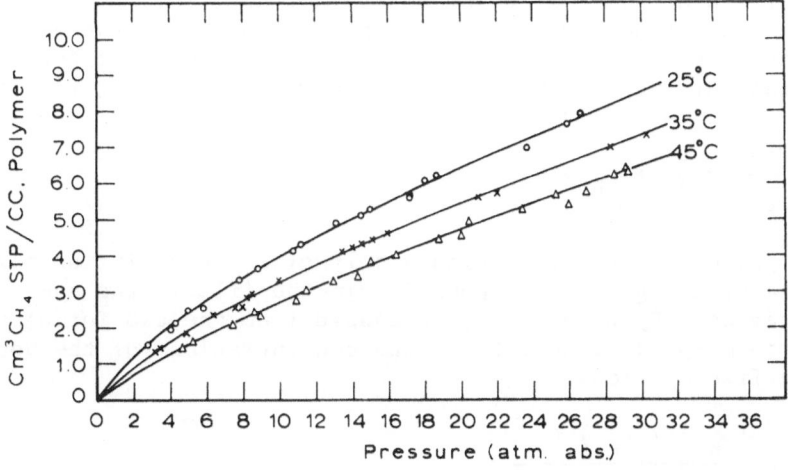

Figure 1.   Solubility of methane in oriented polystyrene [4].

Subtracting the solubility of the diffusing gas from the overall
solubility yields the gas solubility in the holes.

$$S_h = C - C_d = C - K_d \, p \qquad\qquad\qquad (7)$$

A test of this equation is to plot p/S_h versus p and to check for
linearity. Good correlation has been found for CO_2 in polyethylene terephthalate [3,1], CH_4 in polystyrene [4,5], and CO_2 in
polycarbonate [6]. Phenomenologically then, it would appear that
this kind of a description is valid. Thus, immobilization at
fixed sites in a non-swollen matrix can be used to explain negative
deviations from Henry's law. Positive deviations occur when the
polymer is swollen by the penetrant, exposing more sites and increasing the levels of the apparent solubility coefficient.

Diffusion

A diffusion model is needed in order to describe the sorption
transient of a penetrant in a polymer. Normal diffusion across a
membrane is described by Fick's second law in one-dimension.

$$D \, \frac{\partial^2 C}{\partial x^2} = \frac{\partial C}{\partial t} \qquad\qquad\qquad (8)$$

> where D = the diffusion coefficient for the
> free species.

To analyze the diffusion process for dual mode sorption, a more
complex model is necessary. A differential mass balance is performed for penetrant diffusing across a flat film. One term allows
for the rate of accumulation by simple dissolution; the other for
the fact that the film is also entrapping an amount of the penetrant and holding it. The result is a modification of Fick's law.

$$D \, \frac{\partial^2 C}{\partial x^2} = \frac{\partial}{\partial t} \left(S_h + C_d \right) \qquad\qquad\qquad (9)$$

For sorption of a gas in a glassy polymer the equilibrium partial
pressure of the gas at any position in the film is the same for
both species. By equating the pressure from the two formulations,
a relationship is found between the concentrations of the bound
and the free species.

$$S_h = \frac{(S_h' \, b \, / \, K_d) \, C_d}{1 + (b \, / \, K_d) \, C_d} \qquad\qquad\qquad (10)$$

Substituting this expression for S_h into equation 9 yields the diffusion equation for dual mode sorption.

$$D \frac{\partial^2 C_d}{\partial x^2} = \frac{\partial C_d}{\partial t} \left[1 + \frac{S_h' \, (\, b \, / \, K_d \,)}{[\, 1 + (\, b \, / \, K_d \,) \, C_d]^2} \right] \tag{11}$$

At low pressures, equation 10 reduces to a linear relationship between S_h and C_d.

$$S_h = R \, C_d \tag{12}$$

$$\text{where } R = b \, S_h' \, / \, K_d$$

The diffusion equation can then be expressed in a modified form of Fick's law, where an effective diffusivity is defined.

$$D_{eff} \frac{\partial^2 C}{\partial x^2} = \frac{\partial C}{\partial t} \tag{13}$$

$$\text{where } D_{eff} = \text{the effective diffusivity}$$

$$= D \, / \, (\, 1 + R \,) \tag{14}$$

The measured diffusion coefficient is less than the true diffusion coefficient; that is, the attainment of sorption equilibrium is slowed down by the immobilization process.

At higher pressures, however, where the sorption process is nonlinear, the partial differential equation for diffusion (equation 11) is also nonlinear; it is no longer a simple modification of Fick's law. No analytical solution has been found for this nonlinear equation. Any attempt to linearize the equation has resulted in intractable boundary conditions. Numerical solutions have been found, however, using finite difference techniques.

For normal sorption processes occuring in finite baths of penetrant, the dimensionless pressure decay (fractional equilibration), Φ, is a simple function of the square root of dimensionless time, Θ [7].

$$\Theta = D \, t \, / \, L^2 \tag{15}$$

$$\Phi = (\, p_i - p \,) \, / \, (\, p_i - p_f \,) \tag{16}$$

where 2 L = the thickness of the membrane
p_i = the initial pressure
p_f = the final pressure

For the dual sorption model, a modification of this correlation is made. The pressure decay is plotted versus the square root of Θ', which is Θ multiplied by a factor similar to the nonlinear term in the partial differential equation.

$$\Theta' = \Theta \left[\frac{1 + S_h' \, b \, / \, K_d}{(1 + b \, p)^2} \right]^{-1} \tag{17}$$

Three numerical solutions were found by varying the sorption parameters, b, S_h', and K_d over a wide range. Expressed graphically, all of the data can be correlated within a narrow band, encompassing routine calculational errors not exceeding a few percent of

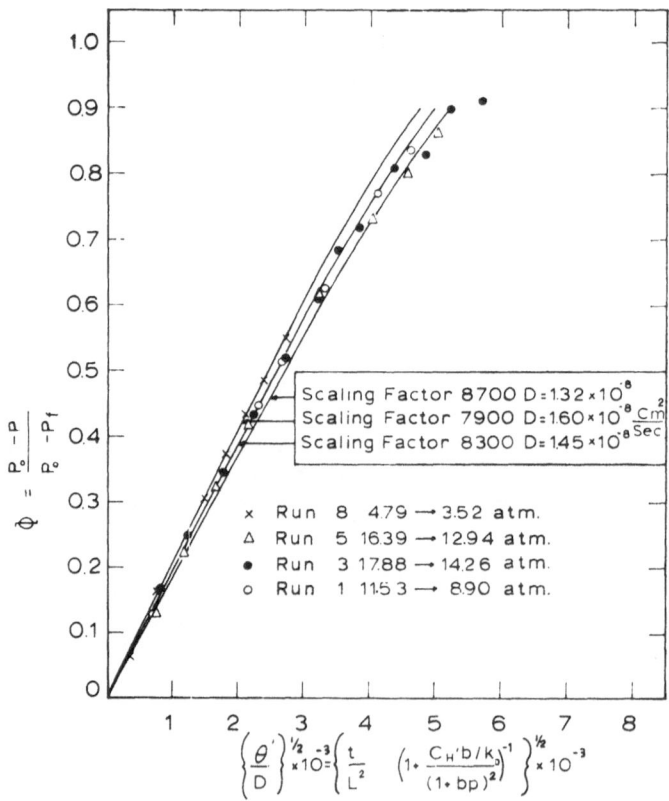

Figure 2.   Methane sorption in polystyrene at 45°C [1].

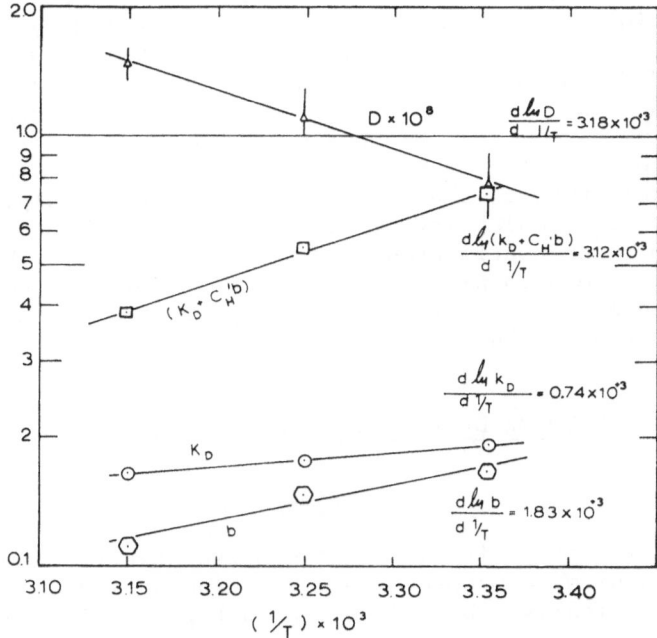

Sorption Parameters K_D, b, $(K_D + C_H'b)$ and
Diffusion Coefficients, D, vs. $\frac{1}{T}$ for Methane
in Oriented Polystyrene

Figure 3.   Temperature dependence of the diffusion coefficient and
sorption parameters for methane in polystyrene [5].

the mean of the band at any point.   Therefore, the mean of the band
was taken as the correlation.

The correlation was tested against actual pressure decay data.
The data were fitted to the generalized correlation with a scaling
factor $(1\sqrt{D})$ from which the diffusion coefficient can be found.
Diffusion coefficients found by this method for methane sorption
in polystyrene [5] fall within the narrow range of 1.3 to 1.6 x
10^{-8} cm^2/sec at 45°C (Figure 2) and between 1.1 and 1.3 x 10^{-8}
cm^2/sec at 35°C.   A very good correspondence of the theory and
data was also found for carbon dioxide in Mylar* [1] and polycarbonate [6].   Other critical tests of the correlation demonstrate
that dual sorption theory is dependable for predicting diffusion
coefficients when negative deviations from Henry's law are encountered.

*registered trademark of E. I. du Pont de Nemours and Co., Inc.

The temperature variation of the true Henry's Law coefficient
for dissolution of methane in polystyrene (Figure 3) yields a
normal enthalpy of dissolution of -1.5 kcal. This would ordinarily be found for such a process in a rubbery polymer when unaccom-
panied by the immobilization mode. Uncorrected or spuriously
large negative enthalpies can thus be explained and corrected by
accounting for this second mode of sorption; for example, for
methane in polystyrene, the overall uncorrected enthalpy of sorption is -6.2 kcal, which may be attributed chiefly to the hole
filling term which contributes -3.7 kcal. Thus, when the sorption
parameters can be uncoupled and distributed, the data seem to make
much more sense and to correlate better with what is known about
the sorption process in rubbery polymers.

POSITIVE DEVIATIONS FROM HENRY'S LAW

Sorption

In the case of positive deviations from Henry's law, the polymer network swells to expose more sites, increasing the sorption
level synergistically. Data is shown in Figure 4 for the sorption

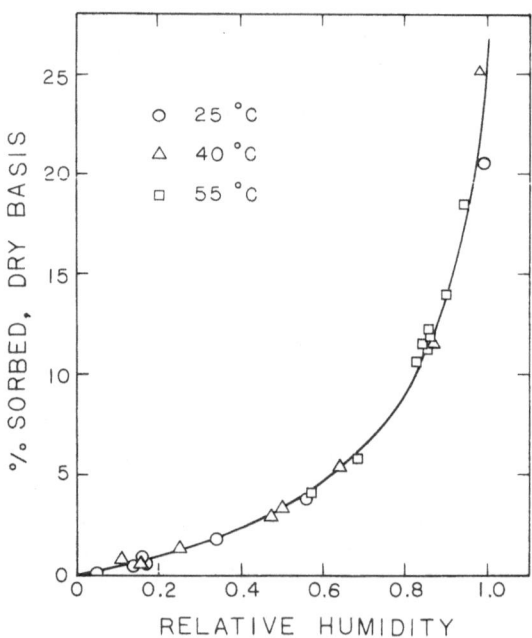

Figure 4.   Sorption of water in a polyurethane membrane. Poly-
            ethylene oxide, mol wt 350, 2% cross-linked [2].

of water in a polyurethane membrane. The curve is nonlinear and
displays a positive deviation from Henry's law. The amount of
water sorbed exceeds 10% by weight on a pure polymer basis. This
is a much higher level of sorption than in cases where negative
deviations are observed. For example, carbon dioxide sorption in
glassy polymers falls at a level far below even 1% on a pure polymer basis. The data for three different temperatures fall on the
same curve, indicating that the only heat effect accompanying the
process is the condensation of the vapor. There are evidently no
strong interactions involving the solvent and the polymer. Thus,
the driving force for sorption is entropic in kind.

In this case, because of the swelling effect, the concentration of available sites for solvent aggregation changes with the
extent of sorption. Under the random conditions corresponding to
Henry's law behavior, the sorbed penetrant molecules occur singly;
however, under the non-random conditions corresponding to the case
in point, these molecules are aggregated. A statistical mechanical
analysis of the phenomenon of clustering [8] allows the calculation
of the mean positive cluster size as a function of the volume fractions of polymer and solvent, and the activity of the solvent at
constant pressure and temperature. All that is needed for this
analysis is the sorption isotherm (the relationship between the
volume fraction of water and the water activity). The amount from
which the system departs from Henry's law is a measure of the aggregation of the solvent. A plot of the sorption of water vapor
in non-crosslinked hydroxyethylmethacrylate [2] is very nonlinear
at high values of relative humidity. Near saturation the polymer
network opens up and the water molecules aggregate. The average
size of a cluster approaches five molecules. This is a significant
deviation from Henry's law, which predicts no cluster formation at
all. (When this analysis is formally applied to the Langmuir model
used in the first section, the calculated mean cluster size is a
small negative number of order -2 which is indicative of site binding.)

Diffusion

The Langmuir model used to describe negative deviations to
Henry's law cannot be used for positive deviations because the site
density is not constant when swelling occurs; it changes with the
extent of sorption. These non-ideal effects will be included in
the activity coefficient.

In order to characterize a positive deviation, Fick's law is
once again modified for the transient case.

$$D \frac{\partial^2 C_d}{\partial x^2} = \frac{\partial C_d}{\partial t} + \frac{\partial S}{\partial t} \tag{18}$$

Because the clustering phenomenon may frequently display relatively slow kinetics, a rate equation is written postulating that clustering is a first order, reversible reaction with rate constants for the forward (λ) and backward (μ) reactions.

$$\frac{\partial S}{\partial t} = \lambda \, C_d - \mu S \tag{19}$$

$$\text{where} \quad \frac{S}{C_d} = \frac{\lambda}{\mu} = R \qquad \text{at equilibrium} \tag{20}$$

Solution of this set of equations for boundary conditions corresponding to an infinite reservoir shows that the effective diffusivity is again given by equation 14. The immobilization of the penetrant within the polymer retards the diffusional process. The correlation of the data with theory is quite good; not only in yielding the effective diffusion coefficient, but also in predicting the sigmoidal shape of the curve for a series of sorption transients for polyurethane membranes [2].

Considering the steady state case, the flux of the mobile species is given by Fick's first law.

$$J_d = -D_d \, \nabla C_d \tag{21}$$

$$\text{where} \quad \nabla C_d = \frac{\partial C_d}{\partial C} \cdot \nabla C \tag{22}$$

Substitution of equation 20 into equation 2 yields an expression for the solubility.

$$C = C_d \, (1 + R) \tag{23}$$

Note that this is the key relationship here, irrespective of the kinetic effects alluded to earlier which might accompany the transient case. Fick's law can then be written in terms of a diffusion coefficient, D.

$$J_d = -D \nabla C \tag{24}$$

$$\text{where} \quad D = D_d \, / \, (1 + R) \tag{25}$$

The flux can also be related by elementary thermodynamics to the gradient of chemical potential, $\nabla \mu$, and a constant, Ω.

$$J_d = -\Omega \nabla \mu \tag{26}$$

By equating the flux from Fick's law (equation 24) and from the thermodynamic formulation (equation 25), the diffusion coefficient can be found.

$$D = \Omega \, \frac{\nabla \mu}{\nabla C} = \Omega \left(\frac{\partial \mu}{\partial C} \right)_{T,P} \tag{27}$$

$$\text{where} \left(\frac{\partial \mu}{\partial C} \right)_{T,P} = \frac{R_g T}{C} \left[1 + \frac{\partial \ln \gamma}{\partial \ln C} \right] \tag{28}$$

The flux is found by substitution into equation 26.

$$J_d = \frac{-D_d}{1+R} \left[1 + \frac{\partial \ln \gamma}{\partial \ln C} \right]^{-1} \frac{C}{R_g T} \nabla \mu \tag{29}$$

$$\text{where} \ D_{eff} = \frac{D_d}{1+R} \left[1 + \frac{\partial \ln \gamma}{\partial \ln C} \right]^{-1} \tag{30}$$

and R_g is the gas constant.

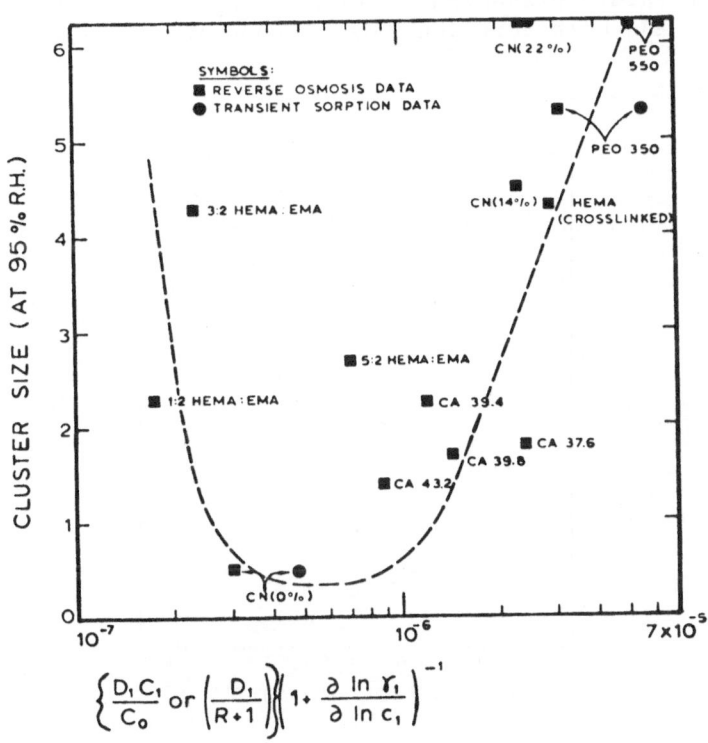

Figure 5.   Dependence of clustering size on effective diffusivity.

In this case, the true diffusion coefficient is modified by two factors: one for the second mode of sorption, which reduces it, and another for network swelling, which increases it.

    Applying the theory to experimental data, a plot of cluster size versus the effective diffusivity yields a hydrophobe-hydrophile balance for water interactions with polymers (Figure 5). Hydrophilic membranes are characterized by high diffusivities and cluster sizes, hydrophobic membranes by low diffusivities and high cluster sizes. Very good agreement is achieved between effective diffusivities extrapolated from transient sorption data at low pressures and calculated from reverse osmosis data at high pressures for the polyurethanes and the cellulosics [2].

    Three of the four major types of sorption isotherms have already been discussed: Henry's law behavior and positive and negative deviations from it. The isotherm of the biopolymer, collagen, is an example of the fourth type of curve (Figure 6). It passes from a negative to a positive deviation from Henry's law, indicating a transition from site binding at low penetrant levels to diffuse network swelling and cluster formation at high penetrant levels. The theory presented here may be applied to the limiting cases of low and high sorption levels, thus bracketing this behavior.

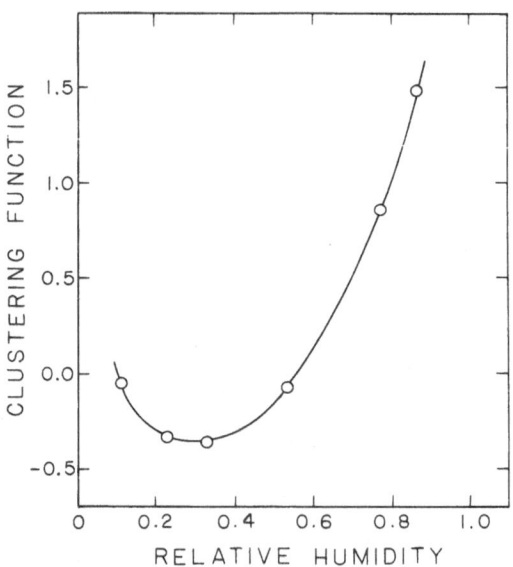

Figure 6. Water clustering of glycerine plasticized collagen (30.99% wt/wt) [9].

APPLICATION

By making use of these rather simple ideas of dual mechanisms of sorption of penetrants in polymers, a very large amount of data can be explained, including transient gas and vapor sorption processes and reverse osmosis solvent transport. Sorption is a key phenomenon underlying the barrier properties of a package. When dealing with non-ideal cases, such as in packaging applications, simple diffusion equations and Henry's law do not describe the process very accurately; more realistic theories such as the dual sorption theory proposed here should be used. For instance, this theory is applicable to studies of carbonated beverages in plastic bottles where the contents are under fairly high pressure and the plastics used have a relatively high affinity for carbon dioxide. The gas not only diffuses through the wall of the bottle, but also tends to be trapped in the walls. As a result, the pressure of the carbonated beverage tends to drop and the shelf-life is shorter than would be anticipated by considering only the diffusion process. Therefore, the generalized dual sorption theory can be applied to achieve a much more refined estimate of this effect.

REFERENCES

1. W. R. Vieth and K. J. Sladek, J. Colloid Sci., $\underline{20}$, 1014 (1965).
2. W. R. Vieth, A. S. Douglas and R. Bloch, J. Macromol. Sci., $\underline{B3}$, 737 (1969).
3. A. S. Michaels, W. R. Vieth, and J. A. Barrie, J. Appl. Phys., $\underline{34}$, 1 (1969).
4. P. M. Tam, "High Pressure Sorption in a Glassy Polymer," S. M. Thesis, M.I.T., Cambridge, Massachusetts (1965).
5. W. R. Vieth, C. S. Frangoulis, and J. A. Rionda, Jr., J. Colloid and Interface Sci., $\underline{22}$, 454 (1966).
6. J. A. Eilenberg and W. R. Vieth, "Transport and Mechanical Properties of Polycarbonate" in "Advances in Polymer Science and Engineering," K. D. Pae, D. R. Morrow, and Y. Chen, ed., Plenum Press, New York, 1972, pp. 145-162.
7. J. Crank, "The Mathematics of Diffusion," Oxford University Press, Oxford, 1956, pp. 52-56.
8. B. H. Zimm and J. L. Lundberg, J. Phys. Chem., $\underline{60}$, 425 (1956).
9. E. R. Lieberman, "Studies on the Permeation of Gas Through Collagen Films," Ph.D. Thesis, Rutgers University, New Brunswick, New Jersey, 1971, p. 137, 322.

ACKNOWLEDGMENT

The authors wish to express their gratitude to the Rutgers Packaging Science and Engineering Program for support of this work.

DIFFUSION IN GLASSY POLYMERS

T. K. Kwei and Tsuey T. Wang

Bell Laboratories

Murray Hill, New Jersey 07974

It is well known that the diffusion of organic vapors or liquids in glassy polymer often fails to obey Fick's Law. Both experimental results and theoretical explanations of the "anomaly" were summarized by Park in 1968.[1] By and large, the interpretations proposed at that time still appear to offer a sound basis, at least qualitatively, for the understanding of the complex phenomena. In the meantime, interest in this subject was further stimulated by the work of Alfrey, Gurnee and Lloyd.[2] In their elegant paper, a limiting case of non-Fickian behavior of liquid diffusion in glassy polymers was characterized by the following features. (1) As the solvent molecules penetrate the polymer, a sharp advancing boundary separates the inner glassy core from the outer swollen shell. The existence of a sharp boundary, however, is not a sufficient criterion for non-Fickian diffusion.[3] (2) At temperatures well below the glass temperature of the unswollen polymer, the distance of penetration, or the weight gain, increases linearly with time, i.e., the boundary between the glassy core and the swollen shell advances at a constant velocity, v. This limiting case was designated by Alfrey, et al., as case II diffusion.[2]

Many studies have appeared since the discovery of case II diffusion. In this brief review, some of the recent results are summarized with emphasis toward the development of a generalized diffusion equation which embodies the characteristics of both Fickian and case II diffusion. Before we discuss the generalized equation, however, some pertinent experimental observations are listed below.

RATE LAWS

When the distance of penetration of the solvent front, ℓ, and the weight gain, Δw, are measured, both Fickian and case II rate laws have been observed.[4] In the former, ℓ, and Δw are proportional to $t^{\frac{1}{2}}$, and in the latter, to t. The $t^{\frac{1}{2}}$ rate law is favored by solvents of high swelling power while case II swelling occurs with a relatively poor solvent. For intermediate solvents, the initial distance of penetration is proportional to neither $t^{\frac{1}{2}}$ nor t. Instead, it is approximately proportional to t^n (for a limited range of t) where $1.0 > n > 0.5$. The trend of decreasing solvent power appears to go hand in hand with increasing value of n.[4] Generally speaking, the initial rate of penetration increases with increasing solvent power.[4]

With a properly chosen mixture of solvents, only one advancing front was observed and the distance of penetration was given by

$$\ell = N_1 k_1 t + N_2 k_2 t^{\frac{1}{2}}$$

where N was the molar fraction of the solvent in the mixture and k the rate constant for the pure solvent.[4] The superposition of case II and Fickian diffusion[2,4] is a crucial observation which leads to the formulation of the generalized equation.

It is possible to alter the swelling characteristics by using polymers of different crosslink densities. For a given solvent, swelling approaches case II diffusion as the crosslink density of the polymer is increased.[4] When blends of polystyrene and poly(2,6 dimethyl 1.4 phenylene oxide) were used in a study of the case II diffusion of liquid n-hexane, there was a maximum in front velocity when blend composition was changed.[5]

CRACK AND CRAZING OF POLYMERS

The swelling of the outer region of the polymer exerts a stress on the remaining glassy core;[2] conversely, the core applies a restraint on the swelling of the outer shell. The stresses developed as a result of differential swelling are often sufficiently large in magnitude to cause crack and crazing of the polymer.[2,4,6-8] When a solvent of high swelling power is used and the swollen layer is relatively soft, fracture of the specimen occurs near the outer surface and small pieces of the swollen material can be seen to break away from the specimen. Where case II diffusion prevails, cracks and crazes appear at or near the advancing boundary in the direction of solvent penetration.

FICKIAN DIFFUSION

The existence of a sharp advancing solvent front is an
unfamiliar concept in Fickian diffusion. However, Crank[3]
has shown that a sharp boundary can result from a discontinuous
change of the Fickian diffusion coefficient, D, with concentration, C. The movement of such a boundary is proportional to $t^{\frac{1}{2}}$.
The experimental observations for diffusion of a good solvent in
a glassy polymer[4] are therefore in accord with Crank's predictions.
The application of Crank's solution of the diffusion problem to
the experimental data can be found in reference (9) and will not
be elaborated here. Rather, we wish to reiterate the physical
reason for the apparent discontinuity of D vs. C in the system
of interest. The discontinuity arises because the swollen polymer,
unlike the unattacked core, is not a glassy material. At the
outer swollen surface, the polymer-solvent mixture may exist in
the rubbery state because of the high concentration of the solvent
and consequently a high diffusion constant, about 10^{-7} cm^2 sec^{-1}
or larger, is expected. As solvent concentration decreases toward
the interior of the polymer, the diffusion constant also decreases.
Finally, a point may be reached where the polymer-solvent mixture
is no longer rubbery and a rapid decrease in the diffusion constant
to about 10^{-9} to 10^{-13} cm^2 sec^{-1} (typical order of magnitude of D
in glassy polymers) occurs when the solvent concentration falls
below a certain critical value. The apparent discontinuity of D
and the attendant sharp boundary are therefore consequences of the
different physical states of the swollen versus the glassy regions.

CASE II DIFFUSION

The constant velocity, v, of case II diffusion is phenomenologically the same as in simple convection. Convection in the
absence of an externally applied force must result from an
internal stress. Indeed, such an internal stress is capable of
contributing to the transport phenomenon, according to the
thermodynamics of irreversible processes.[10]

$$\underline{J} = -\Omega \left[\operatorname{grad} \mu - \frac{1}{c} \operatorname{div} \underline{S} \right] \tag{1}$$

or, in one dimensional flow,

$$J = -\Omega \left[\frac{\partial \mu}{\partial x} - \frac{1}{c} \frac{\partial S}{\partial x} \right], \quad S = S_{xx}. \tag{2}$$

In Eq. (1), $\underline{J}$ is the flux, Ω is the Onsager coefficient, μ is the
chemical potential and $\underline{S}$ is the partial stress tensor of the
penetrant. If the contribution of the internal stress is the

predominant factor, only the second term on the right-hand side of Eq. (2) needs to be considered. If one makes the assumption that the partial stress, S, is proportional to the total uptake of the penetrant, then the flux J is given by:

$$J = vc .$$ (3)

For this to occur, the system cannot be in a state of mechanical equilibrium. A detailed discussion of the partial stress is given in the last section of this paper.

GENERALIZED DIFFUSION EQUATION

Equation (1) forms the basis of our generalized diffusion equation which embodies both Fickian and internal stress contributions. In a one-dimensional flow, the generalized equation becomes:

$$\frac{\partial c}{\partial t} = \frac{\partial}{\partial x} \left[D \frac{\partial c}{\partial x} - vc \right] .$$ (4)

In order to obtain tractable results, it is convenient to assume both D and v to be constant. Solutions of the above differential equation have been obtained for diffusion in a semi-infinite solid and in a plane sheet. These solutions are available elsewhere[11],[12] and will not be reproduced here.

COMPARISON OF EXPERIMENTAL RESULTS WITH GENERALIZED DIFFUSION EQUATION

For diffusion in a semi-infinite medium the solutions for $C(x,t)$ and M_t were obtained in a closed form.[11] Excellent agreement was achieved between the diffusion data of liquid acetone in polyvinylchloride and theoretical predictions.[13] The solution for a plane sheet was compared with the successive vapor sorption data which showed sigmoidal, pseudo-Fickian and two-stage characteristics.[14] Again, the agreement was remarkably good.[12] A point of particular interest is that the value of v remains constant as the successive sorption curves change from sigmoidal shape to pseudo-Fickian to two-stage and again to pseudo-Fickian as vapor pressure increases. At the same time, the diffusion coefficient D increases by three orders of magnitude. It appears that the internal stress contribution does not alter significantly as long as the swollen region is non-rubbery.

SWELLING STRESS

The assumption that the partial stress of the penetrant be proportional to the total solvent uptake is an important step in arriving at Eq. (3). In what follows we shall demonstrate that this assumption is reasonable at least in cases where the degree of swelling is relatively small. Consider a solid sphere of the glassy polymer which is exposed at time $t > 0$ to a solvent capable of swelling the polymer. If the strains due to swelling are small and the relaxation time of the swelling stress in the polymer is either very short or very long compared with the time of experiment, we may, as a first order approximation, calculate the swelling stress on the basis of linear elasticity theory. Since the glassy core contains a negligible amount of solvent it is assumed to have the original elastic property of the polymer. In the swollen region whose property is concentration dependent, the polymer is assumed to have a uniform property corresponding to that averaged over the concentrations taken up by the swollen shell.

Let β be the volume swelling ratio for a given solvent concentration, C. Then under the rule of additivity of volume, β is related to C and the specific volume of the solvent, $\bar{v}$, by,

$$\beta = 1 + C\bar{v} \ . \tag{5}$$

For isotropic swelling the swelling ratio in any direction, γ, is given by

$$\gamma = \beta^{\frac{1}{3}} \cong 1 + \tfrac{1}{3} C\bar{v} \ . \tag{6}$$

In analogy to thermal expansion we define the linear coefficient of swelling, α, by

$$\alpha = \frac{d\gamma}{dC} = \tfrac{1}{3} \bar{v}. \tag{7}$$

Then the swelling stresses in the polymer can be found in the same manner as the thermal stresses if we treat α as the linear thermal expansion coefficient and C as the temperature.[15]

Neglecting the swelling in the glassy core and noting that because of spherical symmetry there are only three nonzero stress components, the radial component, σ_r, and two equal tangential components, σ_t, we have, from the thermal stress analogy,[15]

$$\sigma_{r_1} = \sigma_{t_1} = \frac{E_1 A_1}{1-2\nu_1} \tag{8}$$

for the glassy core and

$$\sigma_{r_2} = - \frac{2\alpha E_2}{1-\nu_2} \frac{1}{r^3} \int_R^r Cr^2 dr + \frac{E_2 A_2}{1-2\nu_2} - \frac{2E_2}{1+\nu_2} \frac{A_3}{r^3} \qquad (9)$$

$$\sigma_{t_2} = \frac{\alpha E_2}{1-\nu_2} \frac{1}{r^3} \int_R^r Cr^2 dr + \frac{E_2 A_2}{1-2\nu_2} + \frac{E_2}{1+\nu_2} \frac{A_3}{r^3} - \frac{\alpha E_2 C}{1-\nu_2} \qquad (10)$$

for the swollen shell having an inner radius R and an outer radius b. Here E is the Young's modulus, ν is the Poisson's ratio and A_1, A_2 and A_3 are integration constants.

$$A_1 = \frac{1}{\Delta} \frac{6\alpha E_2^2}{(1+\nu_2)(1-2\nu_2)R^3} \int_R^b Cr^2 dr \qquad (11)$$

$$A_2 = \frac{1}{\Delta} \frac{2\alpha E_2}{(1-\nu_2)R^3} \left(\frac{E_1}{1-2\nu_1} + \frac{2E_2}{1+\nu_2} \right) \int_R^b Cr^2 dr \qquad (12)$$

$$A_3 = - \frac{1}{\Delta} \frac{2\alpha E_2}{1-\nu_2} \left(\frac{E_1}{1-2\nu_1} - \frac{E_2}{1-2\nu_2} \right) \int_R^b Cr^2 dr \qquad (13)$$

where

$$\Delta = \frac{2E_2}{1+\nu_2} \left(\frac{E_1}{1-2\nu_1} - \frac{E_2}{1-2\nu_2} \right) + \frac{b^3}{R^3} \frac{E_2}{1-2\nu_2} \left(\frac{E_1}{1-2\nu_1} + \frac{2E_2}{1+\nu_2} \right). \qquad (14)$$

The glassy core is therefore in a state of uniform hydrostatic tension which is proportional to the solvent uptake in the entire swollen shell.

Now if the radius of the sphere is large, the transport process along any given radial direction may be treated as one dimensional flow in the early stage of diffusion in which the thickness of the swollen shell is small compared to the radius of the sphere, i.e. (b - R)/b ≪ 1. (For the one dimensional flow considered here the radial stress is the only relevant stress).

From Eqs. (9) and (11)-(14), one finds, after neglecting higher order terms in $(b - R)/b$,

$$\sigma_{r_2} = \frac{2\alpha E_2}{1-\nu_2} \frac{1}{4\pi b^3} \int_r^b c 4\pi r^2 dr$$

or

$$\sigma_{r_2} = \lambda \int_V C \, dV \qquad (15)$$

where

$$\lambda = \frac{\alpha E_2}{2\pi b^3 (1-\nu_2)}$$

is a constant and V is the volume of the swollen shell bounded by the radii r and b. The radial stress at a given r is therefore proportional to the solvent uptake in the shell bounded by the radii r and b.

It remains to relate the partial stress of the polymer to that of the solvent. In the binary system, the requirement that the sum of the two fluxes be zero leads to the following relation,

$$\text{div. } \underline{\sigma} = \text{div. } \underline{S} . \qquad (16)$$

For one dimensional diffusion along the x-axis, Eq. (16) becomes,

$$\frac{d\sigma_{xx}}{dx} = - \frac{dS_{xx}}{dx} \qquad (17)$$

or, upon integration

$$\sigma_{xx} + S_{xx} = \text{constant.} \qquad (18)$$

Since the sum $(\sigma_{xx} + S_{xx})$ is the total stress of the system, the integration constant is zero in absence of the external force. It follows from Eqs. (15) and (18) that the partial stress of the solvent in the one dimensional flow is proportional to the total uptake of the solvent.*

In the foregoing calculation, the polymer is assumed to maintain its integrity without developing any cracks or crazes in the diffusion process. If cracks or crazes are formed at the swelling front as a result of the swelling stress, convective flow through these minute channels may take place in addition to the transport mechanism described above. According to Eq. (8) the tangential stress σ_{t_1}, which is responsible for the cracking or crazing, increases rapidly as the swelling front approaches the center of the sphere (σ_{t_1} is proportional to the solvent uptake in the entire swollen shell divided by the cubic power of the radius of the glassy core). Thus in the thin film diffusion experiment where the solvent is allowed to penetrate to near the center of the film, additional channel flow may enter the transport process in the later stage of experiment if the stress becomes large enough to cause cracks or crazes. The accelerated rate of penetration of the solvent may then result, although other factors may also enter into play as the core thickness becomes small compared to that of the swollen region.

The authors wish to acknowledge Professor H. L. Frisch for many stimulating discussions during the course of preparation of this manuscript.

REFERENCES

1. G. S. Park, Chapter 5 in "Diffusion in Polymers" ed. by J. Crank and G. S. Park, Academic Press, New York, (1968).

2. T. Alfrey, Jr., E. F. Gurnee, and W. G. Lloyd, J. Polym. Sci., C12, 249 (1966).

---

* For strictly one dimensional flow along a particular radius, $\bar{r}$, Eq. (15) becomes $\sigma_{\bar{r}_2} = s \int_{\bar{r}}^{b} C d\bar{r}$, where s is a constant. If we choose an x-axis which originates at the outer boundary of the sphere and increases toward the center along $\bar{r}$, then $\frac{d\sigma_{\bar{r}_2}}{d\bar{r}} = -\frac{d\sigma_{xx}}{dx}$ so that by virtue of Eq. (17), $\frac{dS_{xx}}{dx} = \frac{d\sigma_{\bar{r}_2}}{d\bar{r}} = Cs$ or $S_{xx} = s\int_{0}^{x} Cdx$.

3. J. Crank, Trans. Faraday Soc., <u>47</u>, 450 (1951).

4. T. K. Kwei and H. M. Zupko, J. Polym. Sci., A-2, <u>7</u>, 867 (1969).

5. C. H. M. Jacques, H. B. Hopfenberg, and V. T. Stannett, Polym. Eng. Sci., <u>13</u>, 81 (1973).

6. A. S. Michaels, H. J. Bixler and H. B. Hopfenberg, J. Appl. Polym. Sci., <u>12</u>, 991 (1968).

7. B.R. Baird, H. B. Hopfenberg, and V. T. Stannett, Polym. Eng. Sci., <u>11</u>, 274 (1971).

8. C. H. M. Jacques, H. B. Hopfenberg, and V. T. Stannett, J. Appl. Polym. Sci., <u>18</u>, 223 (1974).

9. H. L. Frisch, T. T. Wang, and T. K. Kwei, J. Polym. Sci., A-2, <u>7</u>, 879 (1969).

10. R. J. Bearman, J. Chem. Phys., <u>28</u>, 662 (1958).

11. T. T. Wang, T. K. Kwei, and H. L. Frisch, J. Polym. Sci., A-2, <u>7</u>, 2019 (1969).

12. T. T. Wang and T. K. Kwei, Macromolecules, <u>6</u>, 919 (1973).

13. T. K. Kwei, T. T. Wang, and H. M. Zupko, Macromolecules, <u>5</u>, 645 (1972).

14. A. Kishimoto, H. Fujita, H. Odani, M. Kurata, and M. Tamura, J. Phys. Chem., <u>64</u>, 594 (1960).

15. S. Timoshenko and J. N. Goodier, <u>Theory of Elasticity</u>, McGraw-Hill Book Co., Inc., Third Ed., (1970), p. 452.

SUPER CASE II TRANSPORT OF ORGANIC VAPORS IN GLASSY POLYMERS

C. H. M. Jacques,* H. B. Hopfenberg and V. Stannett

Department of Chemical Engineering
North Carolina State University
Raleigh, North Carolina 27607

INTRODUCTION

Alfrey, Turner, and Lloyd have defined Fickian transport and Case II transport as the two limiting cases for the transport of an organic penetrant through an amorphous glassy polymer.[1] Fickian transport refers to the interdiffusion of penetrant and polymer described by Fick's equations. In general, Fickian diffusion is only observed in glassy systems when the penetrants are simple gases, solvents with small molecular diameters, or partial solvents at very low temperatures and penetrant activities. At temperatures and penetrant activities where partial solvents swell the polymer, transport is often controlled by a combination of polymer relaxation and Fickian diffusion mechanisms. Case II transport occurs when the sorption is entirely controlled by stress-induced relaxations taking place at a sharp boundary separating an outer swollen shell, essentially at equilibrium penetrant concentration, from an unpenetrated glassy core. Ideally, this sharp boundary moves through the polymer at a constant velocity during Case II transport.

The kinetics of the sorption process are often defined by the equation:

$$M_t = k\, t^n \tag{1}$$

---
*Present Address: Research Laboratories, General Motors Corp., Warren, Michigan 48090

where M_t is the amount of penetrant absorbed per unit area of the polymer at time t, and k and n are system parameters. For a Fickian system, n = 0.5 and k is related to the diffusion coefficient over the initial half of the sorption experiment.[2] For Case II transport, n = 1.0 and k is directly proportional to the constant velocity of the sorption boundary. When Fickian diffusion and Case II transport occur simultaneously, $0.5 \leq n \leq 1.0$.[1]

Transport kinetics controlled by a simple superposition of Fickian and Case II transport mechanisms may also be described by the equation:

$$M_t = M_I + M_{II} = k_I t \ + k_{II} t \tag{2}$$

where the Fickian and Case II contributions to the total weight gain are given by M_I and M_{II}, respectively.[1,3-5] Equation (2) has been used successfully to describe the sorption kinetics of a single penetrant[6] as well as mixed penetrants by amorphous glassy polymers.[1,7]

The sorption of normal alkane vapors by polystyrene films may obey Fickian kinetics, Case II kinetics, or a simple combination of these two limiting cases depending upon the temperature, penetrant, and penetrant activity.[8] However, at some temperatures and penetrant activities, dramatic increases in the sorption rate occur during the final half of the sorption experiment.[9,10] These portions of the sorption curve exhibit what we call 'super' Case II transport, which cannot be described by equation (2). Super Case II transport occurs when $M_t = k \ t^n$ where n>1.0.

The purpose of this paper is to present specific examples of super Case II transport for the sorption of normal hexane by polystyrene, poly(phenylene oxide) and blended films of these homopolymers. The effects of temperature, penetrant activity, blend composition, and film thickness on the kinetics of super Case II transport are discussed in terms of the mechanism of the super Case II transport process.

EXPERIMENTAL

Materials

Films of Dow grade 66U-26-32 polystyrene, General Electric grade 531-801 poly(2,6-dimethyl-1,4-phenylene oxide), and three intermediate blends of these polymers were kindly prepared by the Celanese Research Company. Polystyrene, poly(phenylene oxide) and mixtures of the resins were dissolved in trichloroethylene and filtered through Celite filter aids to remove the pigments. Films were then prepared by casting the trichloroethylene solutions on glass plates and allowing the solvent to evaporate. The films were immersed in distilled water to aid stripping and subsequently vacuum dried for 3 1/2 days at 50 - 55°C. Each of the films prepared in this manner was approximately 1 1/2 mils thick.

Some 3 mil films were prepared by gluing together two 1 1/2 mil films with an aqueous polyvinyl alcohol solution,[13] which was essentially inert to polystyrene, poly(phenylene oxide), and the blended films. After drying, the 3 mil composite contained less than 5% by weight of the polyvinyl alcohol solution. Films were prepared in this manner to minimize the differences in residual casting solvent and orientation between the 1 1/2 and 3 mil films.

Unlike most polymer pairs, polystyrene and poly(phenylene oxide) are highly compatible.[15] The blended films used in this study were clear, had a single T_g determined by DSC, and exhibited vapor and liquid kinetics and equilibria characteristic of a one-phase homogeneous material.[11,12]

Normal hexane used as a penetrant in the liquid and vapor sorption experiments was 99.5 mole percent pure as supplied by the Phillips Petroleum Company and was used as received without further purification. Dissolved gases were removed prior to vapor sorption studies by repeated freeze-thaw cycles under a vacuum of about 1×10^{-2} mmHg.

Apparatus and Procedure

The kinetics and equilibria of n-hexane vapor sorption were monitored gravimetrically using quartz helical springs supplied by Worden Quartz Products, Inc., Houston, Texas. Spring extension was calibrated as a function of weight gain to $\pm 2 \times 10^{-6}$ grams. A fixed amount of n-hexane sufficient to maintain a specified partial pressure was bled into a temperature controlled evacuated chamber. The consequent spring extension was measured as a function of time to determine the sorption kinetics. Sorption equilibria were determined from the final weight of penetrant absorbed by the PS/PPO films.

The kinetics and equilibria of liquid n-hexane sorption were monitored by placing film samples in a liquid n-hexane bath maintained at a constant temperature. After specific periods of time, each sample was removed from the bath, placed between several folds of tissue paper to remove excess n-hexane adhering to the film surface, placed in a weighing bottle, and weighed to the nearest 0.0001 gram.

RESULTS AND DISCUSSION

    The sorption of normal hexane by polystyrene, poly(phenylene oxide), and blends of these polymers occurs with predominately relaxation-controlled Case II kinetics over the initial half of experiments conducted at penetrant activities of 0.775 to 1.0 over a temperature range of 30 to 55°C.[11,12] Case II transport is characterized by equation (1) when n = 1.0. Accelerated Case II transport or 'super' Case II transport is defined by equation (1) when n>1.0. An example of super Case II transport is presented in Figure 1 for the sorption of n-hexane vapor by a 1 1/2 mil polystyrene film. The normalized amount of penetrant absorbed, M_t/M_∞ where M_∞ is the final weight of penetrant absorbed, is plotted as a function of time in this figure. Penetrant uptake is linear with time over approximately the first 400 hours of the sorption experiment; however, a pronounced acceleration in the uptake rate is observed after 400 hours. The overall shape of the sorption curve is therefore concave upward with respect to the time axis.

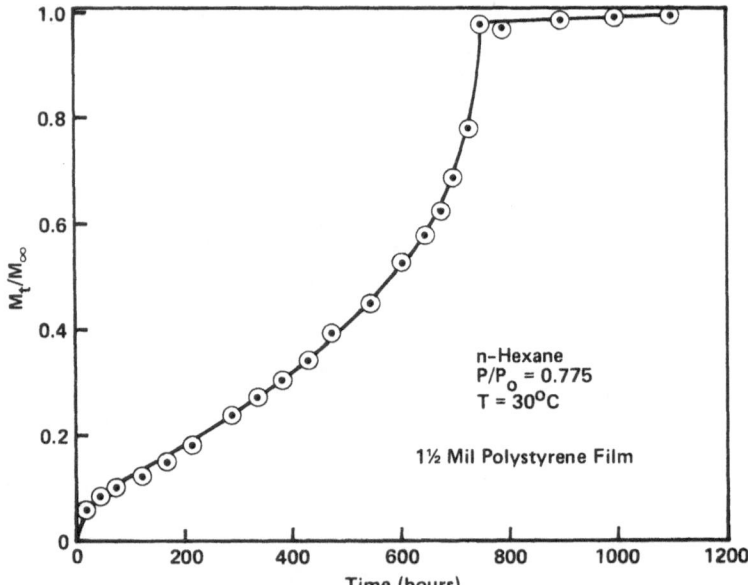

FIGURE 1. Kinetics of n-Hexane Vapor Sorption by a 1.5 Mil Polystyrene Film.

Super Case II transport is fundamentally different from most sigmoid shaped or two-staged non-Fickian transport, where weight gain is normally plotted as a function of $t^{1/2}$. These types of non-Fickian transport can, in many cases, be connected with the transition region between ideal Fickian and ideal Case II transport and are normally characterized by $0.5 \leq n \leq 1.0$.[14] Super Case II transport occurs when n>1.0. For example, a value of n = 2 to 3 is required to fit the data beyond 400 hours in Figure 1 to equation (1).

The kinetics of n-hexane sorption at 30°C and an activity of 0.775 are presented as a function of time for the entire range of blend compositions in Figure 2. Sorption kinetics are predominantly Case II for each polymer composition; however, a distinct increase in the sorption rate characteristic of super Case II transport is observed after approximately 400 hours in the 100% PS film and after about 300 hours in the 75% PS/25% PPO film.

Although accelerated Case II transport only occurs in the polystyrene rich films at this temperature and penetrant activity, super Case II transport is not strictly a function of blend composition. Sorption of n-hexane at 30°C and a penetrant activity of 0.925 by each of the blend compositions is presented in Figure 3. At this higher penetrant activity, the 100% PPO as well as the 100% PS films absorb n-hexane at a constant rate until equilibrium is reached. Only the 75% PS/25% PPO film exhibits acceleration near the end of the experiment. At this temperature and penetrant activity the 75% PS/25% PPO sample absorbs n-hexane at a slower initial rate than the 100% PS film.[12]

For a given film composition, the sorption rate may be increased by increasing either temperature or penetrant activity. The kinetics of n-hexane sorption at 40°C by 75% PS/25% PPO films are presented at four penetrant activities in Figure 4. Sorption of liquid n-hexane occurs with perfect relaxation-controlled Case II kinetics; i.e., a constant sorption rate is observed until sorption equilibrium is reached. However, as penetrant activity is reduced, the sorption rate decreases, and the shape of the sorption curve becomes more complex. At penetrant activities of 0.925 and 0.775, a slight acceleration in the sorption rate characteristic of super Case II transport occurs near the end of the experiments.

The examples of super Case II transport presented in Figures 1 through 4 have three features in common. First, accelerated super Case II kinetics are only observed near the end of the slowest sorption processes, regardless of film composition or experimental conditions. As the initial sorption rate decreases, super Case II kinetics become more pronounced. Second, a constant sorption rate is observed prior to the onset of

C. H. M. JACQUES, H. B. HOPFENBERG, AND V. STANNETT

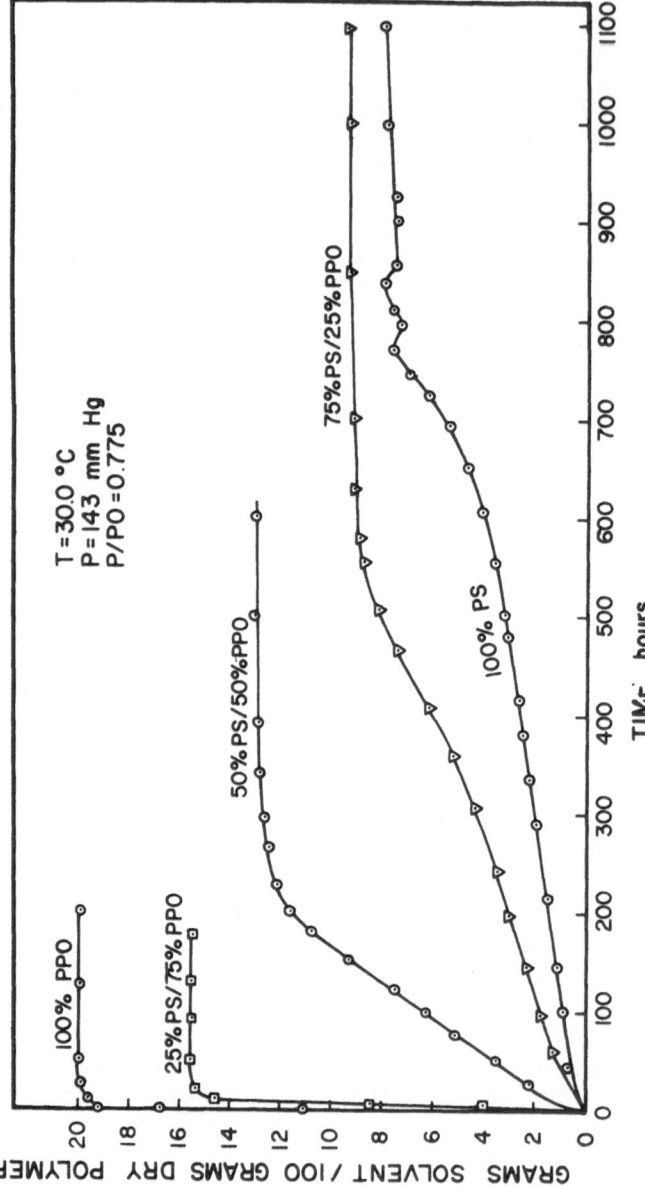

FIGURE 2. Kinetics of n-Hexane Vapor Sorption by 1.5 Mil PS/PPO Polyblend Films.

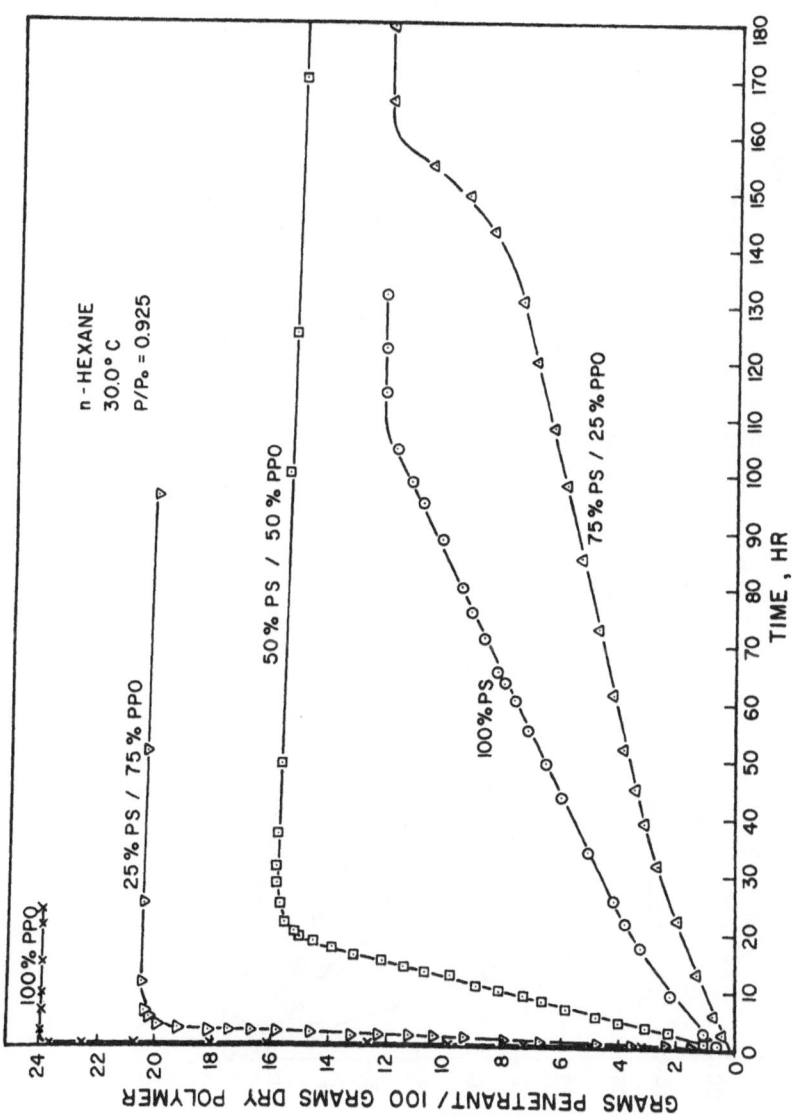

FIGURE 3. Kinetics of n-Hexane Vapor Sorption by 1.5 Mil PS/PPO Polyblend Films.

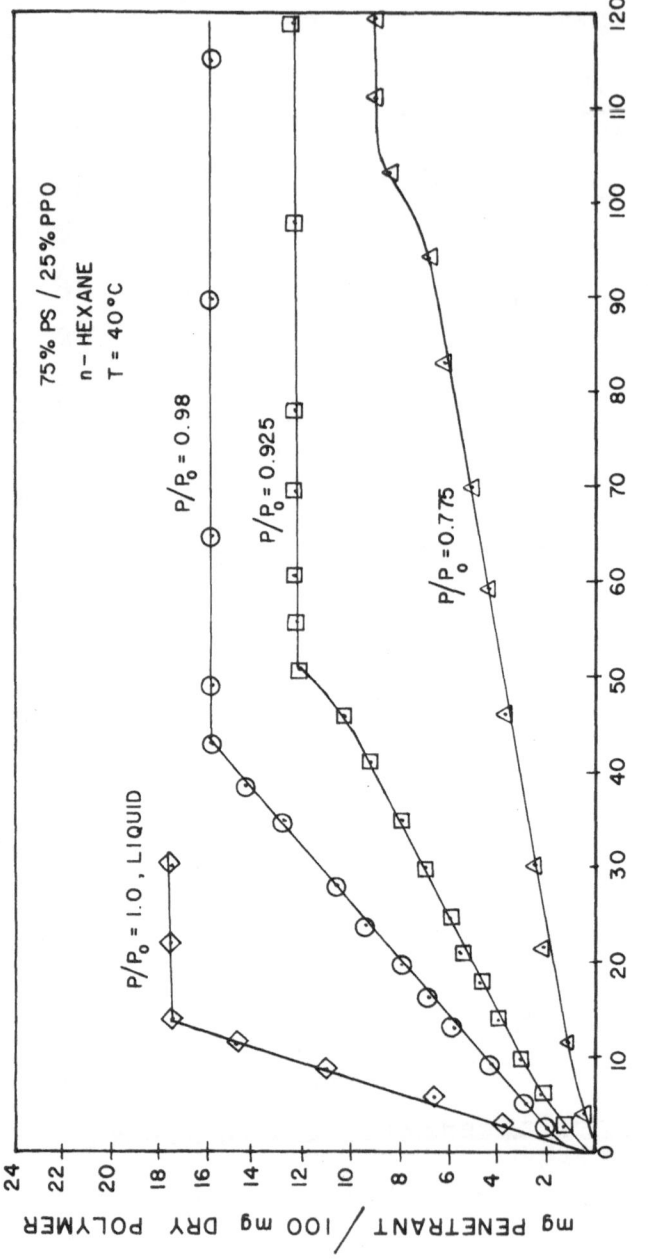

FIGURE 4. Kinetics of n-Hexane Sorption at Four Activities
by 1.5 Mil Films of 75% PS/25% PPO.

accelerated transport; i.e., predominately Case II kinetics
precede super Case II kinetics. Third, the transition from Case
II to super Case II kinetics is abrupt in most cases.

    Each of these features of super Case II transport is most
clearly seen when log weight gain is plotted as a function of
log time. A log-log plot for the sorption of n-hexane at a
constant activity of 0.775 by polystyrene films at four
temperatures is presented in Figure 5. Clearly, super Case II
transport becomes more pronounced at lower temperatures where the
initial sorption rates are slowest. At 30°C more than 70% of
the total absorption occurs with super Case II kinetics. The
abrupt transition from Case II to super Case II kinetics is also
clearly seen in this figure.

    The parameter n from equation (1) may be determined directly
from the slopes of the curves in Figure 5. At 40°C, n is
approximately 1.0, and a constant penetrant uptake is observed
until equilibrium is reached. However, at temperatures of 30 to
35°C, the value of n during the constant rate period drops to
about 0.82. The decreasing value of n represents an increasing
contribution of Fickian sorption during the constant rate period,
prior to the initiation of super Case II kinetics.

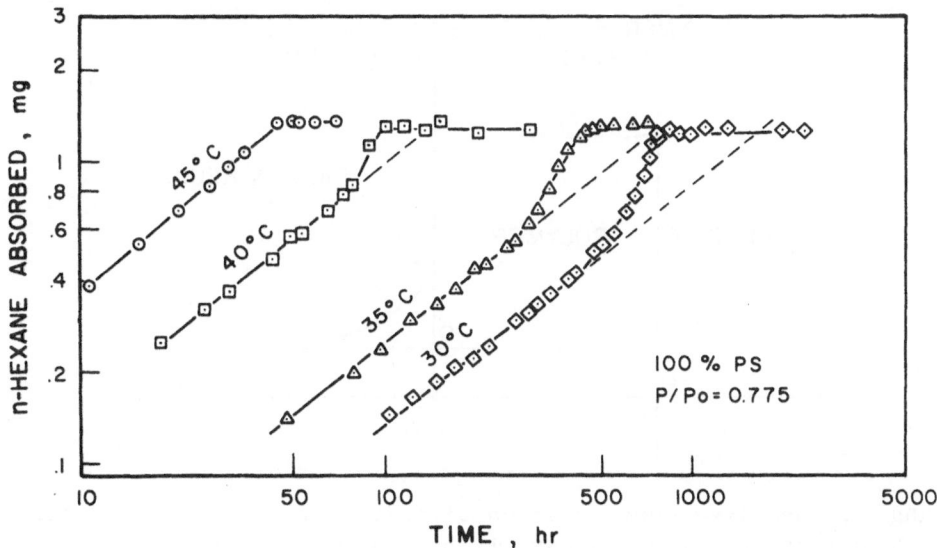

FIGURE 5. Log-Log Plot of the Kinetics of n-Hexane Sorption at
Four Temperatures by 1.5 Mil Polystyrene Films.

    The presence of super Case II kinetics during the latter
portions of slow sorption processes is believed to be a direct
result of the concentration profile established in the glassy film
during the constant rate period. Peterlin has suggested that
concurrent Fickian and Case II transport is, in some cases, a
consequence of a Fickian wave preceding the Case II sorption front
into the film.[3,4] The steady state concentration profile
predicted by Peterlin's model is presented in Figure 6. A discontinuous sorption boundary moves through the polymer at a
constant velocity. Since diffusion in the swollen region behind
this boundary is very rapid compared to the movement of the
front, the swollen region is essentially at a uniform concentration,
c_∞. The size of the Fickian wave ahead of the sorption boundary is
controlled by the rate of advance of the front. If the front
velocity is relatively rapid compared to Fickian diffusion into
the glassy core, the Fickian wave will be small; however, if the
front velocity is slow compared to diffusion, a Fickian tail will
extend far ahead of the advancing sorption boundary.

    The onset of super Case II kinetics presumably occurs when two
Fickian tails, advancing from both sides of the glassy core, meet
at the film midplane. The glassy core is under severe tension
caused by the swelling stresses imposed by the outer swollen
regions of the film. When the Fickian tails overlap, it is believed
that the concentration of penetrant in the highly stressed
central core is increased, and the relaxation processes controlling the rate of advance of the Case II sorption boundary are

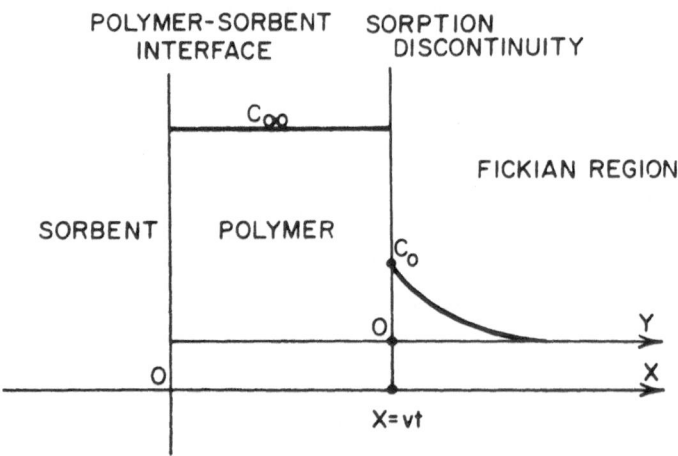

FIGURE 6. Idealized Concentration Distribution Proposed by Peterlin
for Concurrent Fickian and Case II Sorption; where c_∞ is the
concentration of the saturated polymer behind the advancing
boundary (at X = vt), and c_0 is the critical concentration of the
Fickian profile preceding this boundary.[4]

accelerated. The overall effect of an increase in the concentration of penetrant in the central core, therefore, would be to
accelerate the rate of transport near the end of the sorption
experiment.

Since a constant sorption rate is observed prior to the onset
of accelerated transport, the position of the sorption front at
the inception of super Case II kinetics may be determined. The
log of the front velocity is presented as a function of the
normalized front position at the onset of acceleration in Figure 7.
These results for n-hexane sorption over a range of temperatures
from 30 to 45°C at penetrant activities of 0.775 to 0.925 by 1 1/2
mil films of 100% PS, 75% PS/25% PPO, and 50% PS/50% PPO are
surprisingly consistent. When the front velocity is relatively
slow, the Fickian tails extend far ahead of the advancing front,
and accelerated super Case II transport begins early in the sorption
experiment. As the front velocity increases, the size of the

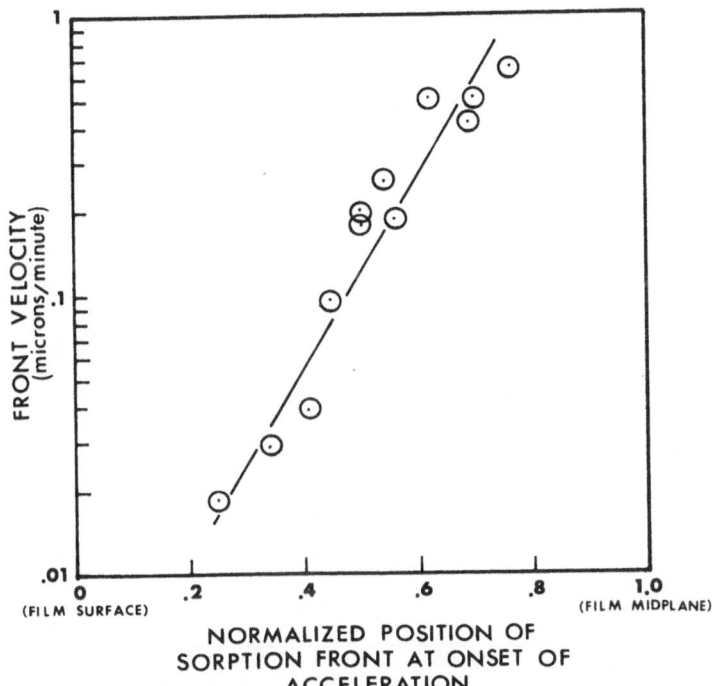

FIGURE 7. The Normalized Position of the Case II Sorption Front at
the Onset of Super Case II Transport for n-Hexane Sorption at
Various Temperatures and Activities by 1.5 Mil PS/PPO Films.

Fickian tails is reduced, and acceleration occurs later during the
experiment. For front velocities greater than about 1 micron per
minute in this system, apparently no Fickian tails develop, and
perfect Case II kinetics are observed until the end of the
experiment.

The duration of the constant rate period prior to accelerated
transport depends upon film thickness as well as from velocity.
The sorption kinetics of two polystyrene films of different
thicknesses exposed to n-hexane vapor are presented in Figure 8.
Both films show an initial, rapid penetrant uptake during the
first 10 hours of the experiment. This initial uptake is essentially Fickian in nature and may be characterized by a diffusion
coefficient of 4.7 x 10^{-12} cm^2/sec. The diffusion coefficient was
calculated from the equation:

$$D = \left(\frac{\pi l}{4}\frac{M_t}{M_\infty}\right)^2 / t \tag{3}$$

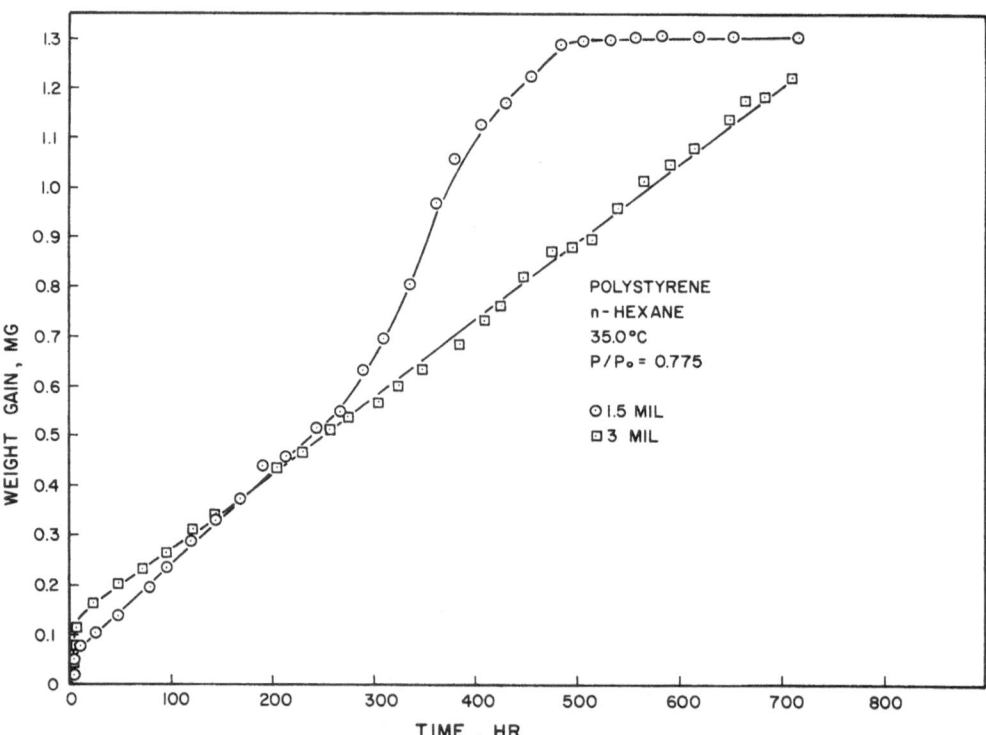

FIGURE 8. Vapor Sorption Kinetics of n-Hexane in Two Polystyrene
Films of Different Thicknesses.

where D is the mean diffusion coefficient, l is the film thickness, M_t is the weight of penetrant absorbed at time t, and M_∞ is the equilibrium penetrant uptake. Holley observed a similar diffusion coefficient of 2.8×10^{-12} cm^2/sec for the desorption of n-pentane at 35°C from cast, annealed polystyrene films.[16] Presumably, the Fickian tails are developed during the initial rapid penetrant uptake. Once established, these tails do not appreciably contribute to the overall uptake; and a constant sorption rate is observed. This constant, relaxation-controlled sorption continues for about 250 hours in the 1 1/2 mil film. At this time, the Fickian tails advancing from both sides of the film meet at the film midplane, and super Case II kinetics are initiated. However, no acceleration is observed in the 3 mil film over the same time period, since the Fickian tails do not reach the film midplane in the thicker film. A constant sorption rate is observed in the 3 mil film up until 750 hours, when the experiment was terminated.

SUMMARY

    The kinetics of n-hexane sorption by films of polystyrene, poly(phenylene oxide), and homogeneous belnds of these homopolymers were studied over a significant range of temperatures and penetrant activities. Although sorption by all blend compositions obeyed predominantly Case II or relaxation-controlled kinetics, super Case II kinetics, defined by the equation $M_t = kt^n$ where n>1.0, were observed during the latter portions of very slow experiments.

    Super Case II transport occurs when the velocity of the Case II sorption boundary is sufficiently slow so that a Fickian tail may develop ahead of the sorption discontinuity. Acceleration is initiated when Fickian tails advancing from both sides of the film meet at the film midplane. The concentration of penetrant in the highly stressed core is then increased, promoting a more rapid relaxation-controlled transport. The position of the sorption front at the onset of acceleration is inversely proportional to the front velocity and is largely independent of film composition, penetrant activity, or temperature for a given film thickness. Super Case II kinetics are initiated earlier in a film 1 1/2 mils thick than in a film 3 mils thick.

REFERENCES

1.   T. Alfrey, Jr., E. F. Gurnee, and W. O. Lloyd, J. Polym. Sci. C, 12, 249 (1966).

2.   J. Crank and C. S. Park, "Diffusion in Polymers," Academic Press, London (1968).

3.   A. Peterlin, J. Polym. Sci. B, 3, 1087 (1965).

4.   A. Peterlin, Makromol. Chem., 124, 136 (1969).

5.   H. L. Frisch, T. T. Wang, and T. K. Kwei, J. Polym. Sci. A-2, 7, 879 (1969).

6.   T. T. Wang, T. K. Kwei, and H. L. Frisch, J. Polym. Sci. A-2, 7, 2019 (1969).

7.   T. K. Kwei and H. M. Zupko, J. Polym. Sci. A-2, 7, 867, (1969).

8.   H. B. Hopfenberg, R. H. Holley, and V. T. Stannett, Polym. Eng. Sci., 9, 242, (1969).

9.   R. H. Holley, H. B. Hopfenberg, and V. Stannett, Polym. Eng. Sci., 10, 376 (1970).

10.   B. R. Baird, H. B. Hopfenberg, and V. T. Stannett, Polym. Eng. Sci., 11, 274

11.   C. H. M. Jacques, H. B. Hopfenberg, and V. Stannett, Polym. Eng. Sci., 13, 81 (1973).

12.   C. H. M. Jacques, H. B. Hopfenberg, and V. Stannett, Polym. Eng. Sci., in press.

13.   H. B. Hopfenberg, Ph.D. Thesis, Mass. Inst. Tech., Cambridge, Mass. (1964).

14.   T. K. Kwei, J. Polym. Sci. A-2, 10, 1849 (1972).

15.   E. P. Cizak (assigned to General Electric Company), U. S. Pat. 3,383,435 (May 14, 1968).

16.   R. H. Holley, Ph.D. Thesis, North Carolina State Univ., Raleigh, N. C. (1969).

CORRELATIONS BETWEEN ENTROPIC AND ENERGETIC PARAMETERS FOR DIFFUSION IN POLYMERS

R. McGregor

School of Textiles, N. C. State University

Box 5006, Raleigh, North Carolina   27607

INTRODUCTION

Although the existence of linear correlations between the logarithms of the pre-exponential factors (Log D_O) and the activation energies (E_D) for the diffusion of simple gases in elastomers has long been recognized [1, 2], there has not been any detailed analysis of the validity of the correlations reported in the literature.

This paper presents a correlation analysis of the data in Stannett's comprehensive tabulation of information from the literature on the diffusion of simple gases in polymers, in chapter 2 of reference [2].

THEORETICAL

Alternative Coordinate Systems for Correlation Analysis

Data on the diffusion of simple gases in polymers often follow the Arrhenius equation; over sufficiently restricted temperature ranges [2]

$$D = D_O \, e^{-E_D/_{RT}} \tag{1}$$

The pre-exponential factor D_O includes entropic terms, whereas the activation energy E_D is an energetic parameter [2]. The logarithmic form of equation (1) may be written as

$$\text{Log } D_O = \text{Log } D + {E_D/_{2.303RT}} \tag{2}$$

If the range in Log D is small by comparison with the range in E_D, then an apparent correlation may be generated between Log D_O and E_D even when no real correlation exists [1]. This observation has apparently never been followed up in detail, on the basis of Exner's analysis of enthalpy -entropy relationships [3, 7].

Suppose that equation (1) is valid over the temperature range $T_1 > T > T_2$. Let D_1 be the diffusion coefficient at $T = T_1$, and D_2 be the diffusion coefficient at $T = T_2$. Then [3]

$$E_D = \frac{2.303RT_1T_2}{T_1-T_2} \left[\text{Log } D_1 - \text{Log } D_2 \right] \tag{3}$$

and

$$\text{Log } D_O = \frac{T_1}{T_1-T_2} \left[\text{Log } D_1 - \left(\frac{T_2}{T_1} \right) \text{Log } D_2 \right] \tag{4}$$

The data in figure (1) for diffusion in polyethyl-methacrylate [4] appear to conform to the linear "entropy-enthalpy" correlation

$$(\text{Log } D_O)_i = A + B \, (E_D)_i \tag{5}$$

where **i** denotes the i'th gas and A and B are constants.

However, equation (5) requires that [3]

$$(\text{Log } D)_i = A + \beta \, (E_D)_i \tag{6}$$

where D is measured at temperature T and

$$\beta = \frac{2.303 \, RBT - 1}{2.303 \, RT} \tag{7}$$

so that

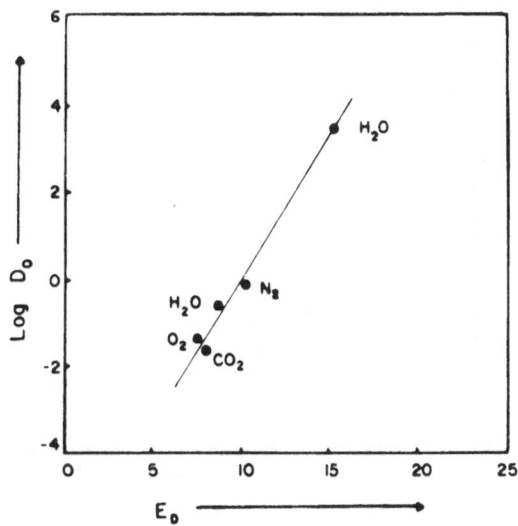

Figure 1. An Apparent Correlation Between Log D_O and E_D.

$$B = \frac{1 + 2.303 \ R\beta T}{2.303 \ RT} \qquad (8)$$

When data in figure (1) are replotted according to equation (6), the linear correlation is no longer so convincing (figure 2). Equation (5) also requires that [3]

$$(Log \ D_1 + Log \ D_2)_i = 2 \ A + \beta^* \ (E_D)_i \qquad (9)$$

in which

$$\beta^* = \frac{4.606 \ RB \ T_1 T_2 - (T_1 + T_2)}{2.303 \ RT_1 T_2} \qquad (10)$$

and

$$B = \frac{2.303 \ R\beta^* T_1 T_2 + (T_1 + T_2)}{4.606 \ RT_1 T_2} \qquad (11)$$

A representation of the data according to equation (9) also suggests that the linear correlation in these data is in fact rather poor (figure 3).

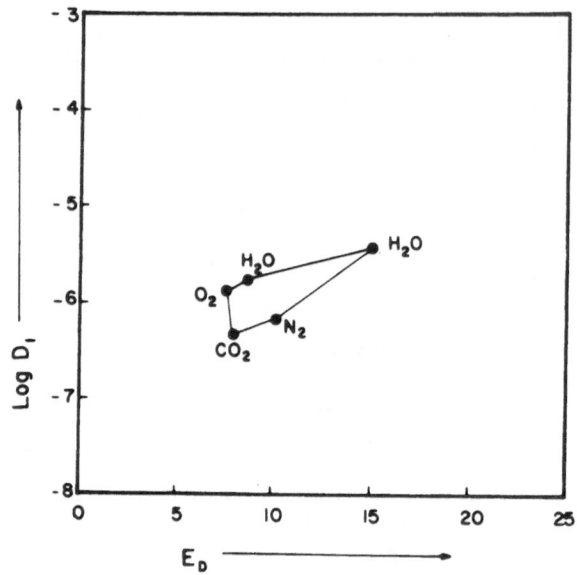

Figure 2.   Transformation of Figure 1.

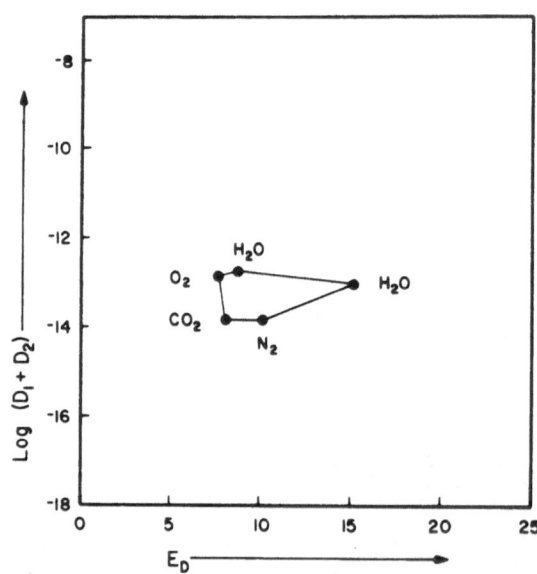

Figure 3.   Transformation of Figure 1.

Equation (5) also requires that [5]

$$(Log \; D_1)_i = \alpha + \nu (Log \; D_2)_i \qquad (12)$$

where

$$\alpha = \frac{A(T_2 - T_1)}{2.303 \; RBT_1 T_2 - T_1} \qquad (13)$$

$$\nu = \frac{2.303 \; RBT_1 T_2 - T_2}{2.303 \; RBT_1 T_2 - T_1} \qquad (14)$$

$$A = \alpha/(1-\nu) \qquad (15)$$

and

$$B = \frac{T_2 - \nu T_1}{2.303 \; RT_1 T_2 \; (1-\nu)} \qquad (16)$$

The data from figure (1) are plotted according to equation (12) in figure (4) and the degree of linear correlation is again shown to be low.

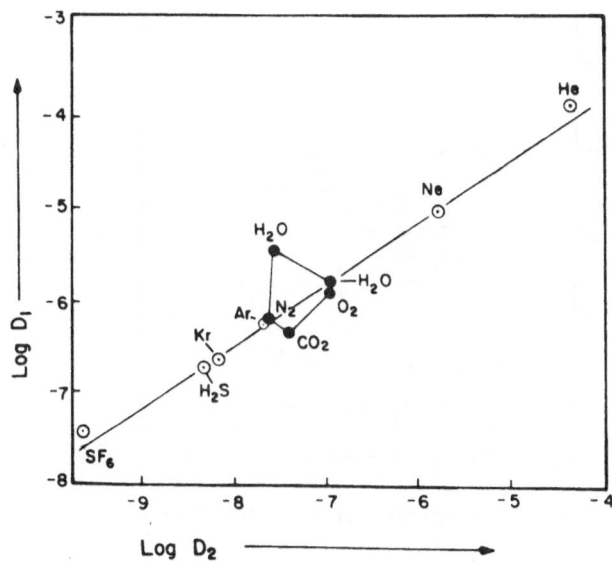

Figure 4.   Transformation of Figure 1.

However, figure (4) contains additional data points [4]
which together establish a good linear correlation. This
linear correlation persists in alternative representations
of the data according to equation (9) (figure (5)), equation (6) (figure (6)), and equation (5) (figure (7)).

The distinction between this valid correlation in
figure (1) is very clear in figures (4) - (6), but is far
from obvious in figure (7). The broken line in figure (7)
has the so-called "error slope" [3, 7] and is generated
by the lack of correlation in the data from figure (1)
(see equations (44) - (47)).

This illustrates some of the problems inherent in
the use of the Log D_O, E_D coordinate system for establishing the validity of such correlations [3].

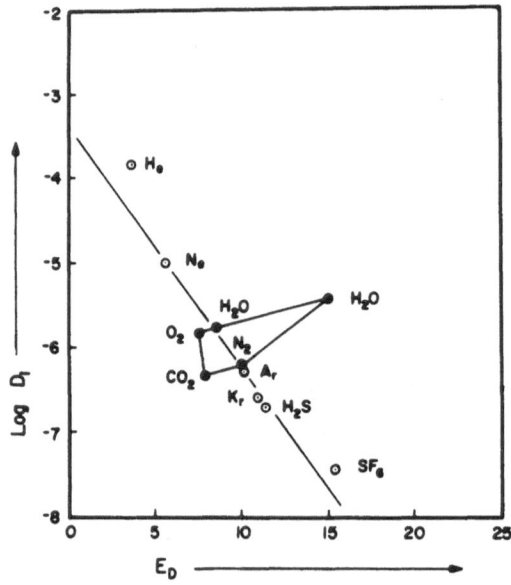

Figure 5. Transformation of Figure 4.

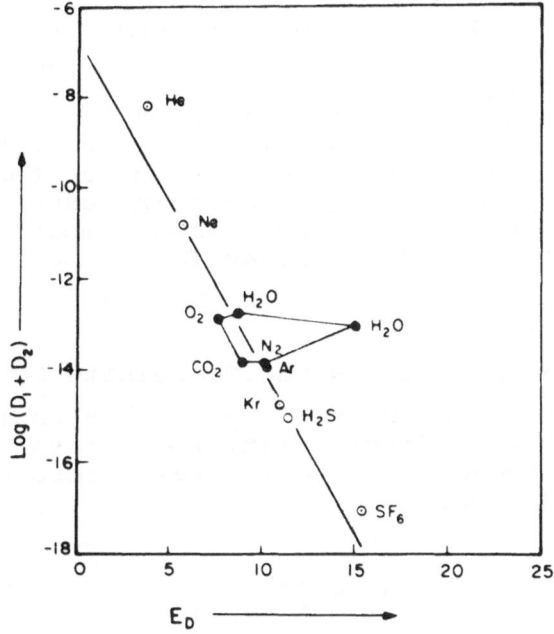

Figure 6.   Transformation of Figure 4.

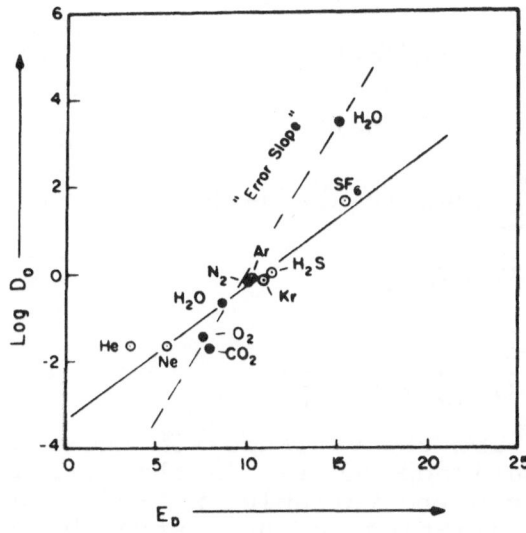

Figure 7.   Transformation of Figure 4.

Figures (8) - (11) are for diffusion in rubbery polyvinylacetate [5] and illustrate a genuine linear correlation between Log D_O and E_D: this linearity persists in
each of the alternative transformations of the coordinates.
When the degree of linear correlation is very high, there
are no strong grounds for preferring one of these coordinate systems over another in representing experimental data.
The choice of coordinate system becomes important when the
degree of correlation is lower, as for the data in figures
(1) - (4).

Correlation Analysis in the Coordinate Systems

The Log D_1, Log D_2 Coordinates. This coordinate system provides the most direct test of the data, and is one
in which the influence of experimental errors can be most
clearly isolated.   Let

$$x = \text{Log } D_2$$
$$\text{for } T_1 > T_2 \qquad (17)$$
$$y = \text{Log } D_1$$

and suppose we have n data points represented by the
coordinates

$$(x_i, y_i); \quad i = 1, 2 \ldots\ldots n \qquad (18)$$

We use the definitions [6]

$$s_x^2 = \sum (x-\bar{x})(x-\bar{x})/(n-1) \qquad (19)$$

$$s_y^2 = \sum (y-\bar{y})(y-\bar{y})/(n-1) \qquad (20)$$

$$\text{cov}(xy) = \sum (x-\bar{x})(y-\bar{y})/(n-1) \qquad (21)$$

$$r_{xy} = \text{cov}(xy)/s_x s_y$$

$$= \frac{\sum (x-\bar{x})(y-\bar{y})}{\left(\sum (x-\bar{x})^2 \sum (y-\bar{y})^2\right)^{\frac{1}{2}}} \qquad (22)$$

where s_x^2 and s_y^2 are the variances in the coordinates, cov
(xy) is the covariance and r_{xy} is the correlation coefficient.   If $r_{xy} = 0$ the variables x_i and y_i are independent.
If $r_{xy} = 1$ the variables define a straight diagonal line
and their relationship is perfect.

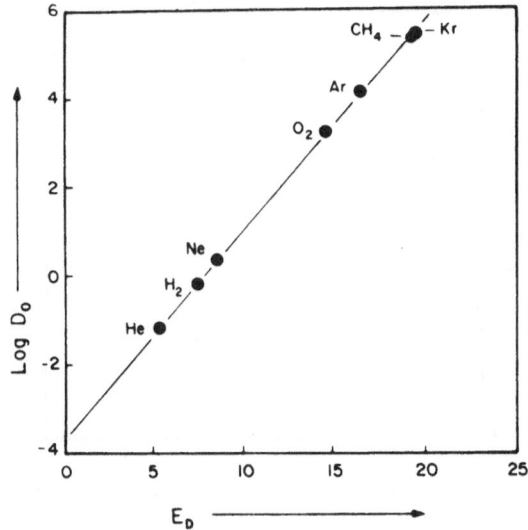

Figure 8. Diffusion in Rubbery Polyvinylacetate.

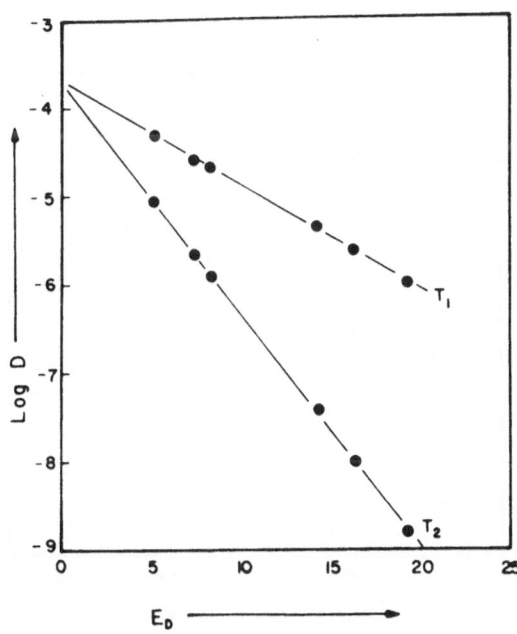

Figure 9. Diffusion in Rubbery Polyvinylacetate.

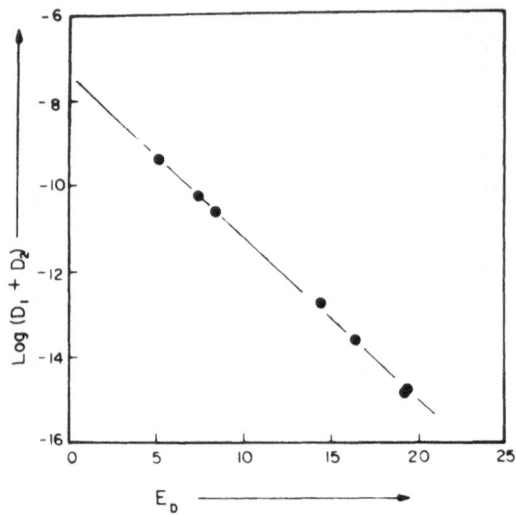

Figure 10. Diffusion in Rubbery Polyvinylacetate.

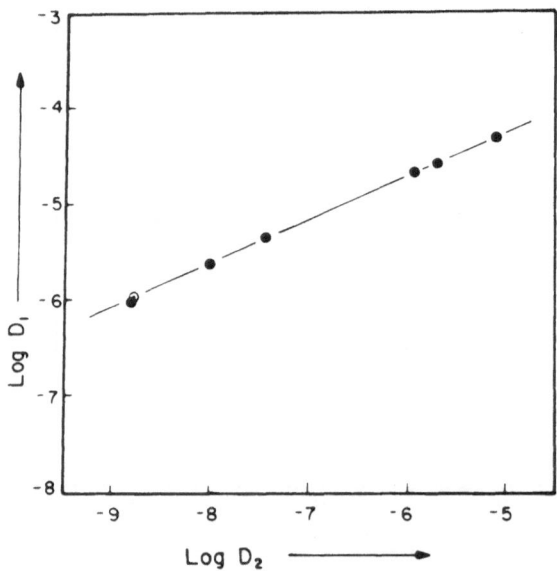

Figure 11. Diffusion in Rubbery Polyvinylacetate.

The correlation line

$$y = \alpha + \gamma \cdot x \tag{23}$$

passes through the point $(\bar{x}, \bar{y})$: $\bar{x}$ is the arithmetic mean of the x_i and $\bar{y}$ is the arithmetic mean of the y_i. Consequently

$$\alpha = \bar{y} - \gamma \cdot \bar{x} \tag{24}$$

and [7]

$$\gamma = \frac{\text{cov}(xy)}{s_x{}^2} = r_{xy} \frac{s_y}{s_x}$$

$$= \frac{\sum (x-\bar{x})(y-\bar{y})}{\sum (x-\bar{x})^2} \tag{25}$$

If this coordinate system is chosen as the primary system, then the linear correlations in the alternative coordinate systems are to be determined by transforming the correlation line of equations (23) – (25) in accordance with equations (6) – (16). It may be noted that the point $\bar{x}$, $\bar{y}$ in the Log D_1, Log D_2 coordinates corresponds to the points defined by the coordinate arithmetic mean values in the other coordinate systems. On this basis, the following results are obtained [3]:

$$\beta(T_1) = \frac{\gamma (T_2 - T_1)}{2.303 \ RT_1T_2 \ (1-\gamma)}$$

$$= \frac{(T_2 - T_1) \ r_{xy} \ s_y}{2.303 \ RT_1T_2 \ (s_x - r_{xy}s_y)} \tag{26}$$

$$\beta(T_2) = \frac{(T_2 - T_1)}{2.303 \ RT_1T_2 \ (1-\gamma)}$$

$$= \frac{(T_2 - T_1) \ s_x}{2.303 \ RT_1T_2 \ (s_x - r_{xy}s_y)} \tag{27}$$

$$\beta* = \frac{(T_2 - T_1)(1 + \gamma)}{2.303 \ RT_1T_2 \ (1-\gamma)}$$

$$= \frac{(T_2 - T_1)(s_x + r_{xy}s_y)}{2.303 \ RT_1T_2 (s_x - r_{xy}s_y)} \tag{28}$$

$$B = \frac{T_2 - \nu T_1}{2.303 \; RT_1 T_2 \; (1-\nu)}$$

$$= \frac{T_2 s_x - T_1 s_y}{2.303 \; RT_1 T_2 \; (s_x - s_y)} \tag{29}$$

$$A = \alpha / (1-\nu)$$

$$= \frac{s_x \bar{y} - r_{xy} s_y \; \bar{x}}{s_x - r_{xy} s_y} \tag{30}$$

Some interesting results emerge when these relationships are compared with those to be deduced from correlation analyses in the alternative coordinate systems [3].

The Log D, E_D Coordinates. Correlation analysis in these coordinates gives

$$\beta(T_1) = \frac{(T_1 - T_2)(s_y^2 - r_{xy} s_x s_y)}{2.303 \; RT_1 T_2 (s_y^2 - 2r_{xy} s_x s_y + s_x^2)} \tag{31}$$

with

$$r(T_1) = \frac{s_y^2 - r_{xy} s_x s_y}{\left[s_y^2 (s_y^2 - 2r_{xy} s_x s_y + s_x^2) \right]^{\frac{1}{2}}} \tag{32}$$

and

$$\beta(T_2) = \frac{(T_1 - T_2)(r_{xy} s_x s_y - s_x^2)}{2.303 \; RT_1 T_2 (s_y^2 - 2r_{xy} s_x s_y + s_x^2)} \tag{33}$$

with

$$r(T_2) = \frac{r_{xy} s_x s_y - s_x^2}{\left[s_x^2 (s_y^2 - 2r_{xy} s_x s_y + s_x^2) \right]^{\frac{1}{2}}} \tag{34}$$

The correlation coefficients $r(T_1)$ and $r(T_2)$ are not the same as r_{xy}. If $r_{xy} = 0$, and there is no correlation in the log D_1, Log D_2 coordinate system, we nevertheless find that

$$\beta(T_1) = \frac{(T_1 - T_2)\ s_y^2}{2.303\ RT_1T_2\ (s_y^2 + s_x^2)} \tag{35}$$

with

$$r(T_1) = \frac{s_y}{(s_y^2 + s_x^2)^{\frac{1}{2}}} \tag{36}$$

and

$$\beta(T_2) = \frac{(T_2 - T_1)\ s_x^2}{2.303\ RT_1T_2\ (s_y^2 + s_x^2)} \tag{37}$$

with

$$r(T_2) = \frac{-s_x}{(s_y^2 + s_x^2)^{\frac{1}{2}}} \tag{38}$$

which implies that some element of linear correlation is generated merely by the process of transformation to the Log D, E_D coordinate system. This coordinate system is therefore inferior to the Log D_1, Log D_2 coordinate system as a means of detecting genuine linear correlations.

The (Log D_1 + Log D_2), E_D Coordinates. Here we have:

$$\beta* = \frac{(T_1 - T_2)\ (s_y^2 - s_x^2)}{2.303\ RT_1T_2\ (s_y^2 - 2r_{xy}s_x s_y + s_x^2)} \tag{39}$$

and

$$r(\beta*) = \frac{(s_y^2 - s_x^2)}{\left[(s_y^2 - 2r_{xy}s_x s_y + s_x^2)(s_y^2 + 2r_{xy}s_x s_y + s_x^2)\right]^{\frac{1}{2}}} \tag{40}$$

If $r_{xy} = 0$, then

$$\beta* = \frac{(T_1 - T_2)(s_y^2 - s_x^2)}{2.303\ RT_1T_2\ (s_y^2 + s_x^2)} \tag{41}$$

and

$$r(\beta*) = \frac{s_y^2 - s_x^2}{(s_y^2 + s_x^2)} \tag{42}$$

and is not zero unless $s_x = s_y$. However, $r(\beta^*)$ is likely to be smaller than either $r(T_1^y)$ or $r(T_2)$ and so this coordinate system is preferable to the Log D, E_D coordinate systems.

The Log D_o, E_D Coordinates. In these coordinates

$$B = \frac{T_1 s_y^2 + T_2 s_x^2 - (T_2 + T_1) r_{xy} s_x s_y}{2.303\ RT_1 T_2 \{ s_y^2 - 2r_{xy} s_x s_y + s_x^2 \}} \tag{43}$$

with

$$r(B) = \frac{(T_1 s_y^2 - (T_1 + T_2) r_{xy} s_x s_y + T_2 s_x^2)}{\{ (T_1^2 s_y^2 - 2T_1 T_2 r_{xy} s_x s_y + T_2^2 s_x^2) \cdot (s_y^2 - 2r_{xy} s_x s_y + s_x^2) \}^{\frac{1}{2}}} \tag{44}$$

If $r_{xy} = 0$, then

$$B = \frac{T_1 s_y^2 + T_2 s_x^2}{2.303\ RT_1 T_2 (s_y^2 + s_x^2)} \tag{45}$$

$$r(B) = \frac{T_1 s_y^2 + T_2 s_x^2}{\{ (T_1^2 s_y^2 + T_2^2 s_x^2)(s_y^2 + s_x^2) \}^{\frac{1}{2}}} \tag{46}$$

so that in general $r(B)$ is not zero.

For systems in which $s_x = s_y$

$$r(B) = \frac{(T_1 + T_2)}{\{ 2(T_1^2 + T_2^2) \}^{\frac{1}{2}}} \tag{47}$$

and

$$B = \frac{(T_1 + T_2)}{4.606\ RT_1 T_2} \tag{48}$$

In most experimental work the ratio T_2/T_1 is unlikely
to be less than 0.8, and for $T_2/T_1 > 0.8$ we find that $r(B)$
> 0.994 for $r_{xy} = 0$. This is the essential origin of the
spurious correlation shown in figure (1). The transformation to the Log D_o, E_D coordinates itself introduces a high
degree of correlation into the data [3]. This particular
coordinate system is the worst of the alternatives available for testing the validity of an "entropy-enthalpy"
correlation [3]. The tendency of random scatter in the
data to propagate along lines of the "error slope" in the
Log D_o, E_D coordinates tends to make such scattered
points merge with the genuinely correlated points, as in
figure (7), so that real deviations from a truly linear
correlation tend to be hidden in these coordinates. When
$r_{xy} = 1$, then equations (26) - (30) become equivalent to
the corresponding equations in the series (31) - (43).
When however $r_{xy} \neq 1$, different results will be obtained
from a correlation analysis, depending on which set of
coordinates is chosen for the analysis. In order of
preference, the coordinates are:

     1. Log D_1, Log D_2

     2. (Log D_1 + Log D_2), E_D

     3. Log D, E_D

     4. Log D_o, E_D

ANALYSIS OF DATA

Correlations for Individual Polymers

Table 1 summarises the results of correlation analyses
in the Log D_1, Log D_2 coordinates, and in the Log D_o, E_D
coordinates, for individual polymers, based on the data
in Table 2 of chapter 2 in reference [2]. For this analysis, $T_1 = 370°$ and $T_2 = 298°$. The linear relationships
between Log D and $1/T$ in the original data have been
extrapolated where necessary, so that the data can be
compared on a common basis: the reference temperatures T_1
and T_2 are used to define the linear relationships in
the original data and do not in all cases correspond to
actual experimental temperatures. The analyses were made
only for those polymers with $n > 4$, and the results are
presented in order of increasing γ.

Table I. Correlation Parameters

Polymer	Correlation of Log D_1 and Log D_2			Correlation of Log D_0 and E_D		
	α	γ	r_{xy}	A	B	r(B)
Polyvinyl acetate (rubbery)	-2.01	0.454	0.999	-3.68	0.471	0.999
74/26 Isoprene-acrylonitrile	-1.96	0.470	0.977	-3.80	0.473	0.994
Butyl rubber	-2.09	0.476	0.892	-4.46	0.508	0.980
Polychloroprene (neoprene)	-1.70	0.514	0.987	-3.57	0.447	0.994
87/13 Vinyl chloride -acetate (Vinylite VYHH)	-1.85	0.517	0.915	-4.50	0.491	0.974
68/32 Butadiene- acrylonitrile (Hycar OR25)	-1.58	0.527	0.998	-3.36	0.432	0.999
61/39 Butadiene-acrylonitrile Hycar OR15)	-1.59	0.530	0.996	-3.41	0.432	0.998
74/26 Isoprene-methacrylonitrile	-1.53	0.543	0.979	-3.53	0.438	0.986
Polyethylene (density 0.914)	-1.45	0.562	0.987	-3.43	0.420	0.988
Polyesteramide diisocyanate (Vulcaprene A)	-1.30	0.574	0.990	-3.17	0.409	0.989
Butadiene-acrylonitrile (Perbunan)	-1.29	0.581	0.997	-3.13	0.396	0.995

Polymer	Correlation of Log D_1 and Log D_2			Correlation of Log D_0 and E_D		
	α	γ	r_{xy}	A	B	r(B)
Natural rubber	-1.29	0.582	0.977	-3.30	0.420	0.975
Polydimethyl butadiene (methyl rubber)	-1.15	0.590	0.996	-2.86	0.390	0.993
80/20 Butadiene-acrylonitrile (Perbunan 18)	-1.28	0.597	0.992	-3.26	0.390	0.986
Polyvinyl acetate (glassy)	-1.83	0.528	0.957	-4.22	0.461	0.980
Excluding data on CH_4	-1.33	0.613	0.994	-3.55	0.374	0.985
Polystyrene	-1.37	0.638	0.999	-3.79	0.339	0.999
Polyethylene (density 0.964)	-1.14	0.651	0.989	-3.53	0.353	0.957
Hydrogenated polybutadiene (Hydropol)	-0.744	0.663	0.962	-3.12	0.410	0.914
Polyethylene terephthalate (rubbery-crystalline)	-1.02	0.671	0.999	-3.13	0.300	0.993
Polybutadiene	-0.899	0.676	0.999	-2.78	0.293	0.996
Polyethylmethacrylate	-0.993	0.675	0.964	-4.10	0.400	0.904

Polymer	Correlation of Log D_1 and Log D_2			Correlation of Log D_0 and E_D		
	α	γ	r_{xy}	A	B	r(B)
Polyethylmethacrylate Excluding data on H_2O II	-0.986	0.685	0.993	-3.39	0.306	0.942
Gutta Percha	-0.550	0.691	0.998	-1.85	0.277	0.977
Polyethylene terephthalate (glassy crystalline)	-0.892	0.723	0.998	-3.37	0.231	0.936
Polycarbonate (Lexan)	-0.753	0.768	0.981	-4.56 / -3.25+	0.304 / 0.116+	0.730
Silicone Rubber (10% filler by wt., vulcanised	-0.464	0.826	0.985	-3.48 / -2.67+	0.242 / -0.091+	0.513

+ calculated from α, γ according to equations (29) and (30).

Each entry in Table I corresponds to correlation analyses of data for a series of n different simple gases diffusing <u>in a given polymer</u>: in this situation the degree of linear correlation is generally high in both coordinate systems. For the $Log D_1$, $Log D_2$ coordinates the average correlation coefficient is $r_{xy} = 0.982$ and the range is $0.999 > r_{xy} > 0.892$. In the $Log D_o$, E_D coordinates the average correlation coefficient is $r(B) = 0.953$ and the range is $0.999 > r(B) > 0.513$.

Correlations for Groups of Polymers

A correlation analysis on all of the data in Table 2 of reference ⌊2⌋, combined as a single group of 148 data points in the $Log D_1$, $Log D_2$ coordinates, gave the result

$$\alpha = -0.922; \quad \gamma = 0.659; \quad r_{xy} = 0.942$$

and these values of α and γ should, on the basis of equations (29) and (30), correspond to A = -2.70 and B = 0.313. A correlation analysis in the $log D_o$, E coordinates however gave (n = 148)

$$A = -3.92; \quad B = 0.448; \quad r(B) = 0.916$$

There is a genuine element of linear correlation in the data as a whole, but analyses in the different coordinate systems lead to different results. The lower degree of correlation for the grouped data suggests real differences between the behavior of the different polymers.

Barrer [1] has published interesting classifications of diffusion processes in different materials, on the basis of the numerical values, and their distributions, of parameters such as $Log D_o$. An inspection of the distributions of the numerical values of the parameters $Log D_o$; E; A; B; $Log D_1$; $Log D_2$; α and γ was therefore made, based on the data in Table I. With one exception, the distributions were all somewhat skewed, and showed only one obvious maximum. The parameter γ however, could be seen to be <u>bimodal</u>, as in figure (12), with a boundary between the two populations of γ-values near $\gamma = 0.6$. On this basis the data were classified into two groups, according to whether $\gamma < 0.6$ or $\gamma > 0.6$.

The group with $\gamma < 0.6$ included 14 polymers, primarily rubbers. The data for butyl rubber and for Vinylite VYHH were characterised by low values of r_{xy}, so that

further analysis was restricted to the remaining 12
polymers. A correlation analysis in the Log D_1, Log D_2
coordinates for the combined data on these 12 polymers
gave (figure 13).

$$r_{xy} = 0.984; \quad \alpha = -1.79; \quad \gamma = 0.498; \quad n = 64$$

These values for α and γ, substituted into equation
(29) and (30), give

$$A = -3.57; \quad B = 0.448$$

in the Log D_O, E_D coordinates. A correlation analysis
on these data, after transformation to the Log D_O, E_D
corrdinates gave (figure (14)) $r(B) = 0.994$; $A = -3.64$;
$B = 0.457$, $n = 64$ which is in reasonable agreement with
equations (29) and (30). It may be concluded that a
genuine, high degree of linear correlation exists between
Log D_O and E_D for this series of polymers, considered as
a single group or as individuals:

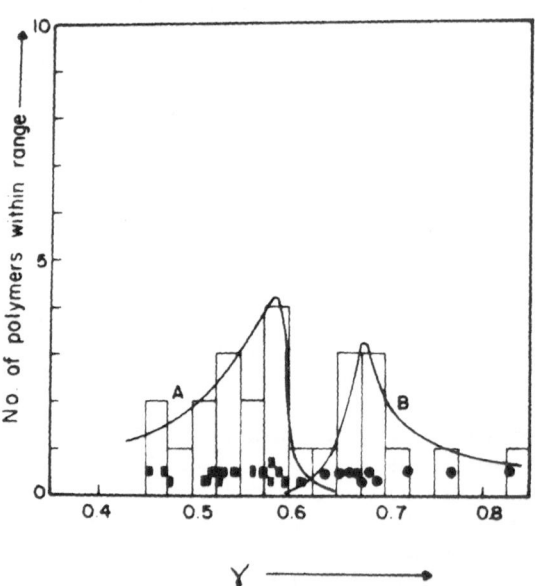

Figure 12.   Distribution of the parameter γ.

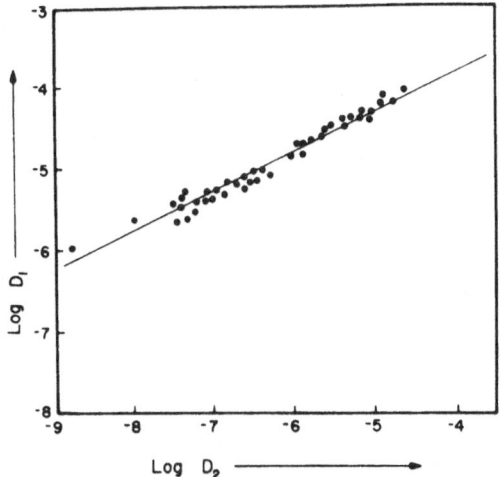

Figure 13.   Correlation for Polymers with $\gamma < 0.6$

diffusion in these polymers must be very similar. Barrer [1] has reported a correlation for rubbery polymers which (using the "average" reference temperature $T = 2T_1T_2/(T_1+T_2) = 333°$) corresponds to

$$A = -3.6; \quad B = 0.42$$

There were 11 different polymers, or polymeric states, in the group with $\gamma > 0.6$. Here a correlation analysis on the combined data (figure (15)) gave

$$r_{xy} = 0.977; \quad \alpha = -0.797; \quad \gamma = -0.708; \quad n = 63$$

These values of α and γ, on the basis of equations (29) and (30) require

$$A = -2.73; \quad B = 0.243$$

but a correlation analysis in the Log D_0, E_D coordinates (figure (16) gave

$$r(B) = 0.865; \quad A = -3.59; \quad B = 0.350; \quad n = 63$$

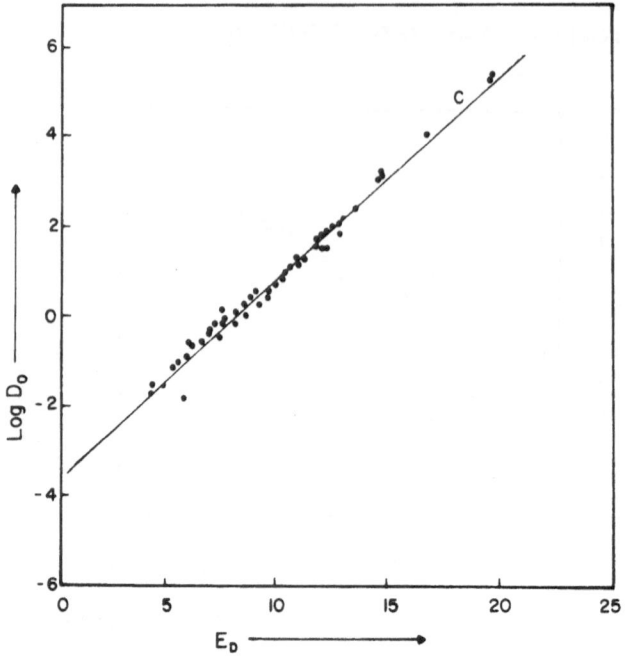

Figure 14.   Transformation of Figure 13.

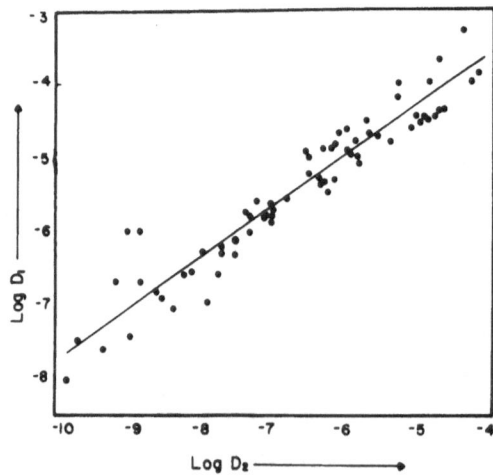

Figure 15.   Correlation for Polymers with $\gamma < 0.6$.

According to equation (48), purely random elements in the original data should be propagated in the Log D_0, E_D coordinates along lines having the "error slope," which in this instance is B = 0.658. The higher slope B obtained by the analysis in the Log D_O, E_D coordinates, and the larger negative intercept A, presumably reflect the effects of the greater randomness in the original data for these 11 polymers, considered as a group. Since in most instances the degree of linear correlation is high for these polymers considered as individuals, we now have a situation in which the individual polymers behave very differently from each other in their diffusion characteristics. This second group contains primarily the glassy and the crystalline polymers, but also contains a silicone rubber.

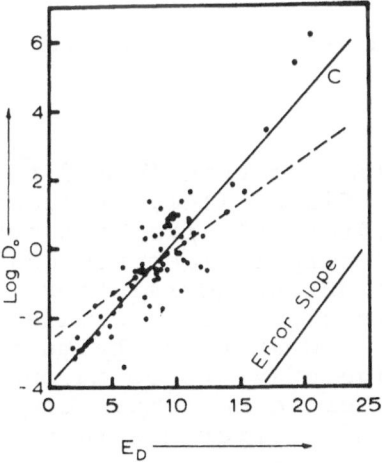

Figure 16.   Transformation of Figure 15.

Advantages of the Log D_1, Log D_2 Coordinates

Since random elements in the original data tend to propagate along lines of the "error slope" in the Log D_O, E_D coordinates, there is a tendency for real deviations from a genuine correlation to be hidden or obscured in the Log D_O, E_D coordinates. The data point for CH_4 in figure (17) does not stand out clearly as an anomalous result.

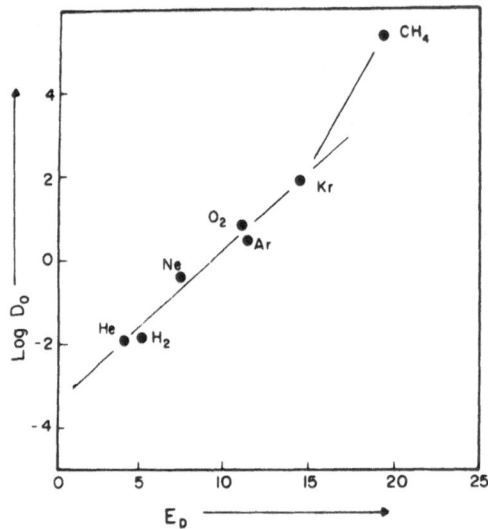

Figure 17.   Correlation for Glassy Polyvinylacetate.

If however these data are transformed to the Log D_1, Log D_2 coordinates, figure (18), it becomes apparent that CH_4 behaves differently from the other gases diffusing in glassy polyvinylacetate [5]. In fact the point for CH_4 falls on the correlation line for rubbery polyvinyl acetate in figure 11. Similarly, the diffusion characteristics for H_2O in polyethylmethacrylate [4] are shown clearly in figure (4) to be in one instance anomalous, and in the other to be comparable with those of the other gases. This difference tends to be obscured in figure (7). In figure (4) the anomalous point corresponds to measurements made above the glass transition temperature T_g; the point for H_2O which is on the correlation line is for measurements below T_g [4]

Interestingly, this 'anomalous' point for H_2O in the Log D_1, Log D_2 coordinates falls on the same line as the data for diffusion in rubbery polyvinyl acetate [5] in figure (11), and the other point for water falls quite close to the correlation line for glassy polyvinyl acetate [5] in figure (18). These relationships are hidden in the Log D_O, E_D coordinates.

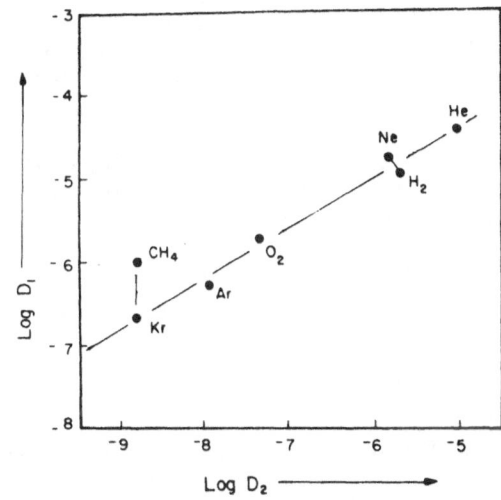

Figure 18.   Transformation of Figure 17.

The Log D_1, Log D_2 coordinates, as Exner has contended [3], provide the best test for the existence of a genuine "enthalpy-entropy" correlation. Since lines of constant activation energy appear as lines of unit slope in the Log D_1, Log D_2 coordinates, i. e.: since

$$\text{Log } D_1 = \frac{E_D(T_1-T_2)}{2.303RT_1T_2} + \text{Log } D_2 \qquad (49)$$

and since the range in E_D is physically restricted, there
will be some element of linear correlation in the Log D_1,
Log D_2 coordinates on purely physical grounds. The degree
of correlation found in this study is however far higher
than this effect alone could produce.

SUMMARY

Genuine correlations exist between the pre-exponential
factors Log D_O and the activation energies E_D for a series
of simple gases diffusing in the polymers examined in this
paper, when the polymers are considered as individuals.
A group of essentially rubbery polymers can be isolated
which shows a high degree of linear correlation when anal-
ysed as a single group. A second group of polymers exhibits large individual differences in behavior: this group
includes, as would be expected, the glassy polymers and
the partially crystalline polymers, but also includes some
rubbery polymers.

The characteristic parameters which describe the
linear correlation are best estimated in the Log D_1,
Log D_2 coordinate system.

Theoretical discussions of the origins and significance of such correlations can be found elsewhere [1, 2]

REFERENCES

1. R. M. Barrer, J. Phys. Chem. 61, 178 (1957)

2. J. Crank and G. S. Park, Eds. "Diffusion in Polymers,"
    Academic Press, New York and London (1968).

3. O. Exner, Coll. Czech. Chem. Commun., 29, 1094 (1964).

4. V. Stannett and J. L. Williams, J. Polym, Sci., C,
    No. 10, 45 (1965).

5. P. Meares, J. Amer. Chem. Soc., 76, 3415 (1954);
    Trans. Faraday Soc., 53, 101 (1957).

6. O.L. Davies, "Statistical Methods in Research and
    Production," 3rd Edn. Revised, Oliver & Boyd, London
    (1967).

7. R. McGregor and B. Milicevic, Nature, 211, 523 (1966).

FORMAL THEORY OF DIFFUSION THROUGH MEMBRANES

R. M. BARRER
Physical Chemistry Laboratories
Chemistry Department
Imperial College, London SW7 2AY
England

INTRODUCTION

Membranes find application in various separations - by dialysis, electrodialysis, ultrafiltration and reverse osmosis. Two large scale separations of gas and vapour mixtures have so far been developed: the preparation of ultrapure hydrogen by permeation through Ag-Pd; and the separation of uranium isotopes using UF_6 and porous septa. At present large scale separations of gases with polymer membranes have not been developed, despite the collection of much basic information. It is the purpose of this communication to outline several advances in the formal treatment of diffusion in membranes. These developments can apply to homogeneous polymeric membranes, polymer-filler composites, laminates and microporous membranes.

THE MEMBRANE EXPERIMENT

The membranes may be slabs, hollow cylinders or spherical shells. Diffusion in a ν-dimensional manner can be written in the transient state as

$$\frac{\partial C}{\partial t} = \frac{1}{r^{(\nu-1)}} \frac{\partial}{\partial r}\left(r^{(\nu-1)} D \frac{\partial C}{\partial r}\right) \qquad (1)$$

where ν = 1, 2 or 3 for slab, hollow cylinder or spherical shell respectively. The boundaries are R_1 and R_2 ($R_2 > R_1$) and the concentrations C_1 and C_2 ($C_1 > C_2$) at these boundaries are constant. Flow in the direction of r increasing is termed 'forward' flow.

113

For 'reverse' flow C_1 and C_2 become the constant values of C at R_2 and R_1 respectively. Forward flow is positive and reverse flow is negative. In the steady state of flow

$$J_\infty = - \omega_\nu r^{(\nu-1)} D \frac{dC(r)}{dr} \tag{2}$$

where $\omega_1 = 1$, $\omega_2 = 2\pi$ and $\omega_3 = 4\pi$ and the flux is through unit area of slab, unit length of hollow cylinder and through the whole surface of the spherical shell. The corresponding amounts of diffusant in the membranes are

$$M_\infty = \int_{R_1}^{R_2} \omega_\nu r^{(\nu-1)} C(r) dr \tag{3}$$

THE STEADY STATE : $D = D_o \varphi(C)$

Where D is a function of concentration, i.e. $D = D_o \varphi(C)$, the function $\varphi(C)$ must be such that $\lim_{C \to 0} \varphi(C) = 1$. For forward flow $(\vec{J_\infty})$ integration of eqn 2 from R_1 to R_2 gives (1)

$$\vec{J_\infty} I_\nu(R_1, R_2) = \omega_\nu D_o \int_{C_2}^{C_1} \varphi(u) du \tag{4}$$

where

$$I_\nu(x,y) \equiv \int_x^y \frac{dr}{r^{(\nu-1)}} \tag{5}$$

so that

$$I_1 = (y-x); \quad I_2 = \ln(y/x); \quad \text{and} \quad I_3 = \frac{1}{x} - \frac{1}{y} .$$

Interchanging C_1 and C_2 alters the sign but not the magnitude of the flux $(\vec{J_\infty} = -\vec{J_\infty}$ for reverse flow). For constant C_2, by differentiation of eqn 4 w.r.t. C_1 one obtains

$$D = D_o \varphi(C_1) = \frac{I_\nu(R_1, R_2)}{\omega_\nu} \left[\frac{\partial \vec{J_\infty}(C_1)}{\partial C_1} \right]_{C_2} \tag{6}$$

Thus the differential value of D and its concentration dependence are readily obtained from a plot of $\vec{J_\infty}$ against C_1 for constant C_2 (Fig.1). If $C_2 = 0$, $\varphi(C_1)$ is found for values of C_1 from zero upwards.

Eqn 2 may be integrated between r and R_1 to give, with eqn 4, the steady state quadrature

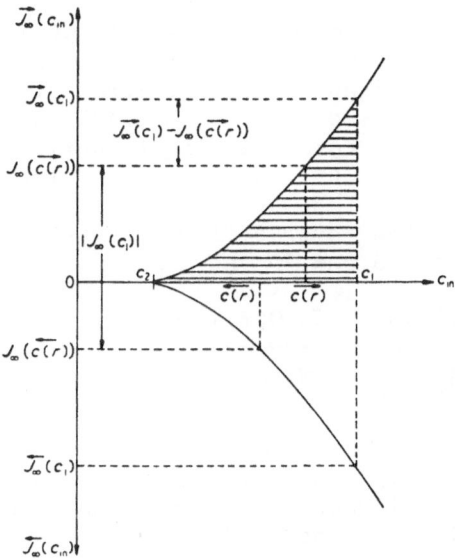

Fig.1. Plots of $\vec{J}_\infty$ and $\overleftarrow{J}_\infty$ against C_{in}, used to find $C(r)$, M_∞ and
$D = D_o\varphi(C)$

$$\frac{I_\nu(R_1,r)}{I_\nu(R_1,R_2)} \int_{C_2}^{C_1} \varphi(u)\,du = \int_{\overrightarrow{C_2}(r)}^{C_1} \varphi(u)\,du \tag{7}$$

where the distribution $\vec{C}(r)$ is for forward diffusion.   Interchanging C_1 and C_2 in eqn 7 gives the quadrature for $\overleftarrow{C}(r)$, the distribution for reverse flow.   Addition of the two quadratures gives

$$\int_{\overrightarrow{C}(r)}^{C_1} \varphi(u)\,du + \int_{\overleftarrow{C}(r)}^{C_2} \varphi(u)\,du = 0 \tag{8}$$

so that when $D = D_o$ and thus $\varphi(u) = 1$

$$\vec{C}(r) + \overleftarrow{C}(r) = C_1 + C_2 \tag{9}$$

In the general case ($\varphi(u) \neq 1$) the distribution $\vec{C}(r)$ can be
found from the master plot of Fig. 1 as follows: From eqn 4 and 7

$$\frac{I_\nu(R_1,r)}{I_\nu(R_1,R_2)} = \frac{\int_{\overrightarrow{C}(r)}^{C_1} \varphi(u)\,du}{\int_{C_2}^{C_1} \varphi(u)\,du} = \frac{\left\{\int_{C_2}^{C_1} \varphi(u)\,du - \int_{C_2}^{\overrightarrow{C}(r)} \varphi(u)\,du\right\}}{\int_{C_2}^{C_1} \varphi(u)\,du} = \frac{\left\{\vec{J}_\infty(C_1) - \vec{J}_\infty(\vec{C}(r))\right\}}{\vec{J}_\infty(C_1)} \tag{10}$$

for a given constant value of C_2. In eqn 10 $\vec{C}(r)$ is the value of C at r for the particular run in which the ingoing concentration, C_{in}, is C_1. The fluxes $\vec{J_\infty}(C_1)$ and $\vec{J_\infty}C(r)$ are identified on Fig.1. Accordingly, from eqn 10 and Fig.1 $I_\nu(R_1,r)/I_\nu(R_1,R_2)$ is evaluated and thence the value of r corresponding to $C(r)$. This gives the steady state distribution $\vec{C}(r)$ as a function of r.

Fig.1 shows also the reverse flows $\overleftarrow{J_\infty}(C(r))$ and $\overleftarrow{J_\infty}(C_1)$ in relation to C_{in}. From eqn 8 and 10 one has

$$\frac{I_\nu(R_1,r)}{I_\nu(R_1,R_2)} = \frac{\int_{C_2}^{\overleftarrow{C(r)}} \phi(u)\,dr}{\int_{C_2}^{C_1} \phi(u)\,du} = \frac{\overleftarrow{J_\infty}(\overleftarrow{C(r)})}{\overleftarrow{J_\infty}(C_1)} \tag{11}$$

so that $I_\nu(R_1,r)/I_\nu(R_1,R_2)$ can again be evaluated, and hence r, at which $C = C(r)$, can be found. Accordingly the concentration distribution for reverse flow is obtained.

From curves of $\vec{C}(r)$ and of $\overleftarrow{C}(r)$ vs r one may also construct those of $r^{(\nu-1)}\vec{C}(r)$ and $r^{(\nu-1)}\overleftarrow{C}(r)$ vs r and thence, using eqn 3, one may by graphical integration, evaluate $\vec{M_\infty}$ and $\overleftarrow{M_\infty}$, the amounts of diffusant in the membrane for forward and reverse flows. Thus for the slab, hollow cylinder and spherical shell information of a basic kind can be obtained.

THE STEADY STATE : $D = D_0 f(r)$ AND $D = D_0 \phi(C) f(r)$

Other situations treated with equal generality (1) were those in which $D = D_0 f(r)$ only, and those in which $D = D_0 \phi(C) f(r)$. In each situation from steady state flow measurements only one may find $f(r)$, $\phi(C)$, concentration distributions across the membranes and amounts of diffusant in them, whether the membranes are slabs, hollow cylinders or spherical shells.

THE TIME LAG

Information obtained from steady state flow measurements can be supplemented by that obtained from time lags (2,3). A membrane in the form of a slab will be considered which is bounded by the planes $x = 0$ and $x = \ell$. The boundary conditions are

(i) $C = C_0$ at $x = o$ for all t

(ii) $C = C_i(x)$ for $0 < x < \ell$ at $t = 0$

(iii) $C = C_\ell$ at $x = \ell$ for all t.

One may take $C_\ell = 0$ and $C_i(x) = 0$ and plot the amounts which have passed through $x = \ell$ and $x = 0$ at time t as functions of t. If their asymptotes are extrapolated to cut the axis of t the intercepts are the 'adsorption' time lags at $x = \ell\,(L_\ell^a)$ and at $x = 0(L_0^a)$ respectively. Similarly, when $C_i(x) = C_0$ and $C_\ell = 0$ these time-lags are the 'desorption' time lags at $x = \ell\,(L_\ell^d)$ and $x = 0(L_0^d)$. L_ℓ^a and L_0^d are positive; L_0^a and L_ℓ^d are negative.

Frisch (4,5,6), using the generalised Fick equation,

$$\frac{\partial C}{\partial t} = \frac{\partial}{\partial x}\left[D(C)\frac{\partial C}{\partial x}\right] \tag{12}$$

obtained an expression for the time lags at $x = \ell$ without requiring an explicit solution of eqn 12. In this equation $D = D_0 \varphi(C) = D(C)$ is a function only of C. Frisch's valuable method has been subsequently developed in several ways (7,8). In particular it has been shown that completely general expressions for the various time lags may be obtained (2,3) without reference to any flow equation such as eqn 12, but using only the conservation of matter condition

$$\frac{\partial C}{\partial t} = -\text{ div } J \tag{13}$$

At any plane $x = X$ within the membrane the time lag L_X was found to be given by

$$\ell\,J_\infty L_X = \int_0^\ell x\big[C(x) - C_i(x)\big]\,dx - \ell\int_X^\ell\big[C(x) - C_i(x)\big]\,dx$$

$$+ \int_0^\infty\int_0^\ell\big[J_\infty - J(x,t)\big]\,dx\,dt \tag{14}$$

In eqn 14 $C(x)$ is the steady state distribution across the membrane and $J(x,t)$ is the transient state flux at time t through the plane $x = x$. If we make $X = 0$ and $X = \ell$ we obtain the time lags for any initial distribution $C_i(x)$ within the membrane at $x = 0$ and $x = \ell$ respectively. If the double integral on the r.h.s. of eqn 8 is termed I, the relations of Table 1 follow between the four particular time lags, L_ℓ^a, L_0^a, L_ℓ^d and L_0^d as already defined above.

Each of the expressions in Table 1 is exact and completely general since it is derived only from eqn 13. For example if flow is expressed in terms of the equation $\frac{\partial C}{\partial t} = \frac{\partial}{\partial x}(D\frac{\partial C}{\partial x})$ the coefficient D can be a function of C, x or t or a function of all three, whether separable or inseparable. In all such situations from

Table 1.   Time lags and related quantities (3)

Eqn No.

15    $\ell J_\infty L_\ell^a = \int_0^\ell xC(x)dx + I^a$

16    $\ell J_\infty L_0^a = \int_0^\ell (x - \ell)C(x)dx + I^a$

17    $J_\infty(L_\ell^a - L_0^a) = J_\infty \Delta L^a = \int_0^\ell C(x)dx = M_\infty$

---

18    $\ell J_\infty L_\ell^d = \int_0^\ell xC(x)dx - \ell M_0/2 + I^d$

19    $\ell J_\infty L_0^d = \int_0^\ell (x - \ell)C(x)dx + \ell M_0/2 + I^d$

20    $J_\infty(L_\ell^d - L_0^d) = J_\infty \Delta L^d = \int_0^\ell [C(x) - C_0]dx = (M_\infty - M_0)$

---

21    $\ell J_\infty(L_\ell^a - L_\ell^d) = \ell J_\infty \delta L_\ell = \ell M_0/2 + (I^a - I^d)$

22    $\ell J_\infty(L_0^a - L_0^d) = \ell J_\infty \delta L_0 = -\ell M_0/2 + (I^a - I^d)$

23    $J_\infty(\delta L_\ell - \delta L_0) = J_\infty \delta \Delta L = M_0$

---

measurements of time-lags and of J_∞ one may determine M_∞ (eqn 17) $(M_\infty - M_0)$ (eqn 20) and M_0 (eqn 23). If the experiments leading to M_0 are conducted at different external pressures of diffusant which give different values of C_0 and with $C_i(x) = C_0$, one may construct the sorption isotherm. The above information can be obtained with the membrane in situ and hence simultaneously with measurements of the kinds described in the previous two sections.

One may consider further the integral

$$I = \int_0^\infty \int_0^\ell [J_\infty - J(x, t)]\, dxdt$$

In the case where $D = D(C)$ only and since then

$$J(x,t) = -D(C)\, dC/dx$$

one may integrate w.r.t. x between O and ℓ and obtain

$$\int_0^\ell J(x,t)\,dx = \int_0^{C_o} D(u)\,du = \ell J_\infty$$

Thus, when $D = D(C)$ only, the integral I vanishes. For this situation therefore eqn 21 and 22 also serve to give M_o.

However, when D is a function of x or of t, I is not necessarily zero. It has been shown that by compaction in increments one may minimise the influence of x dependence of D, arising from physical inhomogeneity, for membranes made by compression of fine powders (2). Also time-dependence in D can arise if there are blind pores in a microporous membrane in which the pore geometry and hence diffusion characteristics differ from those in the through channels (2,9). Thus in the transient state the diffusional properties of the membrane differ from those in the steady state. These considerations have been developed into a procedure for investigating blind pore character in membranes (2,3,10) (the 'Δ-procedure').

THE LAMINATE ABC...

The most general laminate comprises the sequence ABC.... where A, B, C etc total n layers of different thicknesses and in which diffusion coefficients, D, and solubility coefficients of the diffusant also differ. When the solubility in each layer obeys Henry's law and all the D are constants an exact and general treatment has been given for multiple laminates in the form of slabs, hollow cylinders or spherical shells (11). A constant concentration of diffusant is maintained at the ingoing face, R_o, of the first layer, and that at the outgoing face, R_n, is zero. Since Henry's law is obeyed, for the material comprising the ith layer $C_i = k_i' C_g$, C_i being the equilibrium concentration for any gas phase concentration C_g' and k_i' being the relevant Henry law constant. At the phase boundary between the ith and $(i + 1)^{th}$ layer the concentrations λ_{2i} and $\lambda_{2i + 1}$ must accordingly also be related by a Henry's law type of relation:

$$\lambda_{2i} = \lambda_{2i + 1}\,(k_i'/k_{i + 1}') = k_i \lambda_{2i + 1} \tag{24}$$

The derivations will not be given, but some results are summarised below.

$$J_\infty = {}^{\omega}{}_\nu \lambda_1 / \sum_{i=1}^{n} \left\{ H_i \right\} \tag{25}$$

where λ_1 is the concentration just within the ingoing face at R_o.

The functions H_i are given by

$$H_i = \frac{I_\nu(R_{i-1}, R_i)}{D_i} \prod_{j=0}^{i-1} k_j$$

where D_i is the value of D in the ith layer bounded by surfaces R_{i-1} and R_i. The concentration $C_i(r)$ at the surface r in the ith layer is

$$C_i(r) = \lambda_1 \left[\sum_{i=i}^{n} \{H_i\} + \frac{I_\nu(r, R_{i-1})}{D_i} \prod_{j=0}^{i-1} k_j \right] \Bigg/ \left(\prod_{j=0}^{i-1} k_j \right) \sum_{i=1}^{n} \{H_i\} \quad (27)$$

The general expressions for the time lags (11) for $\nu=1$ (slab) $\nu=2$ (hollow cylinder) and $\nu=3$ (spherical shell) will not, on account of their length, be reproduced here. Eqn 27, by graphical integration w.r.t. r across each layer and addition of the result for each layer, also gives the total amount, M_∞, of diffusant in the membrane.

Experimental time-lag and steady flow studies have been made on laminates in the form of slabs (12) (2 and 3 layers), and in hollow cylindrical laminates (13) (2 layers). The latter workers introduced the idea of membrane conductances which will be considered further. A single hollow cylindrical membrane is bounded by inner and outer surfaces of radii R_a and R_b respectively. At $r = R_a$ the gas pressure is p_{in} and is zero at $r = R_b$. In terms of pressure p the Henry's law relation is $C = \sigma p$. The membrane conductance, Y, for the diffusing gas was defined as

$$Y = \frac{J_\infty}{p_{in}} = 2\pi D\sigma / \ln(R_b/R_a) \quad (28)$$

If at $t = 0$ the membrane is free of diffusant, the time lag L at the boundary $r = R_b$ is

$$L = \frac{(R_a^2 + R_b^2)\ln(R_b/R_a) - (R_b^2 - R_a^2)}{4D \ln(R_b/R_a)} \quad (29)$$

For the hollow cylindrical laminate (inner sheath nylon, outer one polyethylene) the conductance Y_c was shown to be

$$\frac{1}{Y_c} = \frac{1}{Y_1} + \frac{1}{Y_2} \quad (30)$$

where Y_1 and Y_2 are the conductances of the inner and outer sheaths whose surfaces are bounded respectively by $r = R_0$ and R_1 and $r = R_1$ and R_2. In terms of the time lags L_1 and L_2 for the individual sheaths that of the laminate was also found to be

$$\frac{L_c}{Y_c} = \frac{L_1}{Y_1} + \frac{L_2}{Y_2} + \frac{B_1 L_1}{Y_2} + \frac{E_2 L_2}{Y_1} \tag{31}$$

where

$$B_1 = \left[\frac{2R_1^2 \ln (R_1/R_0) - (R_1^2 - R_0^2)}{(R_1^2 + R_0^2) \ln (R_1/R_0) - (R_1^2 - R_0^2)}\right] \ln (R_1/R_0) \tag{32}$$

$$E_2 = -\left[\frac{2R_1^2 \ln (R_2/R_1) - (R_2^2 - R_1^2)}{(R_2^2 + R_1^2) \ln (R_2/R_1) - (R_2^2 - R_1^2)}\right] \ln (R_2/R_1)$$

Measurements of Y_c, Y_1, Y_2, L_c, L_1 and L_2 confirmed the above relations to the degree of accuracy illustrated in Table 2.

THE LAMINATE ABAB....

The laminate ABAB.... is a special case of the general laminate ABC.... of the previous section, and has some interesting properties (14). Only two diffusion coefficients D_A and D_B are involved and each was considered to be constant. In the complete slab considered there are m pairs of layers AB, and Henry's law governs equilibrium relations of gas and dissolved gas: $C_A = k_A C_g$; $C_B = k_B C_g$. Thus also $C_A = C_B(k_A/k_B) = kC_B$. All layers of A have thickness a and all those of B have thickness b. Thus the volume fractions v_A of A and v_B of B are $v_A = ma/\ell$ and $v_B = mb/\ell$ where ℓ is the total thickness of the laminate. Several of the results obtained are given below.

The steady state flow, where C_1 is the concentration just within the first layer, is

$$J_\infty = \frac{C_1}{\ell} \cdot \frac{D_A D_B}{v_A D_B + k v_B D_A} \tag{33}$$

The amount of diffusant within the membrane is

$$M_\infty = C_1\left(\frac{\ell}{2}\right)\left\{(v_A + \frac{v_B}{k}) + \frac{v_A v_B(k^2 D_A - D_B)}{mk(v_A D_B + k v_B D_A)}\right\} \tag{34}$$

Table 2. Calculated and Measured Conductances and Time-lags for nylon-polyethylene laminate (13)

(a) Conductances (10^8 x cm^3 at s.t.p./sec. cm atm.)

Gas	Temp.($^{\circ}$C)	Y_1	Y_2	Y_c(expt.)	Y_c(calc.)
Ne	30	3.1_2	41.9	2.8_6	2.9_0
	40	4.8_6	68.0	4.4_5	4.5_3
	50	7.1_7	103	6.5_7	6.6_9
	60	11.4	168	10.5_3	10.7_0
H_2	20	10.9	99.4	9.6_4	9.8_3
	30	16.1	163	14.2_4	14.6_3
	40	23.8	261	21.4_0	21.8_4
	50	35.3	396	31.6_3	32.4_0
	60	51.8	597	46.7_1	47.6_2

(b) Time-lags (mins.)

Gas	Temp.($^{\circ}$C)	L_1	L_2	L_c(expt.)	L_c(calc.)
Ne	30	60.3	9.5	95.6	95.1
	40	39.6	6.4	65.0	62.6
	50	27.6	4.9	45.0	44.7
	60	18.6	3.5	30.9	30.5
H_2	20	40.5	7.2_5	71.5	68.8
	30	26.8	4.9_8	46.5	45.4
	40	17.7	3.3_0	32.0	29.8
	50	13.4	2.6_5	23.5	22.8
	60	9.2	1.8_0	16.2	15.6

and the time lag, L, is

$$L = \frac{v_A D_B + k v_B D_A}{6 D_A D_B} (v_A + \frac{v_B}{k}) \ell^2 (1 - \frac{1}{m^2})$$

$$+ \frac{\ell^2}{m^2} \frac{\left\{ v_A^3 D_B^2 + k v_B^3 D_A^2 + 3 v_A v_B D_A D_B (k v_A + v_B) \right\}}{6 D_A D_B (v_A D_B + k v_B D_A)} \qquad (35)$$

The dependence of the quantities J_∞, M_∞ and L upon the number m of pairs of layers is of interest. The number of pairs may be in-creased by keeping ℓ constant and making each pair AB thinner while keeping v_A and v_B constant ($v_A = (1 - v_B)$); or by keeping each pair AB of constant thickness and increasing ℓ. In the former case, if M_∞ if plotted against $1/m$ a straight line is obtained having the slope and intercept taken from eqn 34. Similarly if L is plotted against $1/m^2$ the slope and intercept can be taken from eqn 35. In the latter experimentally more practical case $\ell = m(a + b)$ where (a + b) is now constant. The linear plots are then those of M_∞ against m and of L against m^2. If a and b are known and therefore v_A and v_B, the slope and intercept, together with eqn 33 for J_∞, are sufficient to determine D_A, D_B and k without making separate experi-ments on the individual membranes A and B.

CONCLUSION

Measurements of steady state flows and of time lags can give more information about the properties of diffusant-membrane systems than has so far been fully explored experimentally. This has been shown by reference to some developments of the formal theory of diffusion and flow through membranes.

REFERENCES

1. R. Ash and R.M. Barrer, J. Phys. D:Appl. Phys., 1971, 4, 888.
2. R. Ash, R.W. Baker and R.M. Barrer, Proc. Roy. Soc., 1968, A, 304, 407.
3. R.M. Barrer, in "Surface Area Determination", 1970, p.227, Butterworth, London(IUPAC Internat. Symp., Bristol, 1969)
4. H.L. Frisch, J. Phys. Chem., 1957, 61, 93.
5. H.L. Frisch, J. Phys. Chem., 1958, 62, 401.
6. H.L. Frisch, J. Phys. Chem., 1959, 63, 1249.
7. J.H. Petropoulos and P.P. Roussis, J. Chem. Phys., 1967, 47, 1491 and 1496.
8. J.H. Petropoulos and P.P. Roussis, J. Chem. Phys., 1968, 48, 4619.

9.   R.M. Barrer and T. Gabor, Proc. Roy. Soc., 1960, A, <u>256</u>, 267.
10. R. Ash, R.M. Barrer, J.H. Clint, R.J. Dolphin and C.L. Murray,
         Phil. Trans., 1973, A, <u>275</u>, 255.
11. R. Ash, R.M. Barrer and D.G. Palmer, Brit. J. Appl. Phys.,
         1965, <u>16</u>, 873.
12. J.A. Barrie, J.D. Levine, A.S. Michaels and P. Wong, Trans.
         Faraday Soc., 1963, <u>59</u>, 869.
13. R. Ash, R.M. Barrer and D.G. Palmer, Trans. Faraday Soc., 1969,
         <u>65</u>, 121.
14. R. Ash, R.M. Barrer and J.H. Petropoulos, Brit. J. of Appl.
         Phys., 1963, <u>14</u>, 854.

POLYETHYLENE-NYLON 6 GRAFT COPOLYMERS I. MECHANICAL PROPERTIES AND
PERMEABILITY CHARACTERISTICS OF COPOLYMERS PREPARED BY ANIONIC AND
AND CATIONIC INITIATED PROCESSES

M. Matzner, D. L. Schober, R. N. Johnson, L. M. Robeson
and J. E. McGrath
Research and Development Department
Union Carbide Chemicals and Plastics
P.O. Box 670, Bound Brook, N. J. 08805

INTRODUCTION

A graft copolymer of polyethylene and nylon 6 combines two
materials each of which possesses good properties. As is well
known, polyethylene is a tough flexible polymer which is insensitive
to moisture. Oxygen permeability is intermediate and its crystalline melting point (Tm) is relatively low. Nylon 6, on the other
hand, has very low oxygen permeability (dry) coupled with a high
Tm of 218°C. However, nylon 6 displays strong sensitivity to moisture. Therefore its utility is significantly impaired under conditions of high humidity.

Thus the present investigation had as its goal the preparation
and characterization of a graft copolymer based on the two homopolymers discussed above. Excellent mechanical properties combined
with controlled transport properties were expected.

A particularly interesting aspect of polymer permeability is
its relationship to the morphology of the material[1]. First of all
the degree of crystallinity has a pronounced effect on permeability[2].
The phenomenon is well documented for low density and high density
polyethylene (LDPE and HDPE).

When two materials are combined, phase relationships will
govern the transport properties. Different behavior is predicted
for a one phase system as opposed to a two phase system. Recent
work on compatible PPO-polystyrene blends has shown the latter to
behave as if they were homogeneous (e.g., statistical) copolymers[3].
The effects of micro-heterogeneity have also been described[4]. This
work pointed out another important factor, namely, that it is

125

important to differentiate between compositions in which one
component is discrete while the other continuous, and those where
both are semi - continuous.   Clearly, if one could devise a system
wherein subtle synthetic variations would enhance or inhibit phase
separation one would have an ideal model for study. A situation
of this type is achievable under certain conditions (e.g., anionic
polymerization of styrene and dienes). However, it is more for-
midable in the case of polyolefins combined with polyamides. Moreover, the dissimilarity of the two constituents of such a system
should allow for a very wide variation in permeability.

The attempt to vary the structure and morphology of polyethylene-nylon 6 graft copolymers has been made and some preliminary
results are presented in this paper.

RESULTS AND DISCUSSION

A. General Considerations

In block and graft copolymers phase separation is a function
of segment molecular weight, crystallizability and differential
solubility parameter[5-7]. For instance, much longer blocks are required in the styrene-butadiene-styrene series[8,9] than in the poly-
sulfone-polydimethylsiloxane systems[10] for phase separation to
occur. Thus, in a graft copolymer composed of two thermodynamically
incompatible components, the degree of miscibility will increase
when the length of either the backbone or of the side chains is decreased. In the present work, it was chosen to vary the length of
nylon 6 side chains on a polyethylene backbone, in order to control
the degree of microphase separation.

B. Synthetic Approach

ε-Caprolactam may be polymerized by a variety of mechanisms
which include hydrolytic, anionic, and cationic[11,12]. Of particular
interest for the present discussion are the cationic and anionic
methods. It is well known that anionic polymerization results in
high molecular weight nylon 6. Moreover, the polymerization proceeds
to its equilibrium conversion[13]. On the other hand, under cationic
conditions both conversion and molecular weights are low. This has
been attributed to the formation of highly basic amidines[14]. Thus
the utilization of these two techniques afforded an easy method of
side chain length control.

The anionic grafting of nylon 6 onto polyethylene was accomplished as shown in equation I. The method takes advantage of the

〰〰〰 CH_2-CH_2 〰〰〰〰 CH_2-CH 〰〰〰
                                        |
                                        CO
                                        |
                                        OC_2H_5

ethylene-ethyl acrylate (E/EA) copolymer
                    (1)

$$NaH \atop Heat \Bigg\downarrow \quad HN<{CO \atop (CH_2)_5} \qquad\qquad (I)$$

〰〰〰 CH_2-CH_2 〰〰〰 CH_2-CH 〰〰〰
                                      |
                                      CO
                                      |
                                      NH
             (2)   〰
                                      〰   Nylon 6
                                      〰

fact that esters are initiators in the anionic lactam polymerization[15] and is analogous to the one previously reported for the
synthesis of polystyrene-nylon 6 copolymers[16]. Note that the ethyl
acrylate content was critical and levels below about one weight
percent were necessary to retain thermoplasticity. The compositions
were translucent, displayed a relatively high degree of crystallinity and were shown to contain two phases by modulus-temperature
measurements. Grafting efficiencies were determined as before[16]
and ranged from moderate (∽50%) to high (∽80%). High molecular
weight was easily achieved as judged by melt index measurement at
250°C and reduced viscosity in m-cresol (at high nylon 6 contents).
These data coupled with the fact that the acrylate content was low
allows the conclusion that long nylon 6 side chains were obtained.

The cationic process is shown in equation II.

〜〜〜〜 CH$_2$-CH$_2$ 〜〜〜〜 CH$_2$-CH 〜〜〜〜
                                      |
                                      COOH

ethylene-acrylic acid copolymer (E/AA)

(3)

anhydrous
24 hrs/230°C   HN〈 CO
                    |
                    (CH$_2$)$_5$

〜〜〜〜 CH$_2$-CH$_2$ 〜〜〜〜 CH$_2$-CH 〜〜〜〜   (II)
                                      |
                                      CO
                                      |
                                      NH
        (4)   ⌇
                                      ⌇   Nylon 6
                                      ⌇

   In contrast to the materials formed by the anionic process
these copolymers derived from E/AA backbones (AA=19 wt. %) were
transparent and exhibited a partial miscibility as judged by
modulus-temperature measurements. Moreover, the degree of conversion was generally low at the preferred polymerization temperature of 230°C (e.g., < 50%), some nylon homopolymer was formed and
was shown to be of low molecular weight. This fact, coupled with
a depressed Tm of the copolymer indicates that the nylon branches
are rather short. This is further supported by the transparency
of the copolymer and the high acrylic acid content in the polyethylene backbone (up to 20 wt. %).

                    C. Mechanical Properties

   The mechanical properties of the graft copolymers prepared by
each route are shown in Table I. Examination of the data in Table I
shows the very significant effect of synthetic conditions on the
mechanical properties. It is believed that this reflects the morphological differences in the two materials. The anionic samples
tend to have high modulus and low elongation characteristics due
probably to the well separated nylon phase. On the other hand,
the cationic copolymers display high elongations and lower moduli

TABLE I

MECHANICAL PROPERTIES OF POLYETHYLENE-NYLON 6 GRAFT COPOLYMERS

Expt. No.	Backbone %EA[1]	Backbone %AA[2]	% N6	Mechanical Properties[4] Tens. Mod. psi	Tens. Str. psi	Elong. @ Break %	Pendulum Impact ft.lbs/in[3]	Tm[3] °C
1	5.2	–	59	156,000	5,200	14	114	225
2	2.3	–	79	204,000	5,100	4.3	18	220
3	–	19	20	13,500	4,000	384	>200	–
4	–	19	31	27,000	4,700	330	190	–
5	–	19	49	49,000	4,400	152	227	–
6	–	19	55	73,000	4,600	87	185	–

1. Polymers prepared via anionic route.
2. Polymers prepared via cationic route.
3. The cationically prepared polymers had depressed melting points; for example for a nylon 6 content of 40 wt. % the Tm was 170°C. See Figure 1.
4. Compression molded; annealing the test specimen generally resulted in significantly enhanced moduli and somewhat decreased elongation and pendulum impact.

TABLE II

PERMEABILITY BEHAVIOR OF PE/N6 GRAFT COPOLYMERS TO OXYGEN, NITROGEN, AND CARBON DIOXIDE

Method[1]	% N6	Permeability $(cc\text{-}cm/cm^2\text{-}sec\text{-}cm\ Hg) \times 10^{11}$ O_2	N_2	CO_2
	Nylon 6	0.13	0.035	0.45
	LDPE	32	10.5	171
	HDPE	7.1	2.3	29
	E/AA[2]	15.5	3.5	50
A	76	0.125	0.035	–
A	74	–	0.053	.54
C	7.2	10.6	2.1	39.4
C	24.9	10.3	1.9	32.4
C	30.5	6.9	–	–
C	41.0	4.8	1.2	18.5
C	54.0	3.6	1.0	14.9
B	90/10	13.4	3.2	49.0
B	80/20	9.1	–	–
B	70/30	5.2	–	–
B	60/40	4.0	0.98	11.3
B	50/50	1.8	–	–
B	40/60	0.95	–	–
B	30/70	0.62	–	–

1. A denotes anionic; B and C denotes the blends and the cationic graft, respectively.
2. Contains 19% AA.

FIGURE 1

MODULUS — TEMPERATURE RELATIONSHIP
FOR PE/N$_6$ GRAFT COPOLYMERS

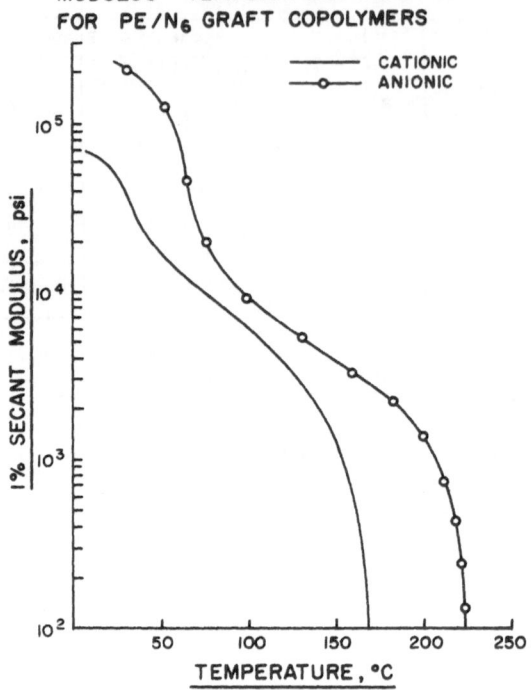

FIGURE 2

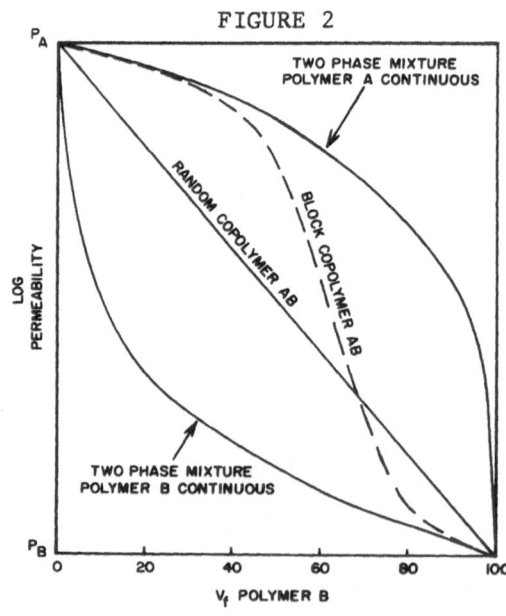

FIGURE 3

DYNAMIC MECHANICAL BEHAVIOR
OF A ETHYLENE-NYLON 6 GRAFT COPOLYMER
(40 WT PERCENT NYLON 6)
PREPARED VIA THE CATIONIC PROCESS

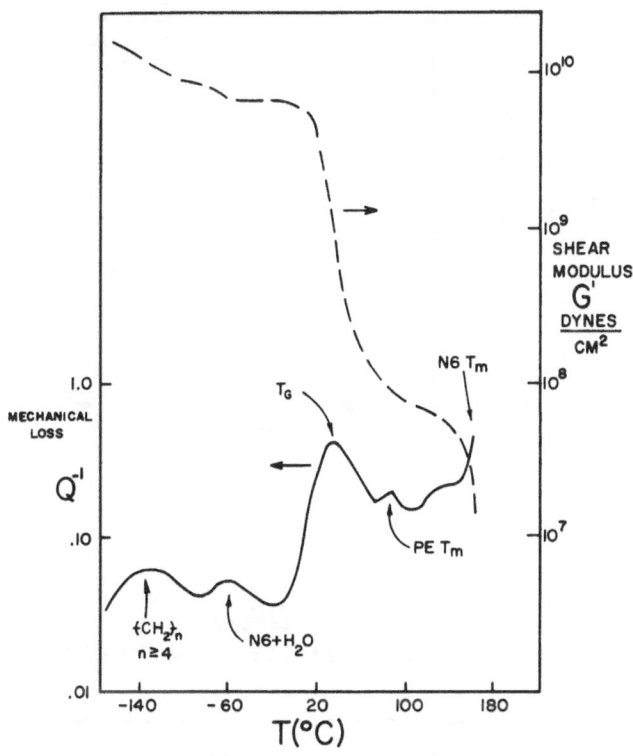

TABLE III

SOLVENT PERMEABILITY OF
POLYETHYLENE/NYLON 6 GRAFT COPOLYMERS

Solvent	Permeability[1]		
	HDPE	Nylon 6	PE/N6 (76% N6)
Benzene	60,000	3,500	2,100
Heptane	70,000	4,400	7,300
Ethyl Benzene	420,000	15,000	21,000

1. In cc. mil/100 in^2, 24 hours, atm.
2. UCC high density polyethylene (d=0.95 gm./cc.).
3. Prepared by the anionic grafting process.

FIGURE 4

EFFECT OF NYLON CONTENT ON
PERMEABILITY OF COPOLYMERS AND ALLOYS

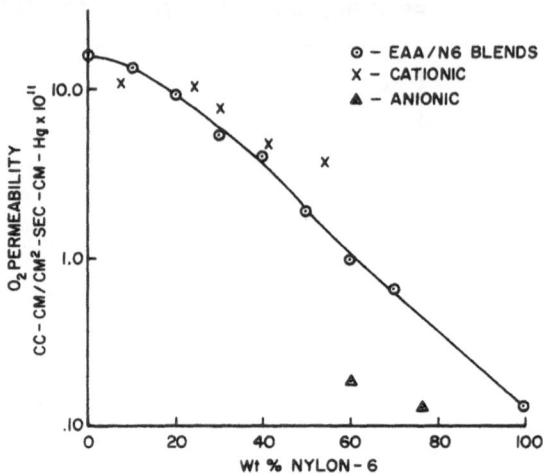

FIGURE 5

PERMEABILITY OF LDPE/NYLON 6
BLENDS

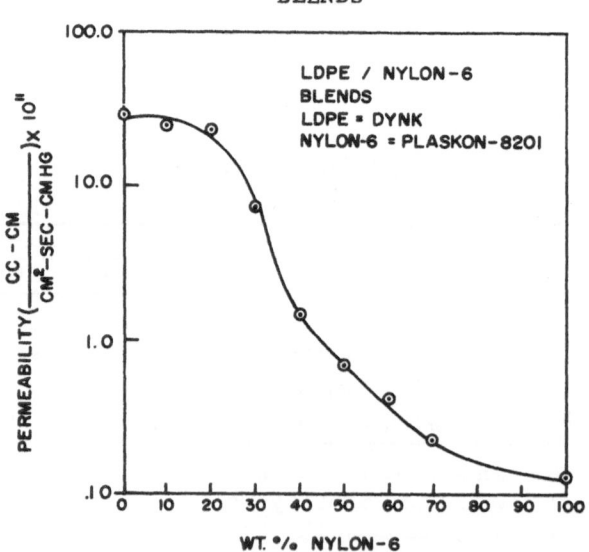

even at comparable nylon contents. This is attributed to partial
miscibility and the concomitant averaging of properties. These
differences are reflected in the modulus-temperature curves shown
in Figure 1. Further examination of the mechanical loss and shear
moldulus-temperature data in Figure 3 illustrates the phase miscibility of the cationically prepared copolymer. Note that only one
Tg at 40°C is observed, and that this value is midway between that
of E/AA (20°C) and dry nylon 6 (60°C).

D. Permeability Characteristics

Permeability data for nitrogen, oxygen and carbon dioxide are
summarized in Table II. They were obtained by ASTM method D-1434-66.
For comparison purposes the corresponding values for the backbones,
e.g., E/AA as well as for low density and high density polyethylene
(LDPE and HDPE) and dry nylon 6 are also shown. Since the E/EA
backbone used contained very low wt. % of ethyl acrylate, permeability
values of LDPE are believed representative. It has been shown that
a linear relationship exists between the logarithm of permeability and
volume fraction in a one phase random copolymer[17]. In a two phase
system the compositional dependence takes the form of a S-shaped
curve[17]. A typical curve is shown in Figure 2. Therefore, with
these facts in mind, we would expect the following: (a) the anionic
graft copolymer should display low gas transport properties even with
significant polyethylene concentrations at volume fractions where
nylon 6 is the continuous phase (e.g., > 65%), (b) the reverse, namely
a polyethylene-like behavior is anticipated even at substantial nylon
contents with the cationic graft copolymer.

Independent experiments have been performed on alloys of ethylene-acrylic acid copolymers (E/AA, 19 wt. % AA) and nylon 6. Mechanical loss data have shown that these blends have one Tg intermediate between the two constituents. However, the peak was relatively broad which indicates a more pronounced heterogeneity than
for the graft copolymer (cf. Figure 3). Nevertheless, the permeability data plotted in Figure 4 closely followed the behavior expected for a one-phase system. Therefore, the characteristics observed on the two systems (the alloys and the cationic grafts) are
believed comparable. Minor differences may be due to either different molecular weights or variation in degree of crystallinity
(the cationic grafts are lower in crystallinity).

With these assumptions, the results shown in Table II and
Figure 4 are in agreement with the postulates (a) and (b) above.
Additional permeability data obtained on blends of LDPE and nylon 6
(a two-phase material) are plotted in Figure 5. The expected S-
shaped behavior is observed. The relatively "low" permeability measured at nylon contents in the range of 20-50 wt. % may be due to
some secondary features such as orientation during compression molding.

Along related lines the resistance to hydrocarbon penetrants should show considerable improvement upon introduction of the crystalline and polar nylon phase in the polyethylene. This phenomenon was well documented in the case of polysulfone block copolymers[6]. Pertinent data on the subject system is shown in Table III.

EXPERIMENTAL

The preparation of the anionic graft copolymers was conducted as reported previously[16]. A typical preparation of a cationic copolymer follows.

Preparation of Polyethylene/Nylon 6 Graft Copolymer from Polyethylene/Acrylic Acid Backbone via the Anhydrous Process

$\mathcal{E}$-Caprolactam (50 g., 0.44 m.) and a polyethylene-acrylic acid copolymer (19.2 wt. % AA) (50 g.) were placed in a 500 ml., round-bottom, 3-neck flask equipped with a mechanical stirrer, gas inlet, thermometer, and a distilling head. Chlorobenzene (100 ml.) was added to the flask. The mixture was then heated in an oil bath to $\sim$200°C. The chlorobenzene distilled off at $\sim$165°C (inside temperature). After the chlorobenzene was removed, the temperature was raised to 230°C (inside temperature) and the distilling head replaced with a reflux condenser. Stirring under argon was continued for four hours. At this point, the stirring rod was lifted above the molten mass and heating was continued for an additional twenty hours under argon. The resulting mixture was cooled, broken into chips, and ground on a Wiley mill. Methanol extraction yielded 71 g. of polymer (40% conversion). The dried polymer (80°C, 15 mm., 24 hours) analyzed for 3.71% N, indicating 29.9% nylon 6. The product exhibited a melt index of 34.1 (1P) at 250°C. The polymer was transparent (haze of 12%). By contrast, nylon 6 has a haze of 71%.

Trifluoroethanol extraction by the Soxhlet method indicated that 13% of the product was soluble. This extract analyzed for 12.12% N, corresponding to nylon 6 homopolymer (12.39% N). The reduced viscosity (0.2% in m-cresol at 25.0°C) of this nylon 6 homopolymer was 0.31.

Extraction with THF indicated no soluble portion after forty-eight hours. Thus, there is no free polyethylene/acrylic acid copolymer in the product.

CONCLUSIONS

Graft copolymers of nylon 6 onto polyethylene were synthesized. Those prepared via the anionic technique were two phase systems.

Their mechanical properties and permeability characteristics reflected the presence of a distinct nylon 6 moiety. By contrast the cationic copolymers showed partial miscibility and concomitant averaging of mechanical and transport properties.

REFERENCES

1. H. B. Hopfenberg and V. Stannett, Chapter 9, in "The Physics of Glassy Polymers", Edited by R. N. Hayward, Wiley (1973).

2. N. L. Zutty, J. A. Faucher and S. Bonotto, Encyclopedia of Polymer Science and Technology, $\underline{6}$, 387 (1967).

3. C. H. M. Jacques, H. B. Hopfenberg and V. Stannett, Polymer Engineering and Science, $\underline{13}$, 81 (1973).

4. L. M. Robeson, A. Noshay, M. Matzner and C. N. Merriam, Die Angew, Makromol. Chemie, $\underline{29/30}$, 47 (1973).

5. M. Matzner, A. Noshay and J. E. McGrath, Polymer Preprints, $\underline{14}$, No. 1, 68 (1973).

6. J. E. McGrath, L. M. Robeson and M. Matzner, Polymer Preprints, $\underline{14}$, No. 2, 1032 (1973).

7. L. M. Robeson, M. Matzner, L. J. Fetters and J. E. McGrath, Polymer Preprints, $\underline{14}$, No. 2, 1063 (1973).

8. D. J. Meier, J. Polym. Sci., Part C, No. 26, 81 (1969).

9. M. J. Folkes and A. Keller, Chapter 10 in "The Physics of Glassy Polymers", Edited by R. N. Hayward, Wiley, (1973).

10. A. Noshay, M. Matzner and C. N. Merriam, J. Polymer Sci., A-1, $\underline{9}$, 3147 (1971).

11. J. Sebenda, J. Macromolecular Sci.-Chemistry $\underline{A6}$, No. 6, 1145 (1972).

12. H. K. Reimschuessel, "Lactams", Chapter 7 in "Ring Opening Polymerizations", Edited by K. C. Frisch and S. L. Reigen, Marcel Dekker, New York (1969).

13. O. Wichterle, J. Sebenda and J. Kralicek, Advances in Polymer Sci., $\underline{2}$, 578 (1961).

14. S. Doubravsky and F. Geleji, Makromol. Chem. $\underline{143}$, 259 (1971).

15.   A. Mattiussi and G. B. Gechele, European Polymer Journal, $\underline{4}$ 695 (1968).

16.   M. Matzner, D. L. Schober and J. E. McGrath, European Polymer Journal, $\underline{9}$, 469 (1973).

17.   M. Matzner, A. Noshay, D. L. Schober and J. E. McGrath, Ind. Chim. Belg. $\underline{38}$, 1104 (1973).

THE INFLUENCE OF DRAWING ON THE TRANSPORT PROPERTIES

OF GASES AND VAPORS IN POLYMERS

Joel Williams

Camille Dreyfus Laboratory, Research Triangle Institute

P. O. Box 12194, Research Triangle Park, N. C. 27709

INTRODUCTION

It is well established that substantial reduction in the equilibrium sorption and even more drastic reduction in the diffusion rates takes place during drawing of polyethylene. The exact influence of the draw ratio on the transport behavior was recently reported by Williams-Peterlin[1] and a more general discussion of this effect was presented earlier by Peterlin-Williams-Stannett.[2] More recently, Williams-Peterlin[3] have demonstrated that drastic decreases in transport behavior do not always accompany drawing. In fact, the extent of decrease in the transport behavior is mainly dependent on the medium in which the material is drawn. Furthermore, under certain drawing conditions the rate of diffusion can be actually increased, i.e., the exact opposite as observed during cold drawing of polyethylene. This increase in transport behavior can be very substantial often several orders of magnitude higher than the undrawn material.

In the case of normal cold drawing, the plastic deformation reduces the number of sorption sites and increases the activation energy of diffusion but does not change the energy conditions at the sorption sites. These changes in the sorption and transport properties were considered as being a direct consequence of the denser packing and ordering of the tie molecules in the amorphous regions brought about by the cold drawing process. This rearrangement of tie molecules results in blocking the easy passage of sorbent

---

Present Address: Becton, Dickinson Research Center, P. O. Box 12016
                  Research Triangle Park, North Carolina 27709

and causes a reduction in the number of unaffected chain folds
where the sorption primarily occurs. Similar effects have been
noted by other investigators using different penetrant-polymer
systems. [4-8] For example, Davis and Taylor[6] have shown that the
diffusion of dyes into nylon 66 is greatly reduced upon drawing,
Takagi [7,8] found that in nylon 6 the radial diffusion increases up
to a draw ratio of 1.6 and subsequently decreases to about 10 percent of the initial value. The axial diffusion shows only a decrease and is always smaller than the radial diffusion. All investigators have also noted that the activation energy increases considerably for diffusion into the drawn material.

In the present paper, the influence of the degree of deformation on the transport properties will be compared and contrasted
for the different drawing conditions.

RESULTS AND DISCUSSION

In Fig. 1 the equilibrium concentration c_∞ of methylene chloride
in undrawn polyethylene and in samples with draw ratios 6, 7, 8, 9,
and 25 are plotted as function of relative pressure p/p_0 of sorbent
vapor. In all cases strong departures from the linear Henry's law
are evident with higher penetrant activities. The departure from a
straight line through the origin is most conspicuous for the undrawn
sample but is still detectable even at the largest draw ratio. The
isotherms obtained are of the normal mixing rather than hole-filling
(Langmuir) type for both the drawn and undrawn material.

There is a drastic reduction in equilibrium concentration of
sorbent caused by the drawing if the draw ratio is 9 or higher.
With the sample of draw ratio $\lambda = 9$ the sorption at p_0 is reduced
to less than 13 percent of the value in undrawn sample. The initial
slope corresponding to the validity of Henry's law is reduced to
about 15 percent. Subsequent drawing to $\lambda = 25$ has very little
additional effect on sorption as the experimental points are on very
nearly the same curve as the points for the sample with $\lambda = 9$.
This fact is in good agreement with former experiments on water
sorption in drawn polyethylene.[9] It was shown that there is a drastic
reduction in equilibrium sorption with drawing but very little change
beyond draw ratios of about nine. For water, the normal Henry's law
relationship for the solubility was found to exist in the drawn and
undrawn material.

The surprising new fact, however, is the very small reduction
of sorption up to a draw ratio 8. In fact, there is even a small
increase of sorption at the low penetrant activities, but it is not
far enough above the error limits of measurement as discussed in the
experimental section at such low sorption and activity to be worthy
of special attention. Following the small reduction in sorption

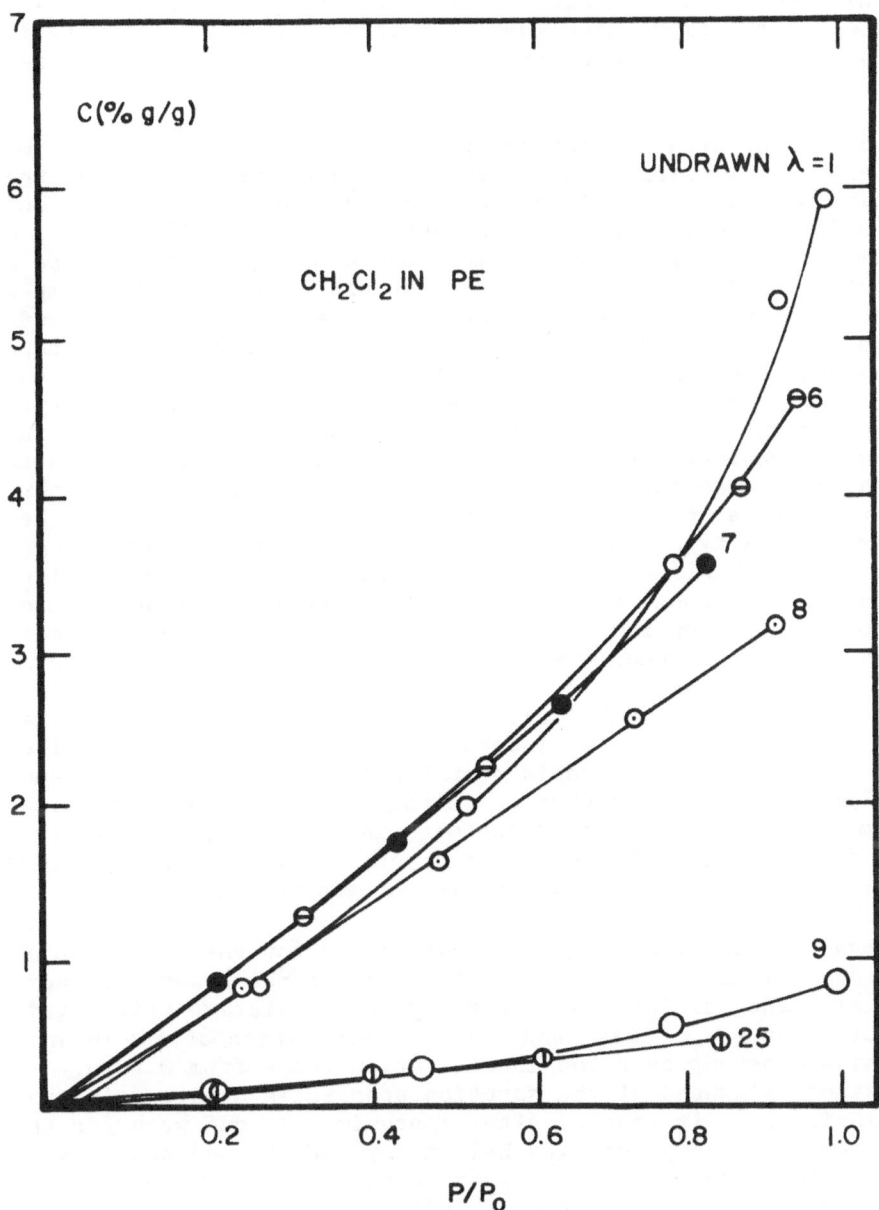

Figure 1.   Methylene chloride sorbed (c in % g/g) at 25°C in
            undrawn quenched polyethylene and in films drawn
            at 60°C to a draw ratio 6, 7, 8, 9 and 25 as
            function of vapor activity p/p_o (p_o =412 torr).

upto a draw ratio $\lambda = 8$ there is an abrupt drop at $\lambda = 9$ to a
new value which hardly changes with increasing λ as seen in Figure
2. One has the impression that at λ between 8 and 9 the material
is transformed into a new structure which does not change appreciably with further drawing.

Since the drawing also changes the density and hence the mass
crystallinity $\alpha = (p_c/p)(p - p_a)/(p_c - p_a)$ of the sample and the
sorption occurs almost exclusively in the amorphous regions, one
has to introduce the specific concentration $c_{sp} = c/(1 - \alpha)$, i.e.,
the amount sorbed per gram of amorphous component. These data are
collected in Table I for penetrant activities 0.2 and 0.9. It was
assumed in these calculations that the density of amorphous and
crystalline components of drawn polyethylene are those of a super-
cooled melt and ideal crystal, respectively. One knows that this
is a rather rough approximation because the density of the crystalline component decreases and that of the amorphous component increases with drawing. There is no drastic change in the draw-ratio
dependence of the specific regain as compared with the directly
measured data at the low draw ratios because the density and hence
the crystallinity of the quenched sample remain very nearly constant
up to $\lambda = 8$. The increase in density at $\lambda = 25$ and the ensuing
decrease in amorphous fraction $1 - \alpha$ very nearly compensates for
the concurrent drop in c so that the specific weight gain remains
nearly constant between $\lambda = 9$ and $\lambda = 25$. The small irregularity
of c_{sp} at $\lambda = 9$ and $p/p_o = 0.2$ is most likely a consequence of the
limited accuracy at such small weight gains. The specific regain
hence shows the same type of dependence on draw ratio as c, i.e.,
and initial increase, a rapid drop between $\lambda = 8$ and $\lambda = 9$, and
very little change at higher λ. Although the density increases and
the amorphous fractions decrease with increasing draw ratio it is
clear that considerably more than crystallinity changes are involved
in the drastic drop of sorption brought about by the drawing process.

Typical sorption-time transient curves for the diffusion of
methylene chloride are shown in Figure 3 for the drawn and undrawn
material. The initial slope of the plot of relative weight gain
$c(t)/c(\infty)$ versus $t^{-1/2}$ was used for the calculation of the mean
diffusion constant over the concentration range from 0 to c_∞.
The extreme slowness of the sorption process is apparent in that
equilibrium is only reached after approximately one week for the
drawn sample as compared with half a day for the undrawn polyethylene in spite of the fact that the thickness of the latter is
three times that of the former ($\lambda=9$).

The diffusion constants during the sorption process are shown
in Figure 4 for the highly drawn materials as function of sorbent
concentration in the polymer. The value extrapolated to zero
concentration of penetrant for undrawn PE is 2.7×10^{-8} cm^2/sec

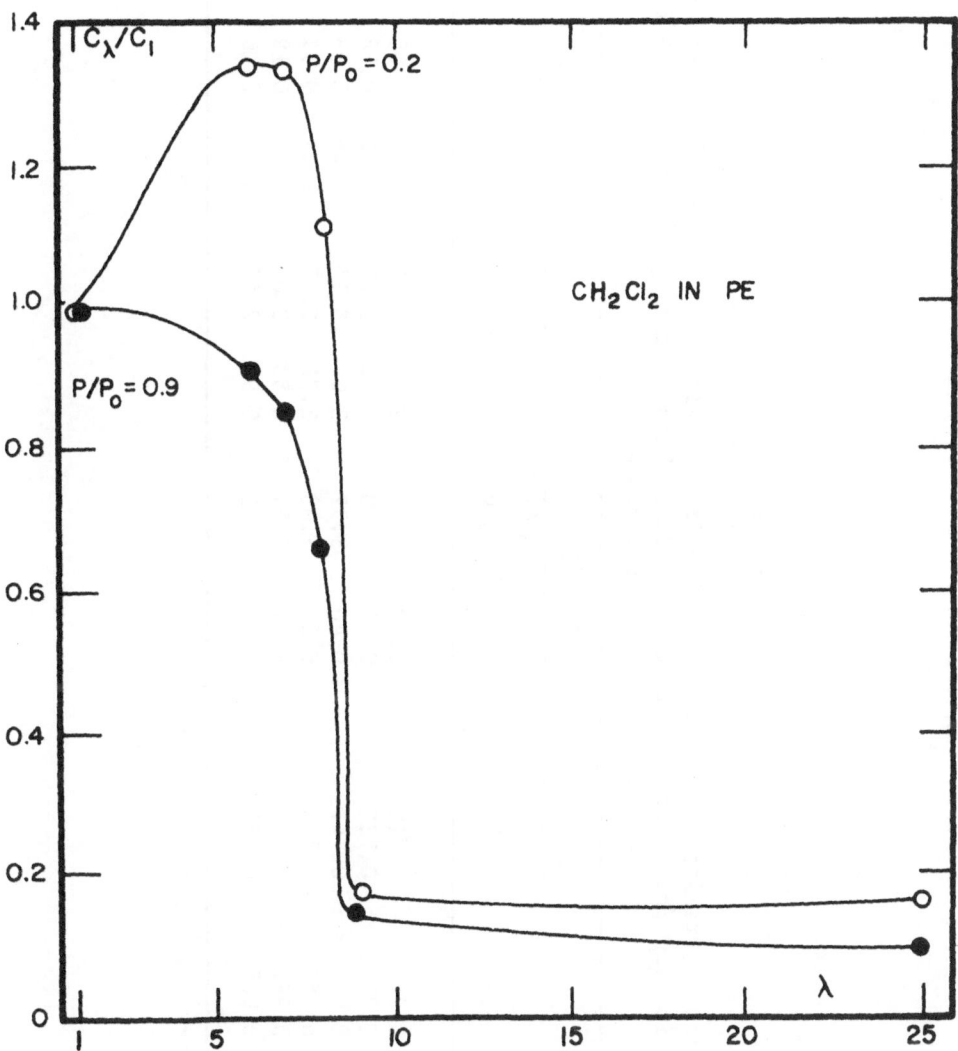

Figure 2. Influence of drawing on the relative sorption
            of methylene chloride in polyethylene at penetrant
            activities of 0.2 and 0.9.

TABLE I

Density ρ, Mass Crystallinity α, Regain c, Specific Regain $c_{sp} = c/(1 - \alpha)$ and Relative Specific Regain $c_{sp\lambda}/c_{sp,1}$ of Methylene Chloride at Vapor Activity $p/p_0 = 0.2$, 0.9 for Polyethylene Films Drawn at 60°C

Draw ratio λ	ρ, g/cc	α[a]	c, g/g		c_{sp}		$c_{sp}/c_{sp,1}$	
			$p/p_0 = 0.2$	$p/p_0 = 0.9$	$p/p_0 = 0.2$	$p/p_0 = 0.9$	$p/p_0 = 0.2$	$p/p_0 = 0.9$
1 (Undrawn)	0.949	0.718	0.60	4.59	2.13	16.21	1.00	1.00
$\lambda = 6$	0.952	0.735	0.80	4.20	3.02	15.85	1.42	0.98
7	0.953	0.741	0.80	3.92	3.09	15.14	1.45	0.93
8	0.955	0.753	0.67	3.08	2.71	12.47	1.27	0.77
9	0.962	0.793	0.11	0.64	0.53	3.09	0.25	0.19
25	0.968	0.826	0.09	0.45	0.52	2.59	0.24	0.16

[a] $\alpha = \rho_c(\rho - \rho_a)/[\rho(\rho_c - \rho_a)]$.

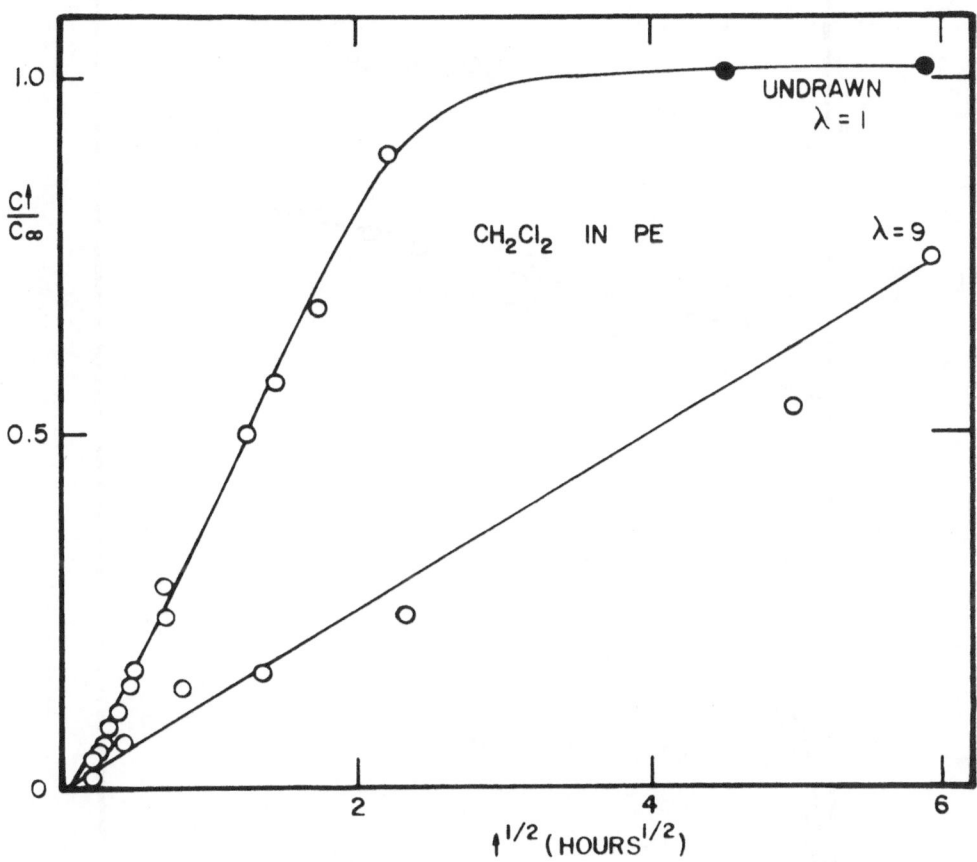

Figure 3. Typical sorption transient curve: $c_t/c_\infty = Q_t/Q_\infty$ VS.
$t^{1/2}$ for methylene chloride in undrawn and drawn
(λ =9) polyethylene at 25°C.

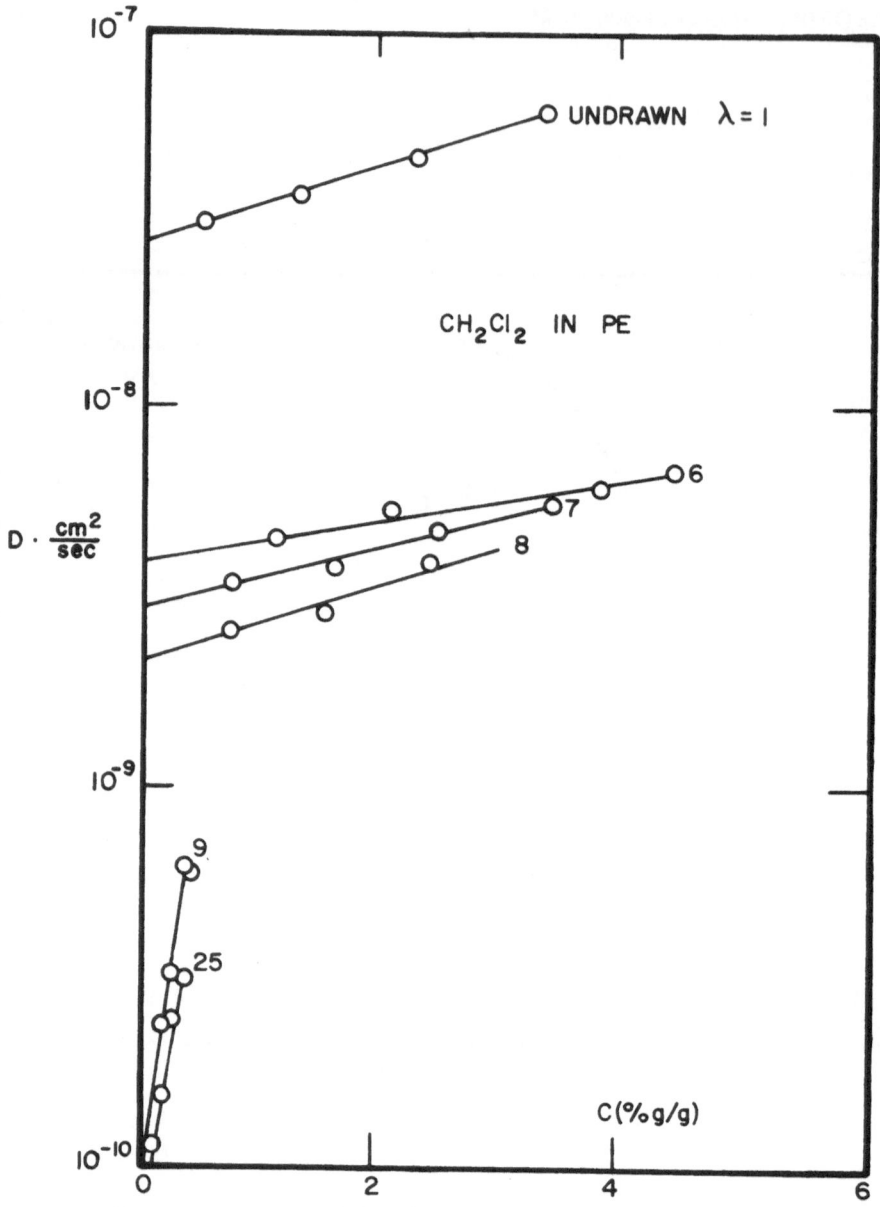

Figure 4. Diffusion constant of methylene chloride at 25°C in
           undrawn and drawn polyethylene film vs. concentration
           of penetrant. The values are derived from the initial
           slopes of sorption transient curves (Fig. 3). The
           extrapolated values D_o are given in Table I.

which is larger by a factor of 270 than the corresponding value
1.0×10^{-10} cm^2/sec for the sample with $\lambda = 9$. Again it can be seen
that upon drawing to $\lambda = 25$ only a small additional reduction of the
diffusion constant, to about 58% of the value at $\lambda = 9$, is obtained.
Also, the dependence of the diffusion constant on concentration is
nearly 20 times higher than with undrawn material and is essentially
the same for samples drawn 900% and 2500%. The diffusion constants
for the drawn material with $\lambda = 9$ and $\lambda = 25$ are 270 and 370 times
lower than for the undrawn material at zero concentration and 50
and 70 times lower for c = 0.45% respectively.

The situation is much less drastic for draw ratios between 6
and 8. The extrapolated zero concentration values for the diffusion
constant are 3.9, 2.94, and 2.13×10^{-9} cm^2/sec for $\lambda = 6, 7$, and 8
respectively. The concentration dependence is practically the same
as with the undrawn material. There is an obvious gap in experimental data between the undrawn sample and the sample with draw
ratios 8 and 9 on the other hand. Both gaps are difficult to
bridge because it is practically impossible to obtain sufficiently
large samples with a draw ratio below 6 and precision of draw ratio
measurements is not good enough for a proper definition of draw
ratios between 8 and 9.

The dependence of the extrapolated diffusion constant D_0 at
zero concentration of penetrant is given in Figure 5 as function of
draw ratio. In the semilogarithmic plot the initial drop seems to
be nearly linear with the draw ratio up to $\lambda = 8$. After that an
abrupt decrease by a factor of 20 occurs in going from $\lambda = 8$ to
$\lambda = 9$. The further decrease in diffusion is small, in fact much
smaller than in the initial stages of drawing. The value of D_0 for
$\lambda = 25$ is still 54% of the value for $\lambda = 9$. The few experimental
points do not allow a detailed curve to be drawn, but one clearly
sees the initial abrupt drop which comes to a stop at $\lambda = 9$ where
a new morphology structure seems to appear and persist with
apparently only small modifications up to the maximum observable
draw ratio.

A linear relationship is obtained for the logarithm of diffusion
constant accompanying the sorption process when plotted versus the
concentration of penetrant which is in line with results reported for
vapors in polyethylene.[2,10] The dependency of the diffusion constants on concentration can be expressed as

$$D = D_o \exp (\gamma c).$$

In this case, the concentration dependence of the diffusion constant
as measured by the coefficient remains constant or even slightly decreases up to $\lambda = 8$, abruptly jumps to a nearly 20 times larger value
at $\lambda = 9$; and remains there up to $\lambda = 25$. There is hardly an indication
of a substantial variation of γ between $\lambda = 9$ and 25 so that one is
permitted to draw a straight horizontal line between the two experi-

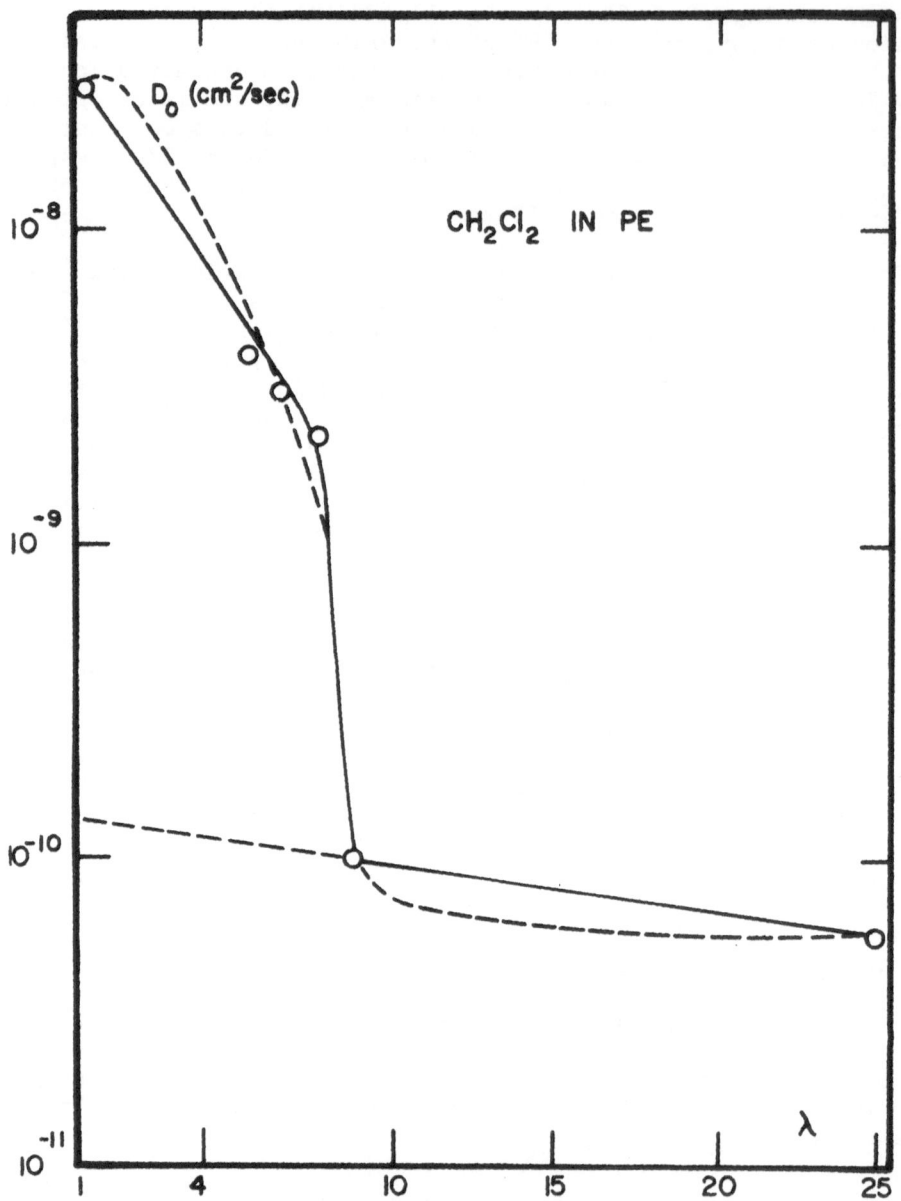

Figure 5.   Extrapolated diffusion constant of methylene chloride
            in drawn polyethylene at zero concentration of sorbent
            as a function of drawn ratio.   The broken line re-
            presents the diffusion constant in the intermediate
            points according to the model for plastic deformation
            of crystalline polymers.

mental points. The main effect is again the abrupt increase of γ between λ =8 and 9 occurring in the same range where also S and D_O exhibit drastic changes. Beyond λ =9 the changes of all three parameters are insignificant.

The drastic decrease of sorption and diffusion constants, along with the concurrent increase in the dependence of the diffusion constant on sorbent concentration, between draw ratios of 8 and 9 and their near constancy at higher draw ratios is sure indication of far-reaching structural changes of the draw material. One can safely conclude that at λ =9 the drawn polyethylene film has, at least as far as material transport phenomena are concerned, fully acquired the fiber structure with no detectable remains of the microspherulitic structure of the starting material. At this draw ratio, the transformation of stacks of crystalline lamellae into microfibrils by micronecking seems to be completed. The subsequent drawing from λ =9 to λ =25, proceeding by longitudinal sliding of microfibrils, does not change the structure of the microfibrils but only changes their mutual location.[11]

The situation becomes quite different, however, when the polyethylene film is drawn in a plasticized state. Instead of obtaining a drastic decrease in diffusion there is an abrupt increase of several orders of magnitude. These results are summarized in Table II for the diffusion of methylene chloride in normal and plasticized drawing of polyethylene. Clearly, the increase in transport properties is in direct contrast to results obtained with cold-drawn samples to date.

The sorption isotherms at 25°C for methylene chloride into PE which had been previously drawn in benzene are presented in Figure 6. The sorption behavior is completely reversible, the same curve representing the sorption and desorption data. Also shown in Figure 6 are the corresponding sorption isotherms for the undrawn material. For comparison, the sorption isotherm is included for a PE sample which had been cold drawn to a draw ratio 9 (ref. 2). It is quite evident that drawing in the presence of benzene has actually increased the sorption capacity from that of the undrawn material. For example, more than a twofold increase over the undrawn material occurs at a relative vapor pressure of 0.80. A slightly smaller increase occurs in the sorption constants S, derived from the the initial slope of the sorption curves of Figure 6 and summarized in Table II. This is in direct contrast to results obtained with the cold-drawn samples, to date. In fact, the cold-drawn sample (λ = 9) shows a reduction in the equilibrium sorption values of more than sixfold at similar vapor activities.

The apparent diffusion constants accompanying the sorption process for the PE sample drawn in benzene and calculated from

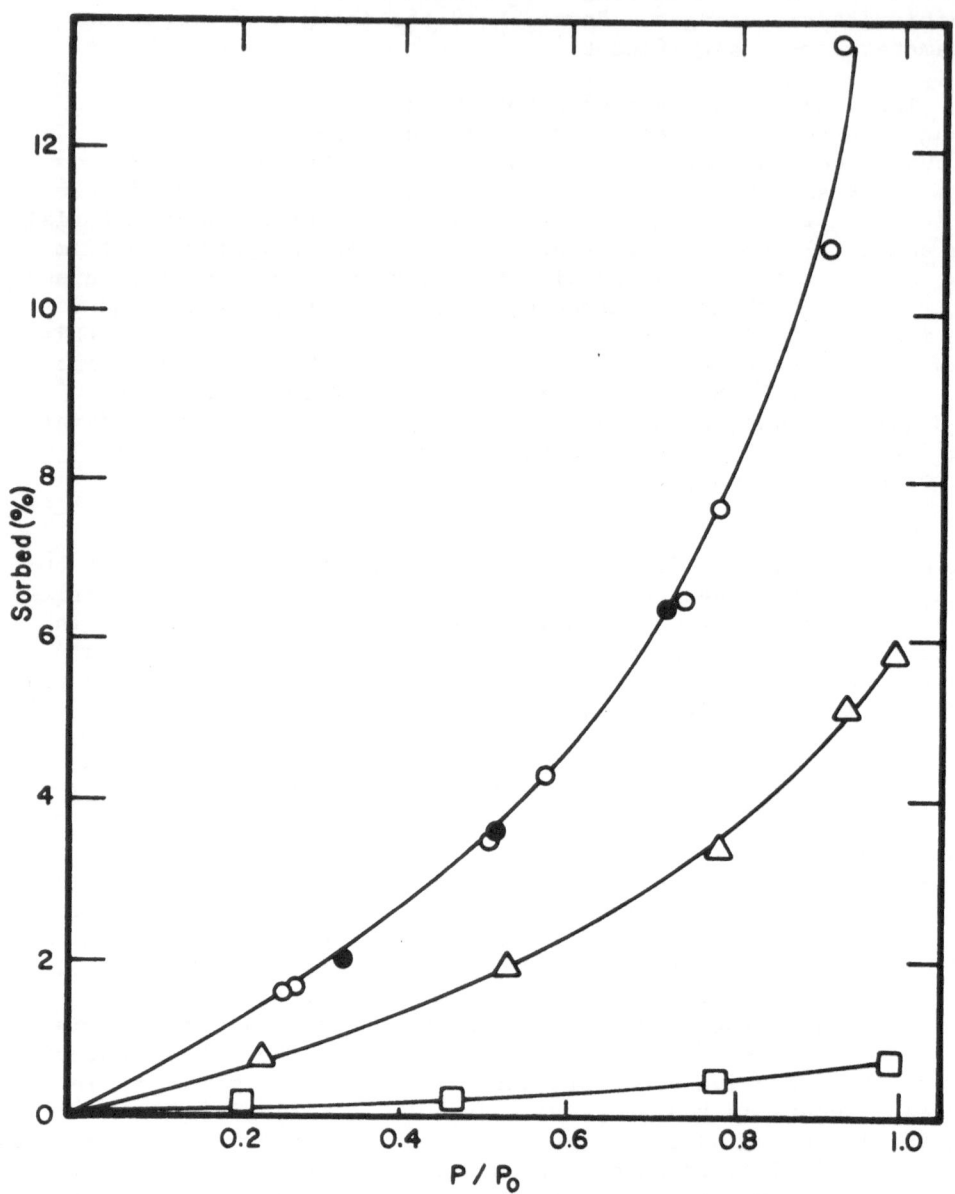

Figure 6.   Sorption isotherms for methylene chloride into various
            oriented samples of polyethylene at 25°C. Filled
            points are for desorption.O●drawn in benzene;
                △ undrawn; □ drawn (λ = 9)

TABLE II

Sorption S, Diffusion Constant D, Permeability P, and the Concentration Parameter λ of D, at Zero Sorbent Concentration, for Methylene Chloride into Polyethylene, at 25°C

Sample Description	S	$D \cdot 10^8$	$P \cdot 10^{10} = SD$	$P_{exp} \cdot 10^{10}$	γ (gm/gm)
Undrawn	0.207	2.66	55.9	– –	9.5
Cold-Drawn ($\gamma = 9$)	0.0324	0.011	0.0356	– –	172.5
Benzene drawn ($\gamma = 2$)	0.410	178.0	7300.0	580,000	0

Units for S: cm^3 (STP)/cm^3 (PE)·cm Hg; D: $cm^2\ sec^{-1}$; P: cm^3 (STP)/cm·cm Hg

the linear increase of sorption with the square root of time are
presented in a semilogarithmic plot in Figure 7 along with values for
the undrawn and cold-drawn material. Again, contrasting behavior
is shown by the solvent-drawn material in comparison to the previous
work with cold-drawn PE[2]. Instead of a 150-fold decrease in
diffusion found with the cold-drawn material, there is a 200-fold
relative increase from that of the undrawn PE in the apparent values
from that of the undrawn PE for the diffusion constants extrapolated
to zero concentration of sorbent (Table II).

From the linear relationship obtained in Figure 7, the dependence of the diffusion constants on concentration can again be
expressed as

$$D = D_0 \exp (\gamma c)$$

The resulting values of γ for the various samples are summarized in
Table II. Once again, the sample drawn in benzene shows consistently
contrasting behavior from the cold-drawn and also from the undrawn
material as evidenced by its near zero value of γ .

In view of the possibility of having a fibrous structure introduced by the solvent-drawing technique, dynamic measurements of
gas permeability were made using methylene chloride as the penetrant
in the benzene-drawn sample. Initial results (Table II) indicated
that the permeability of methylene chloride was 75 times faster than
would be expected based on the calculated value of the permeability
constant from the sorption and corresponding diffusion rates (P=DS).
Although such a discrepancy in the measured and calculated permeability values generally indicates the presence of "holes" a more
definitive measure was made using a number of different penetrants.

The pressure dependence of the apparent permeability constants
for several penetrants ranging in molecular weight from helium (4)
to methylene chloride (85) are shown in Figure 8 for the benzene-
drawn polyethylene samples. It can be seen that there is a considerable pressure dependence for the permeability constants for
most penetrants. For instance, helium shows the highest dependence
of permeability constant with pressure while its permeability behavior is known to be completely pressure independent at such low
pressures with unmodified polyethylene [12].

A plot of the extrapolated permeability constants to zero
pressure as a function of the reciprocal square root of the penetrant
molecular weight is presented in Figure 9. It is readily apparent
that for the range of molecular weights studied there is a linear
relationship when the permeability constants are plotted in such a
fashion for the benzene-drawn material. Such behavior is
characteristic for the KNUDSEN type molecular flow through small
holes at relatively small pressures[13], i.e., as long as the product
of pressure and hold radius is below 10 dyne/cm. At atmospheric

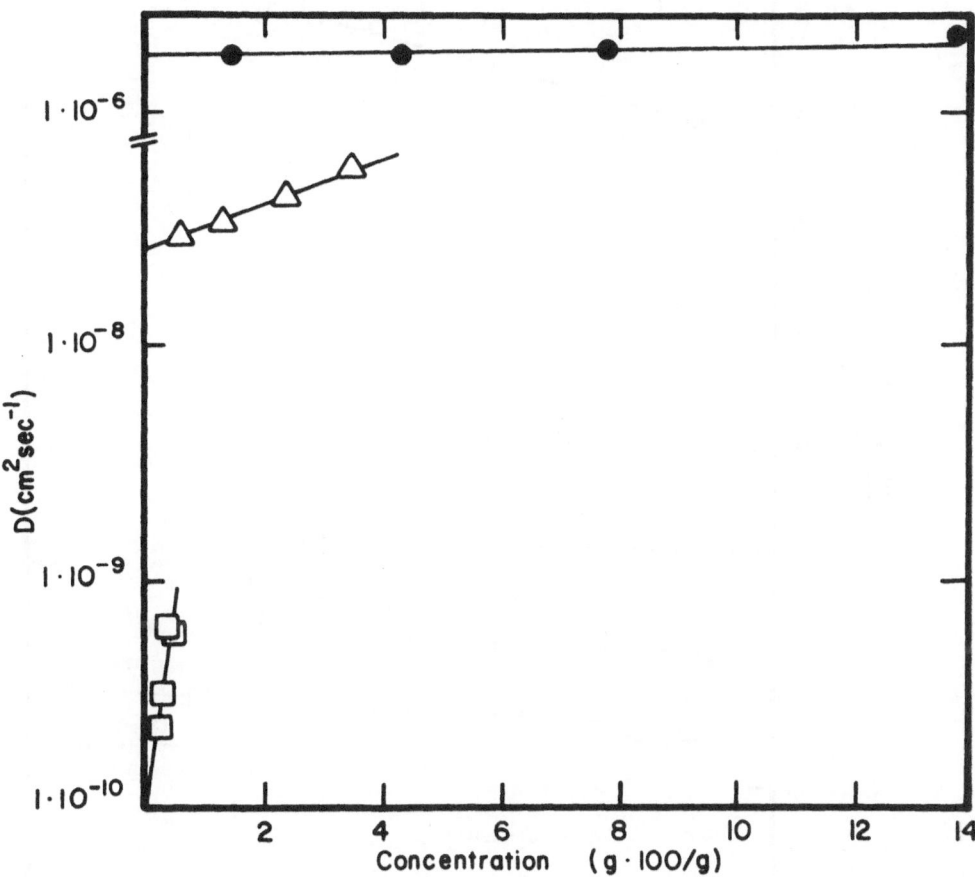

Figure 7.   Concentration dependence of the integral diffusion
            constants for methylene chloride into various oriented
            samples of polyethylene at 25°C.   ● drawn in benzene;
            △ undrawn; □ drawn ($\lambda = 9$)

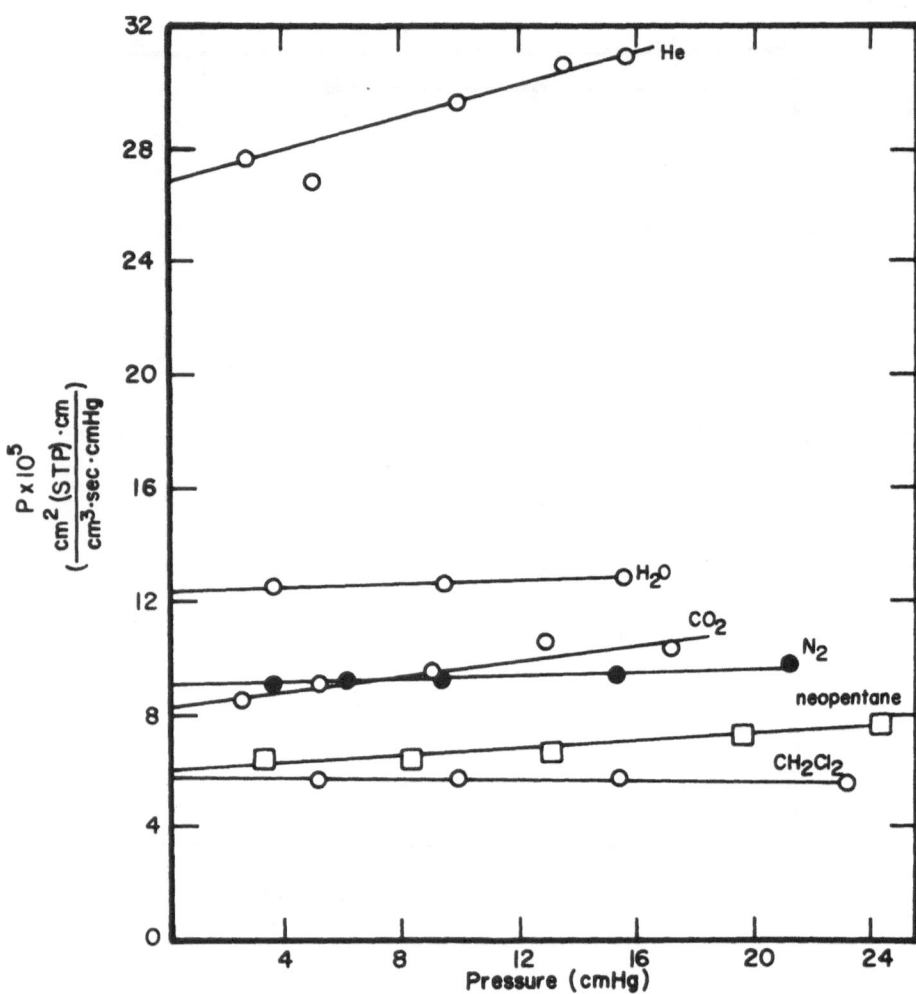

Figure 8.   Pressure dependence of the apparent permeability
            constants for various penetrants in polyethylene
            (drawn in benzene) at 25°C.

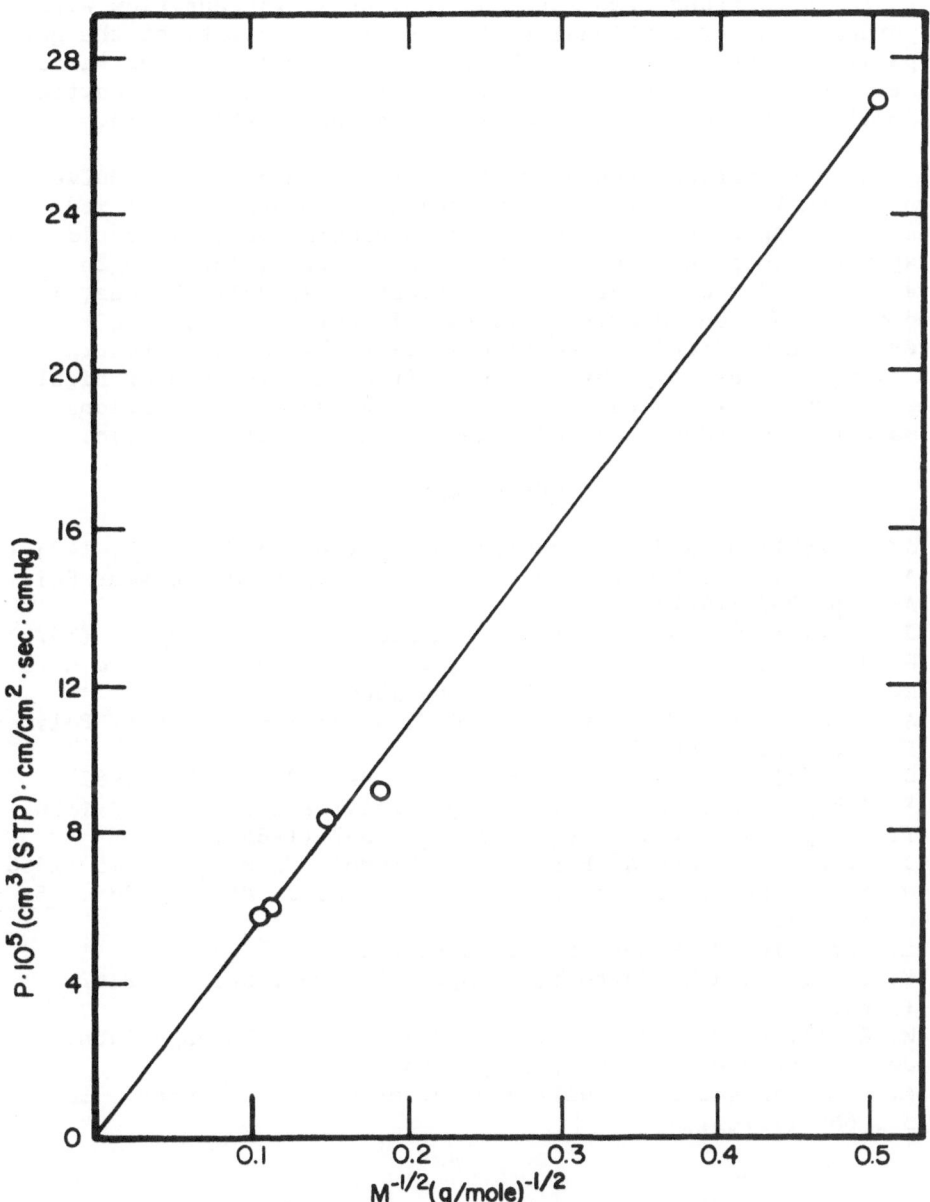

Figure 9. Dependence of the permeability constant on the
reciprocal square-root of penetrant molecular weight
for polyethylene drawn in benzene. All at 25°C

pressure, that means a hold diameter below 2000 Å. Above this
limit, the flow transforms into the convential viscous flow with
the permeability proportional to the inverse viscosity of the gas.
The parameter determining the character of the flow is the ratio
of the mean free path and the radius of the channel; viscous flow
for small and molecular flow for large values of this ratio.

This contrasting transport behavior obtained for the solvent-
drawn material is very interesting and a different or modified
mechanistic approach to the deformation process must be considered
to explain this phenomena. The results, however, indicate that
there is indeed a microporous-like structure in polyethylene, induced by the solvent-drawing process. In any event, by the
proper choice of drawing condition it is truly possible to trans-
form films such as polyethylene into either impermeable or highly
permeable materials by the proper choice of drawing conditions.
The magnitude of these induced changes can be quite dramatic.

REFERENCES

1. J. L. Williams and A. Peterlin, J. Polymer Sci. A-2, 9, 1483 (1971).
2. A. Peterlin, J. L. Williams, and V. Stannett, J. Polymer Sci.
    A-2, 5, 957 (1967).
3. J. L. Williams and A. Peterlin, Makromol. Chem. 135, 41 (1970).
4. H. J. Bixler and A. S. Michaels, paper presented at 53rd Natl.
    Meeting AICE, Pittsburgh, Pa., May 1964.
5. A. S. Michaels, W. R. Vieth, and H. J. Bixler, J. Appl. Polymer
    Sci. 8, 2735 (1964).
6. G. T. Davies and H. S. Taylor, Text. Res. J 35, 405 (1965).
7. Y. Takagi and H. Huttori, J. Appl. Polymer Sci. 9, 2167 (1965).
8. Y. Takagi, J. Apply Polymer Sci. 9, 3887 (1965).
9. J. L. Williams and A. Peterlin, Makromol. Chem. 120, 215 (1968).
10. C. E. Rogers, V. Stannett, and M. Szwarc, J. Polymer Sci. 45,
    61 (1960).
11. A. Peterlin, J. Mater. Sci., in press.
12. Y. Takagi, and H. Huttori, J. Appl. Polymer. Sci. 9, 2167
    (1965).
13. M. Knoll, F. Ollendorff, and R. Rompe, Gasentladungs-Tabellen,
    Julius Springer, Berlin 1935 p. 145.
14. A. Peterlin and J. L. Williams, J. of Coll. and Interf. Sci.,
    32, 654 (1970).

MODIFICATION OF POLYMER MEMBRANE PERMEABILITY BY GRAFT COPOLYMERIZATION

C. E. Rogers, S. Yamada*, and M. I. Ostler

Department of Macromolecular Science
Case Western Reserve University
Cleveland, Ohio 44106

ABSTRACT

The dependence of polymer membrane permeation properties on the nature of grafted polymer chain length, conformation, and domain formation have been elucidated using several membrane materials subjected to controlled graft copolymerization procedures. Improved permeation barrier characteristics of poly(isoprene-g-methylmethacrylate) to inert gas penetrants were found for short chain or densified graft domains as compared with long chain or extended domains. The permselectivity and degradation resistance of polyethylene-g-poly(potassium acrylate) membranes to ionic penetrants were considerably enhanced by surface plus internal grafting of polystyrene. The chemical nature, molecular weight, spatial distribution, and domain conformation of grafted copolymer are factors affecting membrane permeability which can be controlled by feasible variations in polymerization procedures.

INTRODUCTION

The permselectivity and overall transport behavior of polymeric membranes are very dependent on the chemical composition, structure, and morphology of the polymer (1, 2). These factors govern

*Present address:
    Industrial Products Research Institute
    Agency of Industrial Science and Technology
    Tokyo, Japan

the nature and magnitude of possible interactions between polymer
segments and penetrant molecules, the accessibility of sorption
sites within the polymeric matrix, and the mobilities of penetrant
molecules related to component interactions, polymer chain segmental mobility, and the associated free volume content and its
distribution (3, 4). Suitable variation of one or several of
these factors can produce membrane materials with unique and useful transport properties.

Graft copolymerization is a well-recognized general method
for modification of such chemical and physical properties of
polymeric materials. It is essential, however, that the conditions of graft copolymerization procedures be controlled very
carefully if it is desired to prepare materials with specifically
prescribed properties. It has become apparent, on the basis of
the studies of Stannett and his coworkers (5-8) and other investigations (9, 10), that the diffusion and solution behavior of small
molecules in membrane materials is sensitive not only to graft copolymer composition, per se, but also to composition distributions
within the membrane. Changes in the local domain structure of the
membrane can have marked effects on the distribution of a sorbed
penetrant, the nature of the diffusion mechanism, and the resultant
permeation rates.

An appropriate selection of a grafting monomer whose solution
and diffusion behavior in the membrane polymer are known, as well
as its basic polymerization characteristics, allows a degree of
control to be exercised over the graft copolymerization process.
An appreciation of the many factors which may affect the overall
polymerization, such as diffusion control, monomer clustering,
chain transfer control of graft molecular weight, and surface
polymerization, permits a realistic appraisal to be made of the
dependence of membrane permeability on the resultant copolymer
composition and structure.

The utility of suitable graft copolymerization procedures for
the modification of membrane properties has been the subject of
several investigations in this laboratory. Earlier studies (10)
established that a gradient of copolymer composition within a membrane could lead to significant directional permeation rate effects
of some potential advantage for enhanced permselective membrane
applications. The concern of subsequent studies has been on the
effects of variations in the length of graft chains, chain configurations and conformations, surface versus internal grafting,
and the development of effective membranes for use in secondary
battery systems, ion-selective electrodes, and other applications.
The influence of monomer distribution and diffusion during polymerization and the changes in graft composition and overall
polymer structure and morphology induced by radiation treatments

have been significant factors affecting the ultimate properties of the systems which were studied.

ENHANCEMENT OF MEMBRANE BARRIER PROPERTIES

The objectives in this series of studies were to gain insight into changes in transport behavior between short and long side chain polymers and between different domain constitution graft copolymers, i.e., one of which has a domain structure of aggregated branch polymer dispersed in extended chain base polymer and the other with associated backbone polymer chains dispersed in extended side chains. It is known (11) that the physical properties of the system natural rubber grafted with poly(methylmethacrylate) are greatly affected by the conformations of the elastomer and plastomer chains, dependent upon the method of preparation and subsequent treatment of the graft copolymer.

For these purposes, four types of poly(isoprene-g-methylmethacrylate) were prepared under the following types of polymerization conditions:

Type 1. Hexane swollen graft copolymer (rubber chains extended and graft chains collapsed).
Type 2. Acetone swollen graft copolymer (graft chains extended and rubber chains collapsed).
Type 3. Graft copolymer with short side chains by addition of CCl_4 chain transfer reagent (or prolonged radiation degradation).
Type 4. Graft copolymer with long side chains obtained by relatively short, low dose-rate radiation exposure in absence of any chain transfer reagent or solvent.

Typical permeability data for gases in these types of graft copolymer materials are listed in Table I. It is apparent that graft copolymers prepared in the presence of acetone (extended graft domains) have higher permeabilities than those samples prepared in the presence of hexane (collapsed graft domains). The consequent decrease in the rates of permeation are explicable in terms of an anticipated lower permeability of coherent graft regions to the gas flow. The presence of short graft chains acts as relatively inert filler, or excluded volume by chain packing effect, more effectively than does longer graft chains. The short chains are considered to be distributed along the backbone chain allowing a more efficient packing and structural densification than do the relatively isolated compacted long chain domains even though those local domains may be more impermeable, per se. The detailed dependence of solubility and diffusion coefficients on the nature of the graft copolymerization process

TABLE I

Effect of Graft Chain Length and Conformation on Gas Permeation Properties for Poly(isoprene-g-methylmethacrylate)

Polymerization Condition Type	% wt Graft	$P \times 10^9$ (cc cm/cm^{2}sec cmHg)		
		He	Ar	N$_2$
--	0	3.7	2.7	1.1
1-graft collapsed	8	2.3	0.8	0.3
2-graft extended	8	2.9	1.3	0.5
3-short graft chains	5	2.4	0.2	0.6
4-long graft chains	5	2.9	1.9	0.8

and the related variations in polymer structure and composition
shed further light on the mechanism of transport and property-
structure relationships.

Data are presented in Figure 1 for the relative permeability,
P/P_o, defined as the permeability of the grafted specimen divided
by the permeability of the original ungrafted polymer, as a function of graft content for the case of methylmethacrylate grafted

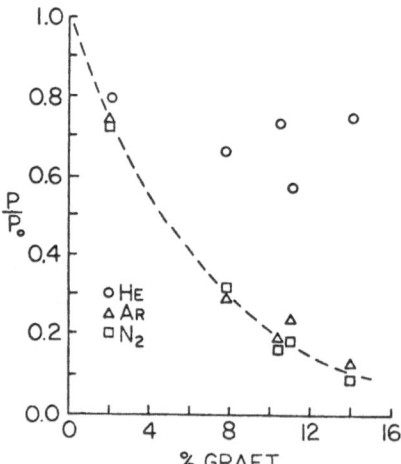

Figure 1. Relative permeability coefficient versus percent polymethylmethacrylate graft content in rubber films with hexane as
swelling agent during polymerization.

into a rubber substrate in the presence of hexane as diluent and swelling agent. Supplementary experiments conducted over the same range of radiation dosage in the absence of monomer confirmed that the level of concurrent crosslinking or related effects was not sufficient to cause significant variation in transport properties, per se.

It is noted that the grafting process has a pronounced effect on the permeation of the test gases in the samples. The permeabilities of the larger size penetrants, argon and nitrogen, are more affected than is helium. The behavior of argon and nitrogen were found to be very similar in the determinations of both permeability and diffusion coefficients under all conditions for all samples. The lesser effect of the presence of grafting on the transport of helium gives an indication of the domain density of the grafted regions within the elastomeric matrix. It is apparent that the densification of local structure is occurring on a packing scale such that there is relatively little change in the void distribution population or ease of hole formation for the migration of helium molecules as compared with argon or nitrogen.

Generally similar effects, although better defined, are obtained for the case of grafting in the presence of acetone, a solvent for polymethylmethacrylate and nonsolvent for rubber. These data are plotted in Figure 2. Again, the relative depression of permeability with increasing graft content is greater for the larger molecules than it is for helium. The decrease in transport properties are nearly linear over the graft content range which was explored.

The diffusion coefficients for these same cases were determined by the transmission time-lag method. The data corresponding to the two cases of polymerization in the presence of either hexane or acetone are shown in Figures 3 and 4. The effects of grafting are very similar to the effects on the permeation process both in the dependence on penetrant size and the general magnitude and linearity of the decrease in transport with graft content. This is an immediate indication that the presence of graft copolymer is affecting the rate of migration of penetrant molecules more than it is causing any change in the number of diffusing molecules, i.e., the solubility.

A measure or comparison of the effects of grafting on the diffusion and solution processes for the various gases in the different samples is given by a plot of relative permeability versus relative diffusion coefficients. If the only effect of grafting was to change the diffusion process, than the plot will represent a one-to-one relationship between the relative permeability and diffusion coefficients in agreement with the definition of perme-

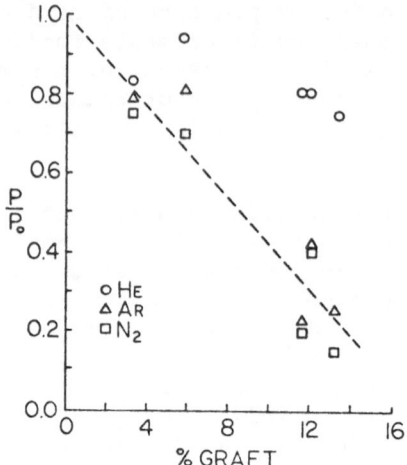

Figure 2. Relative permeability coefficient versus % PMMA graft
content in rubber films(acetone present during polymerization).

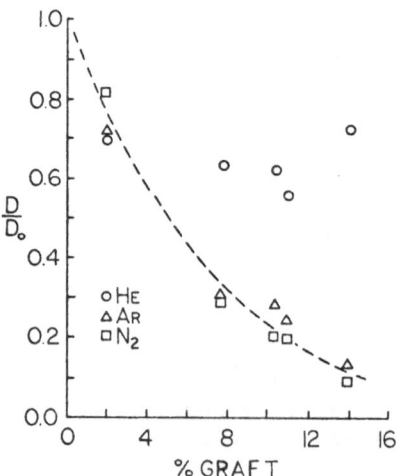

Figure 3. Relative diffusion coefficient versus % graft PMMA graft
content in rubber films (hexane during polymerization).

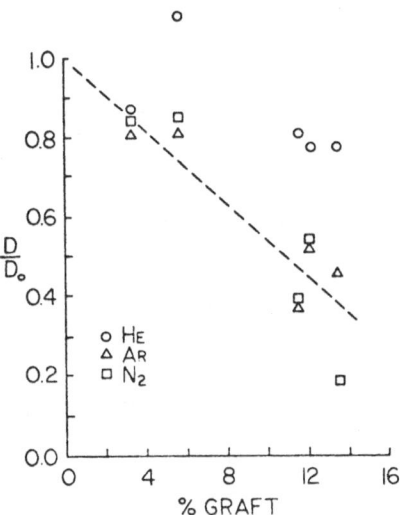

Figure 4. Relative diffusion coefficient versus % PMMA graft content in rubber films(acetone present during polymerization).

ability as the product of diffusion and solubility coefficients;
P = DS. As an example, a decrease in diffusion coefficient by one-
half and no change in solubility would cause a decrease in permeability by one-half. If the solubility coefficient also was affected by the presence of graft copolymer, then the change in
permeability will be greater or less than that for diffusion.

     The data from Figures 1 through 4 are plotted in this fashion
in Figure 5. If only the diffusion process was affected then the
data points should fall along the line drawn for the one-to-one
relationship. Such behavior is effectively obtained. Any
changes in the solubility, indicated by those data points well off
the line, are relatively minor in comparison to the changes in
diffusion coefficients.

     This comparison of the effects of graft content on diffusion
and solution processes suggests that the dominant effect of grafting, in the system under study, is to impede the migration of penetrant rather than cause much variation in the equilibrium amount
or distribution of penetrant molecules within the polymer. Effectively the same number of sorption sites exist but these sites are
less readily accessible. The minor change in sorption level further suggests that sorption is probably taking place in an envi-

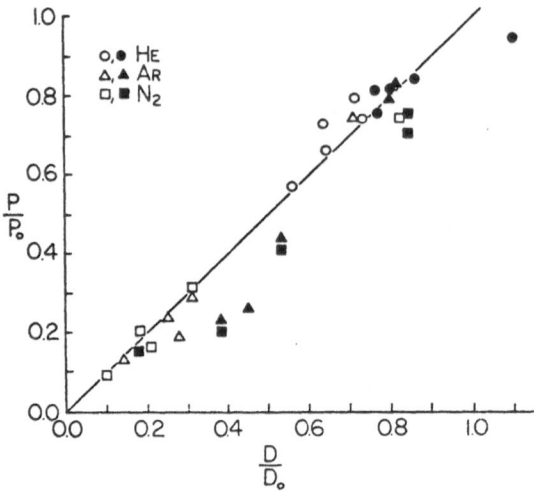

Figure 5. Relative permeability versus relative diffusion in rub-
ber-g-MMA copolymers [hexane(open symbols) or acetone(solid symbols) present during polymerization].

ronment quite similar to that in the original elastomer sample.
The effects of excluded sorption site volume or overall change
in site energies, related to chemical composition and interaction
variations, are not apparent. These considerations lead to the
proposition that the graft polymer exists within domains in the
more readily accessible regions of the original polymer. Their
presence in those regions would be expected to reduce the subsequent rates of penetrant transport without substantial changes
in the eventual sorbed penetrant content.

The presence of acetone during the polymerization leads to a
greater change in solubility than for the case of hexane. The
more extended grafted polymethylmethacrylate domains would be expected to be more effective as excluded sorption volumes than
would be the case for constricted graft domains resulting from hexane nonsolvent treatment. The deviation of the acetone-treated
sample data points from the linear P/P_O versus D/D_O relationship
is consistent with this expectation.

A more uniform distribution of shorter chain graft copolymer
could be expected to show somewhat different behavior. The dis-
persal of bonded chemically dissimilar species throughout the
elastomer matrix would correspond in effect to a change in overall
polymer composition and density. These changes should affect both

diffusion and solution processes. Consequently, a plot of relative permeability and diffusion coefficients should deviate from
a one-to-one relationship.

   This type of short chain graft copolymer may be obtained by
conducting the radiation initiated polymerization in the presence
of an active chain transfer agent. In this study, methylmethacrylate monomer/carbon tetrachloride mixtures were used to prepare
samples with up to 20% graft content. The results of permeation
and diffusion determinations in the samples is shown in Figure 6.
Once again, the effect of grafting on helium transport is much
less than that on argon and nitrogen. The data for argon and
nitrogen would seem to indicate a decided effect on both the diffusion and solution processes as evidenced by the deviations from
linear representation between diffusion and solution. This conclusion is not well established, however, since other limited
experiments, using very low dose rates in the absence of any chain
transfer agent, gave one set of data which also deviated from the
linear relation, Figure 6. A low dose rate, corresponding to a
low rate of initiation of polymerization in an essentially constant
concentration of monomer, should produce isolated local domains of
higher molecular weight graft copolymer. This is the opposite
situation than that obtained using chain transfer agents. These
experiments are complicated further by concurrent swelling action

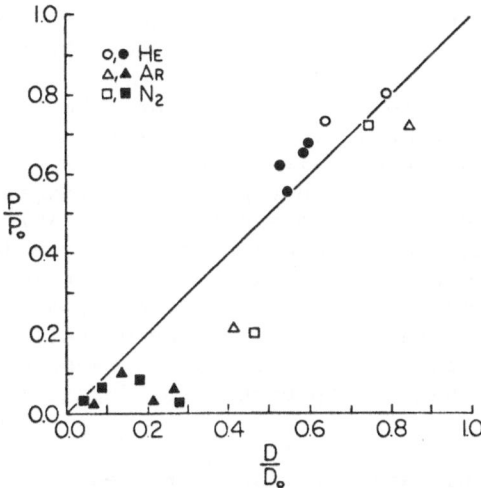

Figure 6. Relative permeability versus relative diffusion in rub-
ber-g-MMA copolymers (Solid symbols: chain transfer agent; open
symbols: low dose rate).

of sorbed monomer, solvent, and chain transfer agents. The sum
of these effects must be resolved before any more detailed
explanation is proposed.

On the basis of the data which have been obtained it seems
reasonable to state that the major effect of grafting on penetrant
transport is to reduce the diffusion rate with generally lesser
effect on the sorption magnitude. The precise domain structure,
as affected by solvent treatments, the chain length and spatial
distribution of graft copolymer, and other factors such as thermal
or mechanical treatments, remains an important aspect for establishing variations in permeation behavior-especially the permselectivity.

ENHANCEMENT OF PERMSELECTIVITY

A great many present and potential applications of polymer
membranes relate to the ability of membrane systems to affect separations of gaseous or liquid mixtures comprised of components of
quite similar chemical composition and/or physical structure, size,
or shape. The sensitivity of permeation processes to rather subtle variations in penetrant characteristics is a primary factor in
these applications. Suitable modifications in polymer composition
and structure, as they affect the solution and diffusion behavior
of low molecular weight materials within them, can enhance considerably the efficiency of such permselective processes.

As an example, certain high current density battery systems,
such as zinc-silver-KOH, require the use of a separator membrane
to prevent contact or contamination of the electrodes. The membrane must allow adequate transport of electrolyte ions for battery operation while preventing, or at least retarding, the permeation of slightly soluble silver salts into the zinc half-cell
compartment. In a recent study (12) an existing separator membrane material, crosslinked polyethylene-graft-poly(potassium
acrylate) was modified by further graft copolymerization using an
inert grafted polymer to obtain a material suitable for critical
applications. A major consideration in this procedure was the
relative effectiveness of internal grafting within or about the
domains of grafted poly(potassium acrylate) compared with the
formation of surface grafts to change the overall permeation behavior and the resistance of the system to redox degradation of
the acrylate graft chains with concurrent deposition of metallic
silver.

The original membrane material contained about 25 weight percent of poly(potassium acrylate). The subsequent grafting of styrene into this material was carried out under conditions which led
to either predominantly internal grafting, or predominantly sur-

TABLE II

Effect of Styrene Graft Modification on Membrane Properties

Grafting Yield(%)	KOH Flux (moles/cm^2-sec $\times 10^6$)	Silver Salt Flux(moles/cm^{2}hr $\times 10^8$)	Silver Deposition Rate(g/cm^2 hr $\times 10^8$)	Flux Ratio (J_{Ag}/J_{OH} $\times 10^8$)
0	3.94	1.40	2.14	9.87
8.5	3.66	0.51	1.48	3.85
14.9	3.62	0.24	1.73	1.80
19.6	2.49	0	(1.1)	0

face grafting, or a prescribed amount of each mode of grafting. These procedures involved control of the amount of monomer sorbed at the time of grafting, its distribution within the polymer and in the ambient phase, the dose-rate and dose of initiating radiation and the presence of different solvent media.

Results of measurements of the fluxes of the KOH electrolyte and silver salt, the silver deposition rate, and ratio of the silver salt to KOH fluxes are given in Table II. Consideration of these data indicates that this method of membrane modification does produce an improved battery separator material. The transport of soluble silver salt is significantly reduced without seriously decreasing the electrolyte flow. Other data indicate that the internally grafted polystyrene acts mainly as a blocking agent to protect the PKA from degradation action of the silver salt. The effect of the surface graft is to present a densified layer to the penetrating species, resulting in an increase in membrane permselectivity.

The results of these, and other, studies of membrane modification by graft copolymerization indicate that more precise control of membrane permeation processes can be obtained by suitable variations in the composition and structure of polymer materials. The feasibility of such modification procedures depends on knowledge of the complex interdependence between transport behavior and the nature of membrane materials and on the effectiveness by which graft copolymerization procedures can be controlled to yield polymeric materials of known and desired composition, structure, and properties. It is anticipated that these methods may prove to be useful procedures for preparation of membrane materials with desirable permeation properties.

REFERENCES

1. J. Crank and G. S. Park, Eds., <u>Diffusion in Polymers</u>, Academic
    Press, London, 1968.

2. C. E. Rogers, in <u>Physics and Chemistry of the Organic Solid
    State</u>, Vol. II., D. Fox, M. Labes, and A. Weissberger, Eds.,
    Interscience, New York, 1965, Chapter 6.

3. D. Machin and C. E. Rogers, "The Concentration Dependence of
    Diffusion Coefficients in Polymer-Penetrant Systems", CRC
    Critical Reviews in Macromolecular Sci., CRC Publ., Cleveland,
    April 1972.

4. C. E. Rogers, J. R. Semancik, and S. Kapur, Polymer Sci. Tech.,
    $\underline{1}$, 297 (1973).

5. A. W. Myers, C. E. Rogers, V. Stannett, M. Szwarc, G. S.
    Patterson, A. S. Hoffman, and E. W. Merrill, J. Appl. Polym.
    Sci., $\underline{4}$, 159 (1960).

6. J. L. Williams and V. Stannett, J. Appl. Polym. Sci., $\underline{14}$,
    1949 (1970).

7. H. B. Hopfenberg, V. Stannett, F. Kimura, and P. T. Rigney,
    Appl. Polym. Symp. No. 13, 139 (1970).

8. V. Stannett, H. B. Hopfenberg, and J. L. Williams, Polym. Sci.,
    Tech., $\underline{1}$, 321 (1973).

9. R. Y. M. Huang and P. Kanitz, J. Appl. Polym. Sci., $\underline{13}$,
    669 (1969); $\underline{15}$, 67 (1971).

10. C. E. Rogers and S. Sternberg, J. Macromol. Sci., Phys., B5(1),
    189 (1971).

11. F. M. Merrett, J. Polym. Sci., $\underline{24}$, 467 (1957).

12. M. I. Ostler and C. E. Rogers, J. Appl. Polym. Sci., $\underline{18}$(6),
    1359 (1974).

THE SORPTION AND DIFFUSION OF WATER IN POLYURETHANE ELASTOMERS

J. A. BARRIE,   A. NUNN   and   A. SHEER

Department of Chemistry, Imperial College of Science

and Technology,   London, S.W.7 2AY

ABSTRACT

The sorption and permeation of water has been measured at
several temperatures for three polyurethane elastomers, namely,
Adiprene CM with a polyether soft segment and Elastothane ZR625
and Genthane S both with polyester soft segments. The
equilibrium sorption behaviour of all three materials is similar
and the heat of dilution is zero, or close to zero, over most of
the concentration range. The results indicate a comparatively
strong interaction between the water and the polymer but the
isotherms themselves do not exhibit significant Langmuir-type
curvature at lower relative pressures. A Zimm-Lundberg analysis
of the isotherms reveals that clustering of the sorbed water tends
to dominate over most of the relative-pressure range. This is
consistent with the observation that the diffusion coefficient, D,
decreases with concentration for all three elastomers. The
concentration dependence of D, bearing in mind the moderate water
solubilities of these materials, is rather weak and the activation
energy for diffusion is virtually constant. The results suggest
that competing processes are operative which tend to oppose the
effect of clustering on D.

INTRODUCTION

The measurement and analysis of the concentration dependence
of the diffusion coefficient of small molecules in polymer films
has proved valuable in the interpretation of the mechanism of the
transport process in these systems. Inert gases are weakly sorbed

at normal pressures and the diffusion coefficient D is generally
constant. For more strongly sorbed penetrants which tend to
plasticize the polymer D normally increases with the concentration,
C, often quite rapidly, and this behaviour can be interpreted in
terms of free-volume theory[1,2]. With water as penetrant both
types of behaviour are observed; in addition for a number of
polymers it is found that D decreases with C when the penetrant,
like water, has the ability to self associate through hydrogen bond
formation.[2,3,4]

The hydrophilic polymers such as nylon, polyvinylalcohol and
the natural fibres, typify systems for which D increases rapidly
with C. The increase in D may be attributed to localized
Langmuir-type sorption in the early stages and to plasticization at
higher concentrations. Generally the permeability also increases
rapidly with relative pressure and the energy of activation for
diffusion tends to decrease with concentration.

For a number of polymers, such as the polyalkylmethacrylates,
ethylcellulose and silicone rubber, the diffusion coefficient
decreases as the concentration of sorbed water increases; this
effect is marked with salt-loaded rubbers. The permeability is
generally not strongly dependent on the relative pressure and in
many instances is constant over the whole of the range. The
activation energy for diffusion normally increases with C but can
be constant as for polymethylmethacrylate. These features have
been interpreted in terms of clustering of the penetrant so as
to render an increasing fraction of it immobile[2,3,4].

With polyvinylacetate and polyethyleneterephthalate, D is
constant even although these polymers have moderate affinities for
water. Similar behaviour is observed for methanol in ethylcellulose
up to sorptions of several per cent. It appears that compensating
effects may be responsible for the constant D behaviour of these
systems. In a few cases maxima are observed in the (D,C) curves
when localized sorption which dominates at lower concentrations
gives way to clustering at higher levels.

The silicone rubbers are representative of materials with
relatively low affinities for water and have been studied in some
detail[4]. In the present investigation the equilibrium sorption
and steady state permeation of water in three polyurethane
elastomers of moderate water solubilities is investigated. A
sorption-desorption kinetic study of water diffusion in a similar
class of materials has also been made[5,6]. The polyurethane
elastomers offer a class of materials for which the water solubility
can vary from moderate to high values depending on the detailed
chemical structure[6].

EXPERIMENTAL

The following polyurethane elastomers were used. Adiprene CM
(E.I. du Pont de Nemours & Co.) with a polyether soft segment is
formed from poly(oxybutylene) glycol reacted with 5 to 10 per cent
of glycerine mono allyl ether and then polymerized with 2,4 toluene
diisocyanate (TDI). Genthane S (General Tire and Rubber Co.) with
a polyester soft segment is prepared from an 80:20 mixture of
ethylene:propylene adipates, further polymerized with methylene-
pp-diphenyl diisocyanate (MDI). Elastothane ZR625 (Thiokol Chemical
Corp.) has a polyester soft segment and is formed from a
caprolactone and T.D.I.; details of the composition were not
made available.

The relative molar concentrations of groups were determined
from proton magnetic resonance spectra and the reported
compositions. For Adiprene CM these were $-CONH.C_6H_3(CH_3).NHCOO-$:
$-(CH_2)_4O-$ as 26:192 and for Genthane S,

$-CONHC_6H_4.CH_2.C_6H_4NHCOO-$: $-CH(CH_3)CH_2O-$: $-CH_2CH_2O-$: $-OC(CH_2)_4COO-$
as 18:65:118:165[7].

The membranes used for both sorption and permeation measurements
were lightly crosslinked with dicumyl peroxide. Film densities
determined by water displacement and glass transition temperatures
using a Du Pont 900 D.S.C. are given in Table 1 along with values
of the membrane thickness.

TABLE   1

DENSITY, THICKNESS AND GLASS TRANSITION
TEMPERATURE OF MEMBRANES

Polymer	Density ρ,gcm^{-3}	Thickness ℓ,cm	T_g $^\circ$C
Adiprene CM	1.08	.082(.075 at 60°C)	-56
Genthane S	1.18	.043	-47
Elastothane ZR625	1.18	.072	-37

Sorption isotherms were measured with a conventional calibrated
silica spiral and strain-gauge, pressure transducer (0-15psi).
Steady-state permeation rates were determined as a function of the
vapour pressure on the ingoing face of the membrane. The pressure
on the outgoing face was always less than 0.5 per cent of that on
the ingoing face. Permeation rates were measured by sorbing the
effluent vapour in a dehydrated aluminosilicate suspended from a
calibrated silica spiral[8]. Alternatively the downstream pressure

was monitored with an electronic micromanometer (MKS Instruments Inc.) and a receiving volume of 6 litres or more to minimise effects arising from sorption of water on the glass surfaces[9]. The sorption of water on the glass surfaces has a more pronounced effect on the time-lag and for this reason accurate measurements of this quantity were not attempted with this apparatus[9,10].

RESULTS

Sorption

Equilibrium sorption isotherms were determined at several temperatures and are shown in Figures 1 to 3. The accuracy of measurement is not sufficient to determine precisely the shape of the isotherm at low relative pressures. However, there is no evidence for pronounced localized or Langmuir-type sorption.

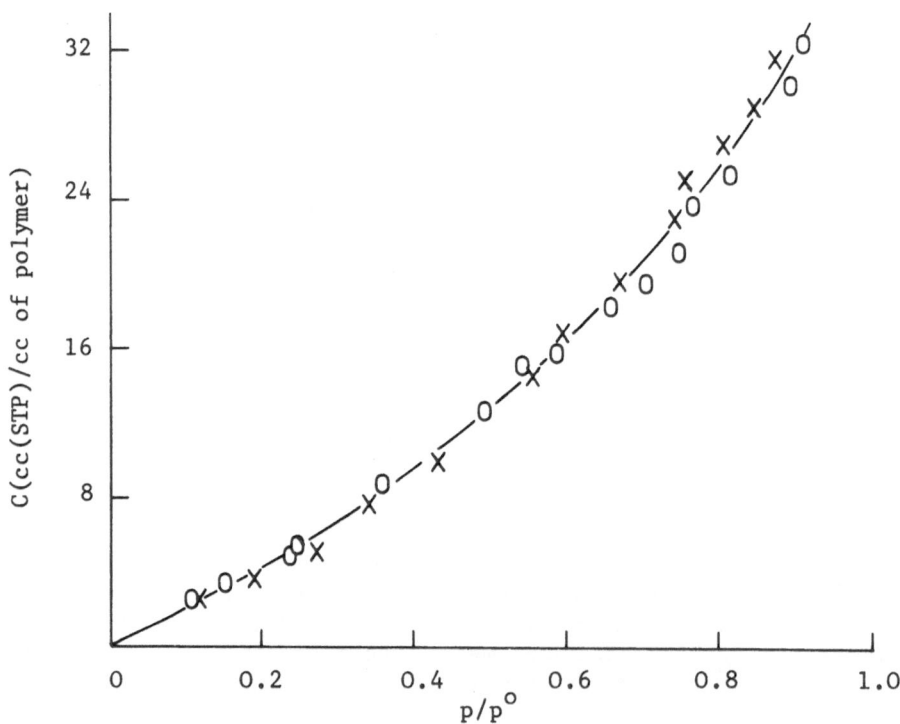

FIGURE 1.   Sorption isotherms for Adiprene CM. (O) 35.8°, (X) 60.2°C.

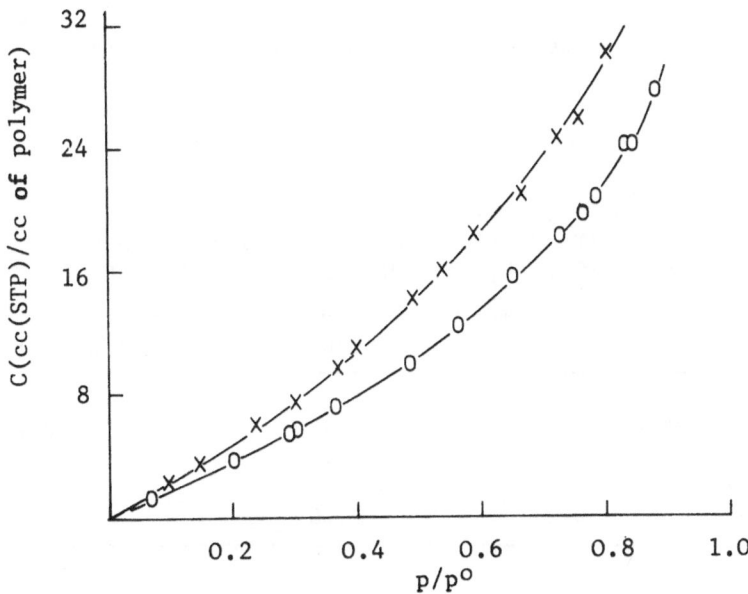

FIGURE 2.   Sorption isotherms for Genthane S. (O) 34°, (X) 60°C.

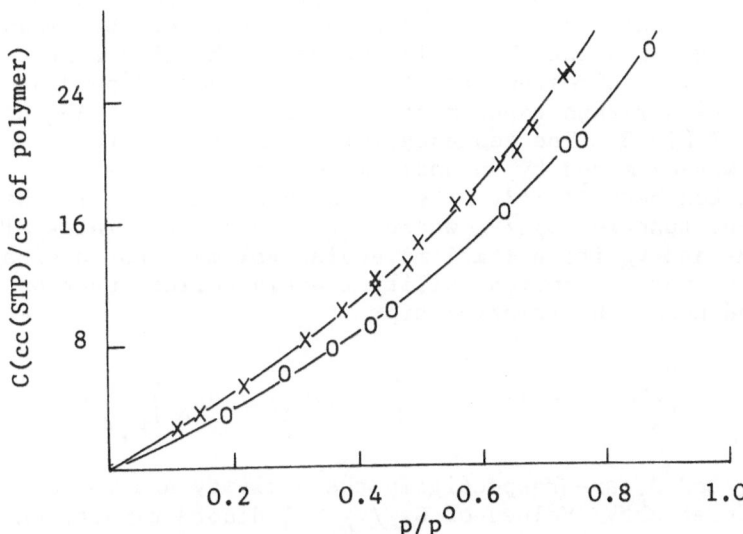

FIGURE 3.   Sorption isotherms for Elastothane ZR 625.
            (O) 35.1°, (X) 60.4°C.

From the limiting values of the slope at zero concentration the Henrys law solubility coefficient, $\sigma_{c=0}$, was estimated. Typical values of $\sigma_{c=0}$ are given in Table 2 along with **values** of the heat of sorption, ΔH_s, obtained from $\sigma_{c=0} = \sigma_0 \exp(-\Delta\overline{H}_s/RT)$.

TABLE   2

SOLUBILITY COEFFICIENTS, $\sigma_{c=0}$, AND HEATS OF
SORPTION, $\Delta\overline{H}_s$, IN THE LIMIT C→0 AT 50°C

Polymer	$\sigma_{c=0}$ cc(stp)/cc.cmHg	$\Delta\overline{H}_s$,kcal mole^{-1}
Adiprene CM	2.3	10.3
Elastothane ZR 625	2.4	8.2
Genthane S	2.1	7.6
Silicone Rubber [4]	0.070	5.8
Fluoro Silicone Rubber [4]	0.104	5.1

Heats, $\Delta\overline{H}_A$, and entropies, $\Delta\overline{S}_A$, of dilution were independent of temperature and were calculated as a function of the amount of water sorbed. For Adiprene CM the isotherms at the different temperatures were virtually coincident and $\Delta\overline{H}_A$ is zero and independent of concentration within experimental error. Values of $\Delta\overline{H}_A$ and $\Delta\overline{S}_A$ are given in Table 3. The tendency for the penetrant to segregate or cluster was examined by an analysis of the isotherms according to Zimm and Lundberg[11,12]. The **quantity** of interest is the clustering function G_{AA}/V_A where G_{AA} is a cluster integral for the **penetrant** and V_A its partial molecular volume. For a binary system of low isothermal compressibility the clustering function can be evaluated using the relationship

$$\frac{G_{AA}}{V_A} = -(1 - \phi_A)\left\{\partial(a_A/\phi_A) \,/\, \partial a_A\right\}_{P,T} - 1 \qquad (1)$$

where a_A and ϕ_A are respectively the activity and volume fraction of the penetrant. Values of $G_{AA}/V_A > -1$ denote clustering and values < -1 segregation of the penetrant. Values of the clustering function at different activities and several temperatures are given in Table 4.

TABLE   3

HEATS, $\Delta\overline{H}_A$, AND ENTROPIES, $\Delta\overline{S}_A$, OF DILUTION

Polymer (35 to 60°C)	C cc(stp)/cm^3 of polymer	$\Delta\overline{H}_A$ kcal mole^{-1}	ΔS_A cal mole^{-1} K^{-1}
Adiprene CM	0	0	∞
	2	0	4.7
	10	0	1.7
	16	0	1.0
Elastothane ZR 625	0	2	∞
	2	2	11
	10	1.6	6.7
	16	1.3	5
Genthane S	0	2.7	∞
	6	2.1	9
	18	1.6	5.7
	24	1.2	4.5
Silicone Rubber[4]	0	3.9	∞
	0.6	1.3	4

TABLE   4

CLUSTERING FUNCTION, G_{AA}/V_A, FOR WATER IN
POLYURETHANE ELASTOMERS

Polymer	T, °C	G_{AA}/V_A (a_A = 0.3)	G_{AA}/V_A (a_A = 0.7)
Adiprene CM	35.8–60.2	12	37
Elastothane ZR625	35.1	26	41
	50.4	14	32
	60.4	14	35
Genthane S	34	31	40
	52	25	34
	60	23	27
Silicone Rubber[4]	30.3	−1	960
	50.0	−1	370

Permeation

Typical plots of the normalized flux, $J\ell$, versus the relative pressure on the ingoing face are shown in Figures 4 to 6. These plots exhibit significant if not pronounced departures from linearity, particularly at the higher relative pressures. Using the isotherm data, curves of $J\ell$ versus C were constructed and differentiated graphically to give D as a function of C, since $D_{c=c_1} = (dJ\ell/dC)_{c=c_1}$. The concentration dependence of the diffusion coefficient is illustrated in Figures 7 and 8. Values of the diffusion coefficient and the permeability $\overline{P}[cc(stp)cm/cm^2(cmHg)sec]$ in the limit of zero concentration are given in Table 5. The activation energy for diffusion, E_D, was obtained as a function of concentration from plots of log D vs 1/T constructed at several values of C.

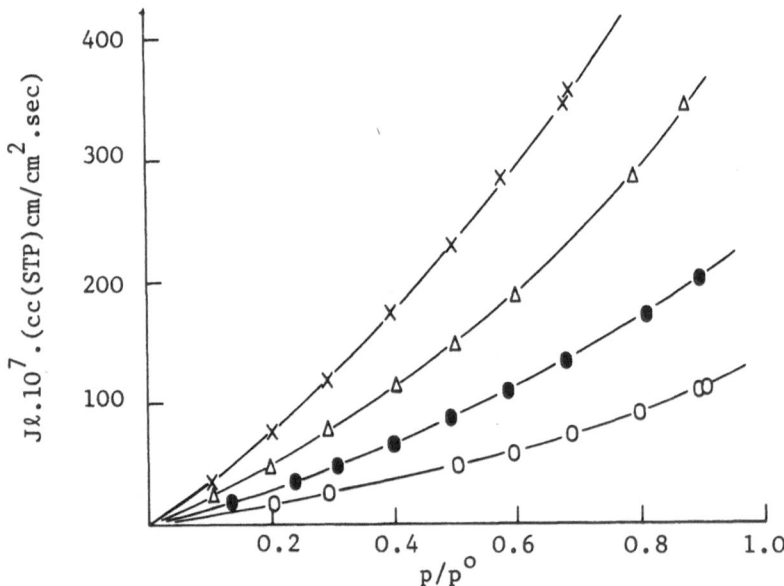

FIGURE 4. Plots of normalized flux $J\ell$ vs relative pressure p/p^o for Genthane S.
(0) 34°, (●) 43°, (Δ) 52°, (X) 60°C.

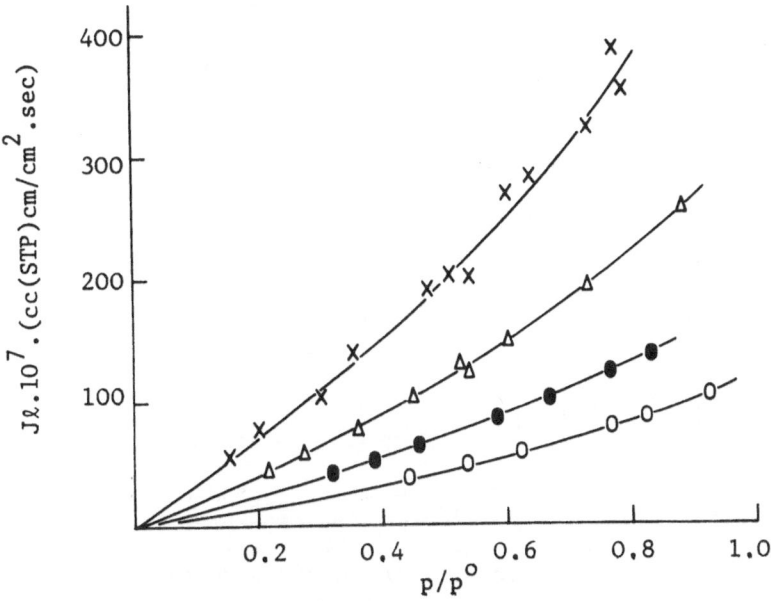

FIGURE 5. Plots of normalized flux $J\ell$ vs relative pressure p/p^o
for Elastothane ZR 625
(0) 35.1^o, (●) 41.9^o, (Δ) 50.3^o, (X) 60.4^oC.

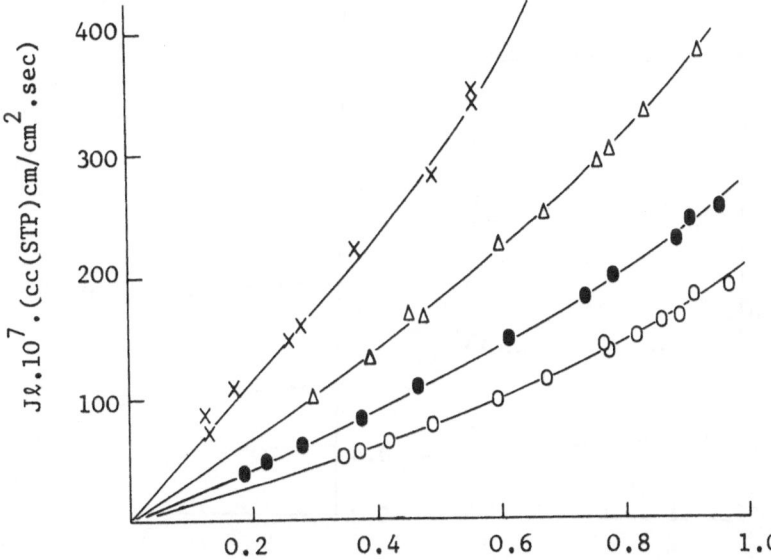

FIGURE 6. Plots of normalized flux $J\ell$ vs relative pressure p/p^o
for Adiprene CM.
(0) 35.8^o, (●) 42.2^o, (Δ) 50.2^o, (X) 60.2^oC

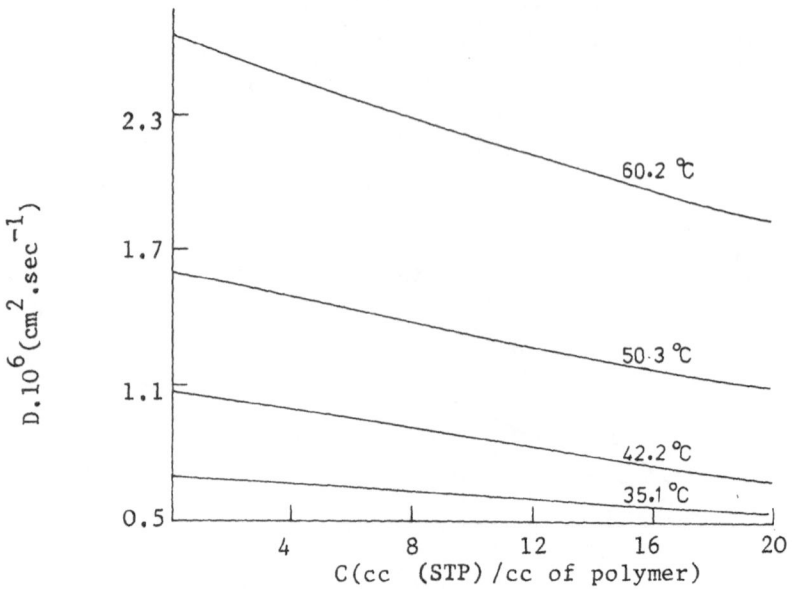

FIGURE 7.   Concentration dependence of D for Adiprene CM.

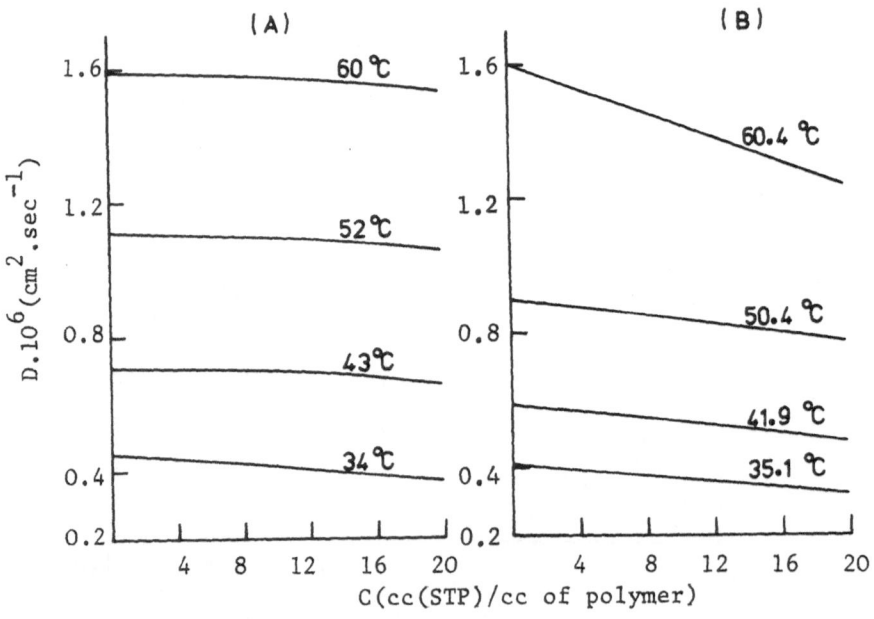

FIGURE 8.   Concentration dependence of D for (A) Genthane S and
            (B) Elastothane ZR 625.

TABLE 5

VALUES OF $D_{c=o}$, $\overline{P}_{c=o}$ AND E_D

Polymer	$T,^{\circ}C$	$D_{c=o}\ 10^6$ (cm^2sec^{-1})	$\overline{P}_{c=o}10^6$ $\left[cc(stp)cm/cm^2(cmHg)sec\right]$	E_D $(kcal\ mol^{-1})$
Adiprene CM	35.8	0.7	3.2	
	42.2	1.1	3.7	
	50.2	1.6	3.8	11
	60.2	2.6	3.7	
Elastothane ZR625	35.1	0.42	1.9	
	41.9	0.59	2.0	
	50.3	0.89	2.1	11
	60.4	1.6	2.4	
Genthane S	34.0	0.47	2.0	
	43.0	0.7	2.1	10 ... 11.5
	52.0	1.1	2.3	
	60.0	1.5	2.4	
Silicone Rubber[4]	50.5	52	3.8	3.1 ..13
Fluorosilicone Rubber[4]	50.3	18	1.8	4 ... 12

DISCUSSION

The differences in chemical structure are not reflected to any
great extent in the affinity for water. For Adiprene CM the heat
of dilution is virtually zero over the whole of the concentration
range indicating that the sorbed water is as tightly bound as in the
bulk liquid state. For the polyester based materials the sorption
process is more endothermic and the sorbed water less tightly bound.
By comparison, the interaction of the sorbed water with the low
solubility silicone rubbers is rather weak at the lower
concentrations. The clustering functions indicate a significant
degree of clustering which increases with relative pressure in
accord with the corresponding decrease in $\Delta\overline{H}_A$ and $\Delta\overline{S}_A$. For the
silicone rubbers, the tendency for water to cluster at the higher
relative pressures is much less even although the level of sorption
is much greater. Because of the polar nature of these materials
it appears likely that some degree of localized sorption occurs

concurrently with the processes of dissolution and clustering. The clustering functions describe the overall behaviour of the system and indicate that clustering is the dominant if not sole mode of sorption over most of the concentration range.

The transport properties reflect more strongly the differences in structure. Adiprene CM is the more permeable with a diffusion coefficient appreciably higher than that of either Genthane S or Elastothane ZR 625. The water permeabilities of the polyurethane elastomers are comparable with those of the silicone rubbers and as such are moderately high. The diffusion coefficient decreases with concentration for all three elastomers, albeit rather weakly for Genthane S and Elastothane ZR, and the activation energy for diffusion, E_D, is practically independent of concentration. Schneider et al[5] from sorption-desorption kinetic measurements on a similar, if not identical, class of material also obtained a D which decreased with C. The concentration dependence was rather more pronounced than in the present investigation and for a polyester based elastomer E_D was found to decrease with concentration.

The decrease in the diffusion coefficient is consistent with some degree of immobilization of an increasing fraction of the sorbed water through clustering[3]. In the absence of other effects, an increase in E_D with C is anticipated if clustering decreases with a rise in temperature[3]. It appears likely that other processes are occurring concurrently and opposing the effect of clustering on both D and E_D with the result that these parameters exhibit at best a weak dependence on concentration.

One may distinguish between two forms of the sorbed water and write for the flux[3]

$$J = -B(C)C \frac{\partial \mu}{\partial x} = J_1 = -B_1(C)C_1 \frac{\partial \mu_1}{\partial x} \qquad (2)$$

$$= -B_1(C) \ RT\left(\frac{\partial \ln a_1}{\partial \ln C_1}\right) \frac{\partial C_1}{\partial x} \qquad (3)$$

where the suffix 1 refers to mobile species only; water sorbed in clusters or on specific sites is treated as relatively immobile. It follows that

$$D(C) = B_1(C) \ RT\left(\frac{\partial \ln a_1}{\partial \ln C_1}\right) \frac{\partial C_1}{\partial C} \qquad (4)$$

$$= D_1(C) \frac{\partial C_1}{\partial C} \qquad (5)$$

The mobility of the free or dissolved water, $B_1(C)$, may be constant or increase with C through plasticization of the matrix. The term $\partial C_1/\partial C$ will increase with C for specific-site sorption of the Langmuir type and decrease with C for clustering. In addition, for systems of moderate or high water solubilities, the thermodynamic term $\partial \ln a_1/\partial \ln C_1$ may also vary with C if the sorption of dissolved species departs from Henry's law. The effect of clustering on D may be opposed at the lower concentrations by localized sorption and at the higher concentrations by plasticization so as to yield a D and E_D which vary weakly with concentration. The increase in the permeability at higher relative pressures is consistent with some degree of plasticization. For a more detailed analysis the fractions of penetrant dissolved, clustered and localized is required as a function of concentration and temperature. It is of interest to examine the effect of specific-site sorption on D and E_D in the limit of zero concentration when clustering will be absent. If k_1 and k denote respectively the Henry's law constants for the dissolution and localized sorption processes then

$$\frac{\partial C_1}{\partial C} = k_1/k \quad \text{and from (5)}$$

$$D_{c=o} = (D_1)_{c=o} \frac{k_1}{k} \tag{6}$$

Since $k > k_1$, the experimental coefficient $D_{c=o}$ will be less than $(D_1)_{c=o}$ for the mobile dissolved species. The temperature dependence of $D_{c=o}$ and k is given respectively by $D_{c=o} = D_o \exp(-E_D/RT)$ and $k = k_o \exp(-\Delta H/RT)$ with similar expressions for $(D_1)_{c=o}$ and k_1 so that

$$(E_D)_{c=o} = (E_{D_1})_{c=o} + \Delta H_1 - \Delta H \tag{7}$$

ΔH is the overall heat of sorption and ΔH_1 the heat of sorption for dissolved species only. The contribution of localized sorption is such that ΔH is more exothermic than ΔH_1 and $(E_D)_{c=o} > (E_{D_1})_{c=o}$. On this basis the experimental activation energy for diffusion is larger than the true value for the dissolved species.

Van Amerongen[13] obtained an almost linear correlation between log D and $(T-T_g)$ for hydrogen and nitrogen diffusing in a series of elastomers. Newns and Park[14] predicted from free volume theory that a plot of $(\log D_{c=o} + A)^{-1}$ vs T_g should be linear at constant temperature and found good agreement for benzene diffusing in a variety of polymers at 25°C. The constant A was always less than 10 per cent of $\log D_{c=o}$. Butyl rubber was a notable exception in both studies with abnormally low values of D whilst $D_{c=o}$ for benzene in polybutylmethacrylate was abnormally

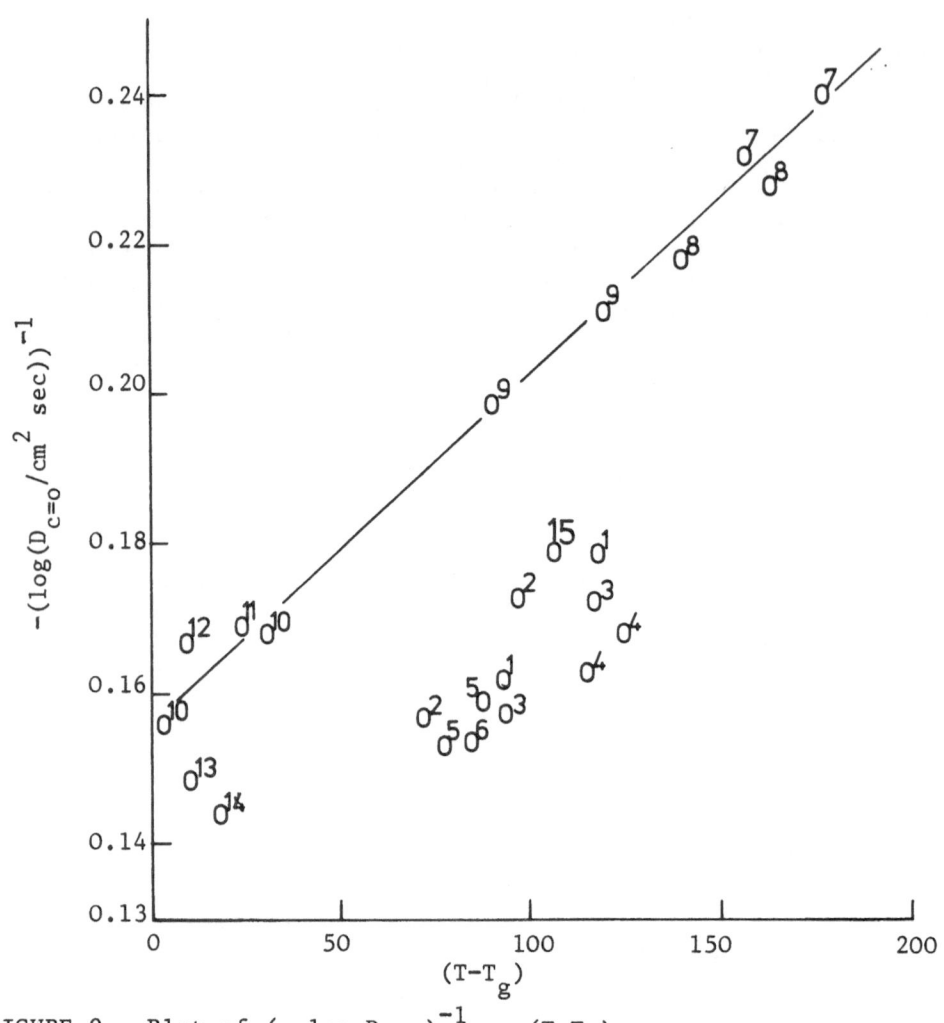

FIGURE 9. Plot of $(-\log D_{c=o})^{-1}$ vs $(T-T_g)$

(1) Adiprene CM, (2) Elastothane ZR 625, (3) Genthane S, (4) Polytetramethylene oxide - polyurethane[6], (5) polybutylene adipate - polyurethane[6], (6) polyethylene oxide-polyurethane[6], (7) polydimethyl-siloxane[4], (8) polydimethyl phenylmethylsiloxane[4], (9) poly-3,3,3, - trifluoropropylmethylsiloxane[4] (10) polybutylmethacrylate[18], (11) polypropylmeth-acrylate[18], (12) polyethylmethacrylate[18], (13) poly vinylacetate[19], (14) polymethylacrylate[19], (15) polypropylacrylate[18].

high. These and similar deviations[15] limit the applicability of
simple correlations between $\log D_{C=O}$ and T_g. Nevertheless a plot
of $(\log D_{C=O})^{-1}$ vs $(T-T_g)$ for water diffusion is revealing in
that the polyurethane elastomers, relative to the silicone rubbers
and less hydrophilic members of the polyalkylmethacrylates, have
low values of $D_{C=O}$ in common with the more hydrophilic polyalkyl-
acrylates and polyvinylacetate (Figure 9). Clough et al [15,16]
demonstrated that domain structures, arising through segregation of
the polyurethane segments, occur to varying degrees particularly
in polyether-polyurethane elastomers. On the assumption that the
domains are impermeable it is argued that a substantial reduction
in the diffusion coefficient may result. However, for polyester
based materials there was little evidence for domain structures.
It appears reasonable to conclude that the relatively low values of
$D_{C=O}$ for the polyurethanes can be attributed, at least in part, to
the effect of specific-site sorption whilst the apparent diffusion
coefficient may be decreased further when domain structures are
present.

ACKNOWLEDGEMENTS

We thank S.R.C. for maintenance awards to A.S. and A.N. and
Dr. D. K. Thomas, R.A.E., Farnborough for preparation of the
membranes.

REFERENCES

1. H. Fujita, Fortschr. Hochpolym.-Forsch., 3, 1 (1961).

2. D. Machin and C. E. Rogers,"The Concentration Dependence of
    Diffusion Coefficients in Polymer-Penetrant Systems", C.R.C.
    Critical Reviews in Macromolecular Science, C.R.C. Publ.
    Cleveland, 1972.

3. J. A. Barrie, Diffusion in Polymers (J. Crank and G. Park eds.),
    Academic, New York, 1967.

4. J. A. Barrie and D. Machin, J. Macromol. Sci. - Phys., B3(4),
    645 (1969).

5. N. S. Schneider, L. V. Dusablon, L.A. Spano and H.B. Hopfenberg,
    J. Appl. Polymer Sci., 12, 527 (1968).

6. N. S. Schneider, L. V. Dusablon, E. W. Snell and R. A. Rosser,
    J. Macromol. Sci. - Phys., B3(4), 623 (1969).

7. Ludbrook, B., Thesis, University of London, (1972).

8. R. M. Barrer and J. A. Barrie, J. Polymer Sci., 28, 377 (1958).

9. H. Yasuda and V. Stannett, J. Macromol. Sci. - Phys. B3(4),589 (1969).

10. J. A. Barrie and D. Machin, J. Appl. Polymer Sci. 12, 2633 (1968).

11. B. H. Zimm and Lundberg, J. Phys. Chem., 60, 425 (1956).

12. J. Lundberg, Pure and Appl. Chem. 31, 261 (1972).

13. G. J. van Amerongen, Rubber Chem. and Tech., 37, 1065 (1964).

14. A. C. Newns and G. S. Park, J. Polymer Sci., C22, 927 (1969).

15. J. B. Alexopoulos, J. A. Barrie, J. C. Tye and M. Fredrickson, Polymer, London, 9, 57 (1968).

16. S. B. Clough and N. S. Schneider, J. Macromol. Sci., B2, 553 (1968).

17. S. B. Clough, N. S. Schneider and A. O. King, J. Macromol. Sci., B2, 641 (1968).

18. J. A. Barrie and D. Machin, Trans. Faraday Soc., 67, 244 (1971).

19. A. Kishimoto, E. Maekawa and H. Fujita, Bull. Chem. Soc. Japan, 33, 988 (1960).

WATER VAPOR TRANSPORT IN HYDROPHILIC POLYURETHANES

J. L. Illinger and N. S. Schneider
Polymer and Chemistry Division
Army Materials and Mechanics Research Center
Watertown, Ma.   02172
and
F. E. Karasz
Polymer Science and Engineering Department
University of Massachusetts
Amherst, Ma.   01002

SYNOPSIS

Water uptake and water vapor transmission rates have been obtained on two series of hydrophilic segmented polyurethanes based on MDI, butanediol, and several block copolymer polyethers with varying proportions of poly(ethylene oxide) PEO and poly-(propyleneoxide) PPO. In series I the ratio of PEO to PPO is varied at fixed hard segment concentration. In series II the hard segment concentration is varied at a fixed 50/50 ratio of PEO to PPO. At constant hard block composition and decreasing amounts of PEO in series I polymers water uptake decreases from 58g/100g of polymer at pure PEO to 2g/100g of polymer for pure PPO. Transmission rates determined with water in contact with the upstream surface and 50% relative humidity downstream show a corresponding decrease. As hard block concentration is increased in II both water uptake and transmission rate decrease. The sorption isotherms are surprisingly simple concave upward curves which rise steeply at saturation. Analysis of the data on the basis of moles of water per EO unit provides a clear picture of the influence of compositional variations on the course of the sorption isotherms and shows that saturation water concentrations are not related in a simple manner to PEO concentration. The diffusion coefficients calculated from steady state transmission rates decrease with increasing water concentration in series I but increase with water concentration in series II. The apparently conflicting trends can be rationalized in terms of water clustering in the first case and the blocking effect of hard segment domains in the second.

Comparisons are also made with results on a series of polyurethane-ureas based on Hylene W and containing a mixture of separate PEO and PPO macroglycols which were studied by Tobolsky and coworkers for reverse osmosis.

INTRODUCTION

There is considerable present interest in hydrophilic polymers for reverse osmosis, biomedical applications and for fabric coatings which can offer high moisture vapor transmission rates. The usual hydrogels suffer the disadvantage of being brittle when dry and may have poor physical properties in the wet state. Hydrophilic polyurethanes based on the incorporation of a poly(ethylene oxide) soft segment remain flexible when dry and can provide very high moisture transmission rates. (1) However the high transmission rates are accompanied by an unacceptably high degree of swelling (up to 100% by weight of dry polymer) and the loss of physical properties. Tobolsky (2) and coworkers have explored one approach to modifying the hydration behavior of polyurethanes based on the use of varied proportions of hydrophobic and hydrophilic polyethers as separate soft segments. The availability of several poly-(propylene oxide) (PPO)-poly(ethylene oxide)(PEO) block copolymers offers another approach to the same end, incorporating both hydrophilicity and hydrophobicity in all of the soft segments of elastomeric polyurethanes. The present study is concerned with elucidating the effect of soft segment composition and urethane concentration on the polymer-water interactions, water sorption levels and permeability behavior of a series of such block copolyether polurethanes.

EXPERIMENTAL

Synthesis and Materials

MDI (4,4' methylenediphenyldiisocyanate) was obtained from Mobay Chemical Co., vacuum distilled at 2 mm Hg and maintained liquid until polymerized to prevent dimerization.

1,4 Butanediol was obtained from GAF Industries and used as received.

Glycols were obtained from Wyandotte Corporation (block polyols and PPO) and Union Carbide (PEO). Hydroxyl numbers were determined and a potentiometric titration performed to measure any residual basic material (3) which was then neutralized with an appropriate amount of etheric HCl. This was necessary to prevent the free base from catalyzing crosslinking reactions which lead to prepolymer gelation.

The polyurethanes were prepared by a two step synthesis. The first step involved end capping the macroglycol with excess dissocyanate at 80°C for one hour under vacuum (approximately 5 mm Hg) with stirring. Butanediol was added under flowing Argon for chain extension. The resultant polymer was poured into a Teflon lined pan and cured under vacuum at 110°C overnight. Weights of components were accurate to + 0.04 g and the OH/NCO ratio was 0.952 to control degree of polymerization. This synthesis produces a polymer of the general structure

$$M \sim\sim\sim [(MB)_x M \sim\sim\sim]_y M$$

where M is MDI

$$O=C=N \langle O \rangle CH_2 \langle O \rangle N=C=O$$

b is 1,4 Butanediol (B'd)
$$HO\ CH_2CH_2CH_2CH_2OH$$

$\sim\sim\sim$ is a polymeric glycol
$$HO(CH_2CH_2O)_a\ (CH_2CHO)_b\ (CH_2CH_2OH_a)$$
$$CH_3$$

Available polyols of $\overline{M}_n$ 2000 are listed in Table I

Characterization data for the resultant polymers are shown in Table II (following page).

In the sample designation the first number represents the poly(ethylene oxide) weight fraction of the block copolyether soft segment, the second number indicates the weight percent of urethane. Thus, 5PE33 is a sample with 33% of MDI and with a soft segment which is 50% PEO. The three intermediate samples in this series have the same soft segment, a 50/50 copolymer of PEO and PPO but increasing urethane concentration. There are five samples with identical hard segment composition (33% MDI) but with a change

TABLE I

Glycol Composition and Structure

Glycol	Composition	a	b
C1540	100% PEO	36	0
L35	50/50 PPO/PEO	11	17
L43	70/30 PPO/PEO	7	23
L61	90/10 PPO/PEO	2.5	33
P2010	100% PPO	0	35

TABLE II

Sample Characterization

Sample	Molar Composition MDI:Bd:Polyol			$M_n \times 10^{-3}$	T_g °K
10PE33	4.20	3	1C1540	22±1	242
5PE28	3.15	2	1L35	31±2	236
5PE33	4.20	3	1L35	38±2	236
5PE40	6.30	5	1L35	20±2	236
3PE33	4.20	3	L43	24±3	230
1PE33	4.20	3	1L61	21±1	---
0PE33	4.20	3	1P2010	23±1	232

in soft segment composition from 100% PEO in the first sample
to 100% PPO in the last sample and all 2000 molecular weight.
All samples are of moderately high molecular weight and in a
range where the properties are expected to be sensitive only to the
differences in molecular composition. The glass transition temperatures in the last column of Table II appear to depend only
on the soft segment composition and are independent of changing
MDI concentration. These T_g values were 25°C higher than the
T_g's of the pure soft segment polyol.

Sorption Measurements

Equilibrium water sorption measurements were performed by
immersing a preweighed sample of polymer in distilled water maintained at 30°C. Samples were allowed to come to equilibrium (up
to 16 hours) then removed from the water, blotted with two sets of
filter paper and immediately placed in a tared weighing bottle.
Weight gain was measured and the samples returned to immersion for
repeat determinations.

Sorption isotherms were measured in a thermostatted vacuum
system incorporating a five liter ballast volume for vapor, a silicone oil manometer to increase sensitivity for measuring pressure,
and a Perkin Elmer AR-1 recording electrobalance to continuously
register weight gain.

Transmission Measurements

Steady State bulk permeabilities were measured in inverted cup
cells, the top being sealed by the polymer film of interest, which
were placed in Aminco Aire cabinet controlled at 30 ± 0.5°C and 50
± 1% RH. Periodic weighings were made to determine weight loss

which was converted to flux and then to bulk permeability. Evaporation rate of water was measured at the same positions and conditions to obtain limiting values required by the flux equation of reference 4.

Scanning Calorimetry

Calorimetric measurements were made using a Perkin Elmer DSC-2 equipped with sub-ambient accessory capable of controlling from 100°K. All scans were run at 20°/min. The wet polymer samples were prepared by placing a weighed sample of polymer in a volatile pan, adding a calculated amount of water and then hermetically sealing the pan. The pan was reweighed to determine the actual amount of water added.

RESULTS AND DISCUSSION

Equilibrium Swelling Ratios

The results of the equilibrium water uptake experiments at 30°C are shown in Table III. The table is organized into two sections. The upper section presents the results for samples of varying PEO concentration at a fixed hard segment length, the lower section of the table shows the results for varying hard segment length and 50% PEO, 50% PPO soft segment composition. The data show two effects of polymer composition: (1) At fixed hard segment length, the increased per cent of ethyleneoxide in the polymer is accompanied by increased sorption; and (2) the increased amount of hard segment decreases the water uptake. We will first discuss the influence of soft segment composition on water sorption referring to the results in the upper section of Table III.

Measurements on the copolymer of MDI and butanediol, repr - senting pure hard segment, showed no measureable water uptake. Therefore all sorption must occur in the soft segment regions of the polymer. Further, if it is assumed that the number of molecules of water per repeat unit bound to the central PPO portion of the segment remains fixed and independent of soft segment composition then the sorbed water can be interpreted in terms of the ratio of molecules of water to ethylene oxide unit. These values are shown in the last column of Table III. It is evident that the ratio is not fixed, but decreases progressively with increasing PPO content, indicating that the water sorption on PEO units is not independent of the composition of the soft segment region. If complete miscibility of hydrated PEO and PPO segments occurred then

TABLE III

Immersion Water Uptake at 30°C

Polymer	% PPO	% PEO	$\dfrac{g\ H_2O}{g\ film}$	$\dfrac{Molecules\ H_2O*}{EO\ unit}$
10PE33	-	54	.58 $\pm$ .03	2.63
5PE33	30	29	.25 $\pm$ .02	2.02
3PE33	41	19	.08 $\pm$ .01	0.85
1PE33	55	6	.03 $\pm$ .005	0.49
0PE33	61	-	.02 $\pm$ .005	0.11[+]
5PE28	34	33	.40 $\pm$ .03	2.88
5PE33	30	29	.25 $\pm$ .02	2.02
5PE40	25	24	.15 $\pm$ .01	1.45

* Calculated assuming constant sorption levels on PPO

+ Molecules H_2O/PO unit

the water solubility behavior could be expected to conform simply
to an additive mixing relation consistent with fixed characteristic
ratios of molecules of water per PPO unit and PEO unit. It is
known that PEO is completely soluble in water whereas PPO is not.
Since PPO is the central block and increases in length with decreasing PEO concentration, these results suggest that phase in-
compatability between PPO and PEO segments may be occurring in
the hydrated state. The resulting association of PPO blocks would
constrain swelling in PEO regions.

It is instructive to compare these results with the data obtained by Tobolsky and coworkers. We select for comparison the
data in their Table 6 for a series of polyurethane-ureas from
Hylene W (the saturated version of MDI) with fixed hard segment
composition and varying proportions of the separate 1000 molecular
weight PPO and PEO segments. The water uptake in the six samples
varies from 22% to 63% with increasing PEO content. Significantly,
when their results are converted to the ratio of molecules of water
per EO unit the values scatter between 3 and 4 and do not show any
defined trend with composition. These results support the proposal
presented above, that simple miscibility conditions do not exist
for the hydrated block copolyether in our polymers.

The influence of increasing hard segment concentration on
water sorption levels at fixed 50/50 PPO/PEO soft segment composition is shown with particular clarity by the sensitive variation in
the ratio of molecules of water to EO unit in Table III. Decreasing the urethane concentration from 33% to 28% allows even higher

local sorption (indicated by the value of this ratio) than elimination of PPO completely from the soft segment. The general dependence of the swelling ratio on composition can be explained in terms of the increasing elastic modulus of the polyurethanes with increasing hard segment length and the corresponding increase in the elastic contribution to the free energy of mixing.

Sorption Isotherms

The sorption isotherms shown in figures 1 and 2 are surprisingly simple being continuously concave upward and rising steeply at high relative humidities. The three isotherms for the samples with 50/50 polyether copolymer soft segment and varying hard segment composition shown in Figure 1, are very close up to a partial pressure of 0.7. The agreement in sorption behavior is more clearly displayed by comparison of the H_2O/EO ratio (Figure 3) where matching occurs up to p/p_0 of 0.9 (see hatched area) corresponding to a maximum swelling of 8%. Apparently at or below this level of swelling, the change in free energy of mixing due to differences in elastic modulus of the samples is too small to affect the sorption behavior.

Figure 2 shows the isotherms for samples with varying "soft" segment composition and fixed hard segment structure. Even at low

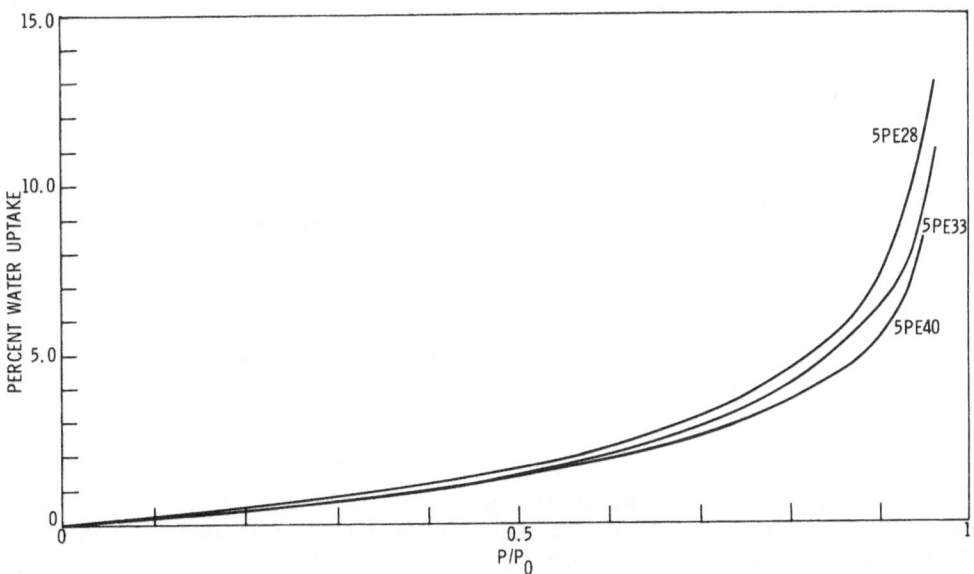

Figure 1. Sorption isotherms for polymers with variation in amount of hard segment at 50/50 PPO/PEO soft segment.

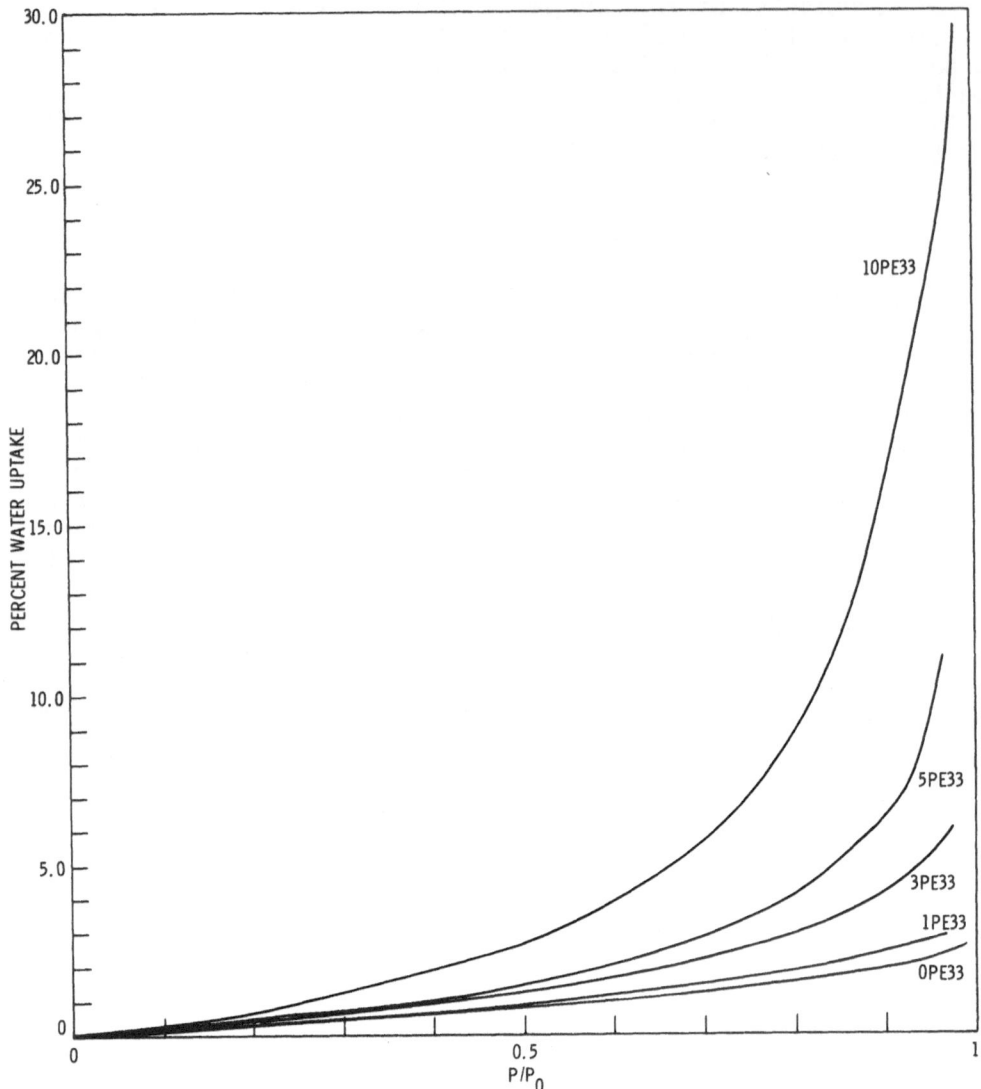

Figure 2. Sorption isotherms for polymers with variation in PPO/
PEO ratio in soft segment at fixed hard segment composition.

humidity, differentiation of the pure PEO polymer becomes apparent.
The differences become more marked in comparing the H_2O/EO ratio.
(Fig. 3) The PPO block affects sorption levels for p/p_0 as low as
0.3. The PPO block shows more effect on the sorptive capacity of
the EO units at lower partial pressures than does hard segment size
or elastic modulus change.

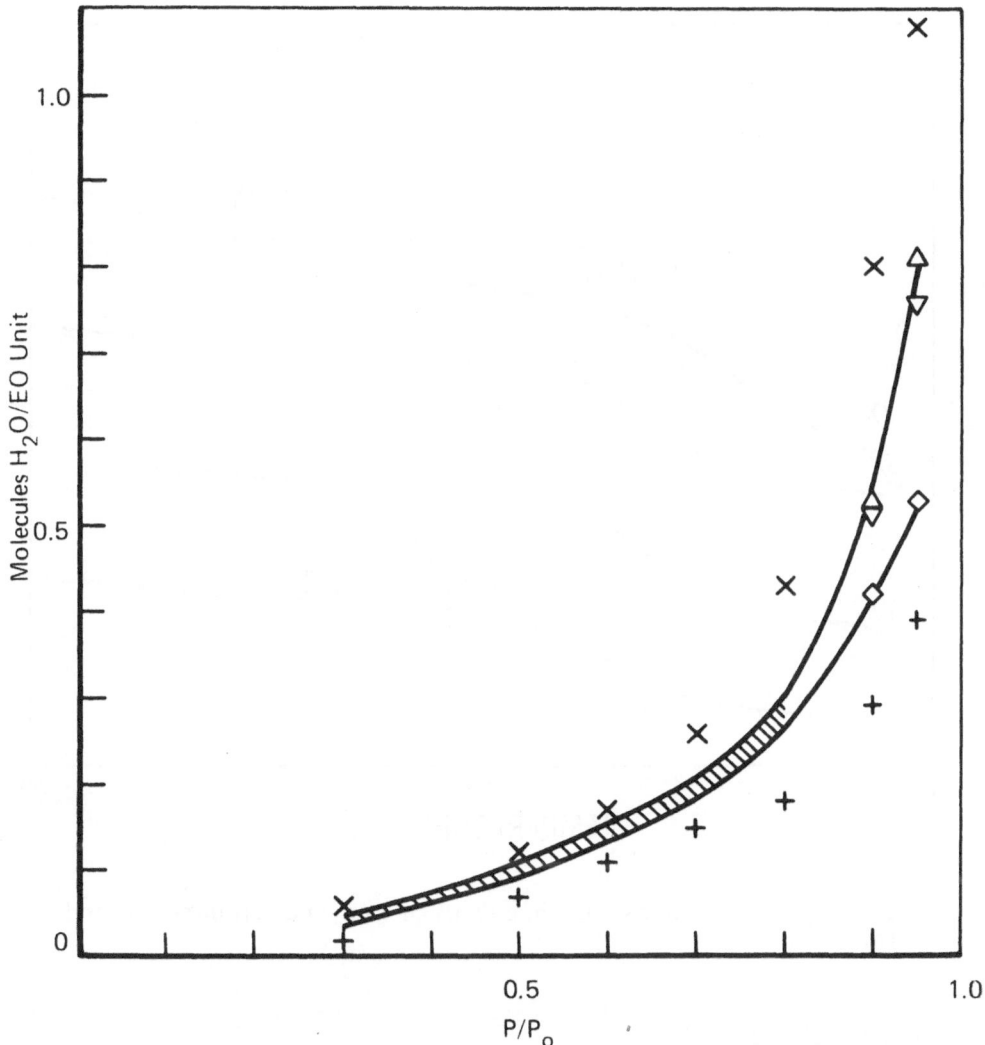

Figure 3. Sorption isotherms plotted as the ratio molecules $H_2O/$ EO unit: ($\times$) 10PE33, ($+$) 1PE33, ($\triangle$) 5PE28, ($\triangledown$) 5PE40, ($\diamond$) 3PE33. Hatched area shows matching of isotherms within limits of error.

The very much higher levels of sorption at very high values of p/p_0 and on immersion are due to the clustering of water in these highly swollen materials.

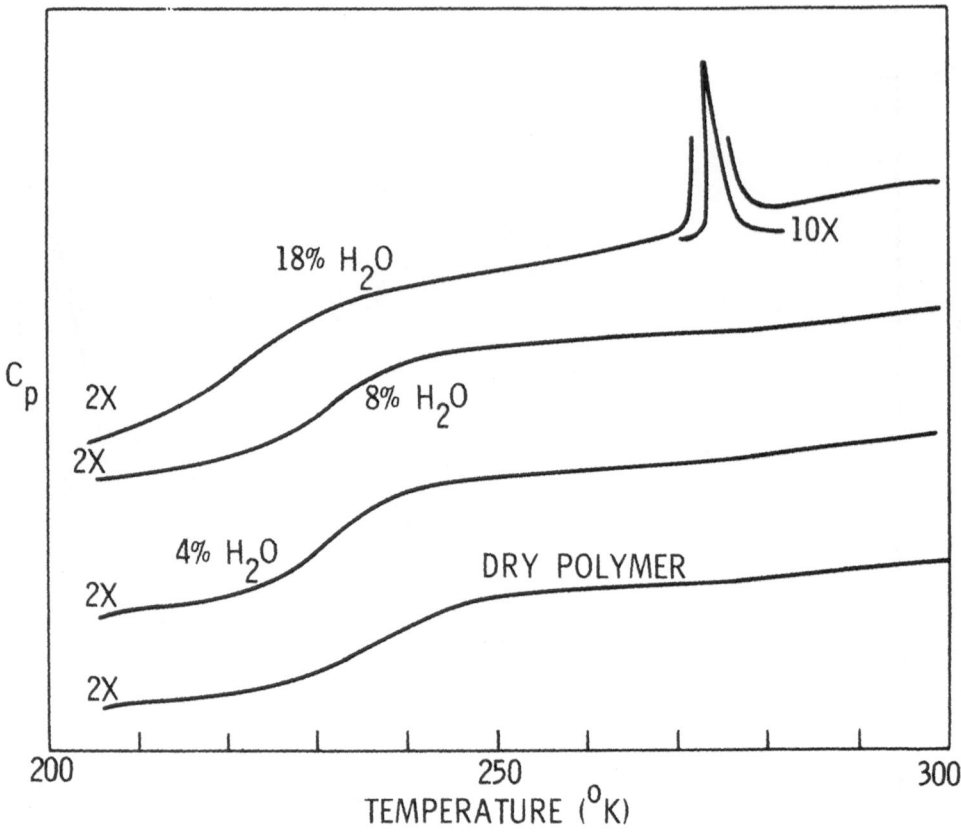

Figure 4. DSC traces for 5PE33 with varying amounts of added water.

TABLE IV

Change in T_g With Water Uptake
(Sample 5PE33)

% H_2O	T_g(°K)
0	237
4	233
8	231
18	226

Scanning Calorimetry

Some DSC studies were carried out to further elucidate the
nature of the interactions between water and the polymer soft segment. One effect which was observed is a progressive decrease in
soft segment glass transition with increasing amount of water.
This is illustrated by the results in Table IV for 5PE33 which
show an overall decrease of 11°C in the glass transition. In addition, it was possible to obtain information concerning the state
of sorbed water. As shown in Figure 4, up to 8% added water there
is no separate contribution of water to the DSC trace, indicating
that it is all bound to the polymer.

At 18% well below the 30°C equilibrium uptake of 25% water,
sufficient "free" or "clustered" water is present to show a sharp
ice melting endotherm. A smaller endotherm was seen at 14% water.
The 8% uptake is found at partial pressure of about 0.9 and is
equivalent to one molecule of water per two units of ethylene oxide
while at 14% uptake this ratio is one to one. Further work is
needed to determine the ratio at which "bulk" water behavior is
first observed.

Bulk Permeabilities

Bulk permeabilities were calculated from the steady state
flux at 30°C measured with liquid water in contact with the film
and 50% relative humidity downstream. Equation 1 (ref. 4) was
applied to correct for the limiting evaporation rate of water

$$J^{-1} = mL + b \qquad (1)$$

Here J is the observed steady state flux, $1/m$ is the bulk permeability, L the film thickness and $1/b$ the evaporation rate of
water. Integral diffusion coefficients D were estimated from the
above determined values of the bulk permeability in units of gm -
mil/100 in 2-24 hours using the relation

$$D = \frac{(1-v_1)}{m(C_2-C_1)} \qquad (2)$$

where (C_2-C_1) is the difference in water concentration across the
film. The additional term $(1-v_1)$, where v_1 is the equilibrium upstream volume fraction of water, is required for highly swollen
films to convert from the fixed polymer frame of reference to the
fixed volume frame of reference appropriate to the calculation of

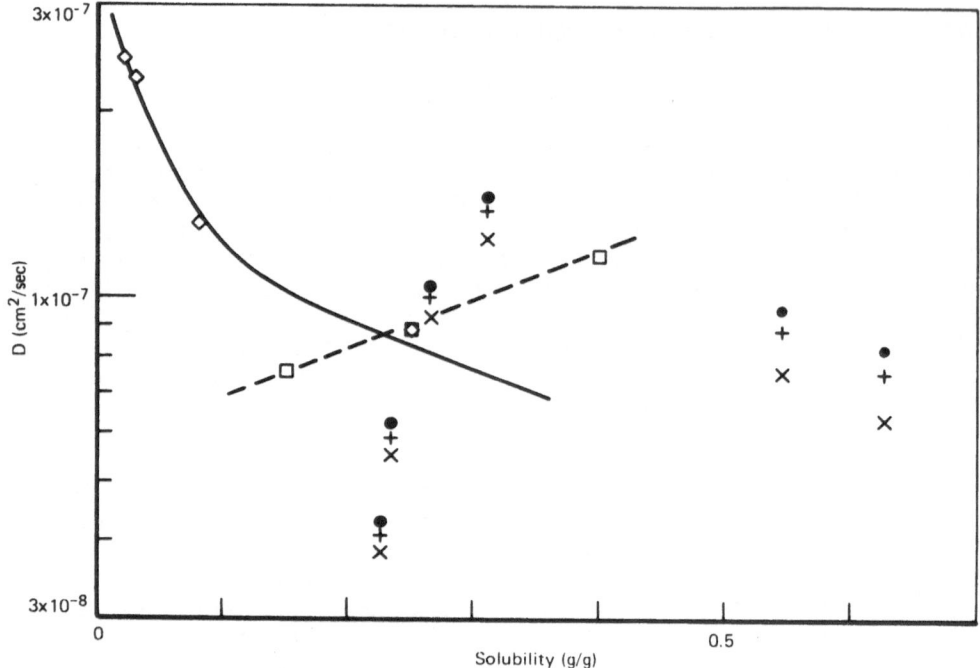

Figure 5. Dependence of diffusion constant on water concentration:
($\diamond$) polymers with soft segment variation, ($\square$) polymers with hard
segment variation. Polymers from Tobolsky and coworkers: D calculated with $x_1 = 0$ ($\bullet$), $x_1 = 0.1$ ($+$), and $x_1 = 0.3$ ($\times$).

TABLE V

Permeation and Diffusion Behavior

Polymer	Solubility (g/g)	Bulk Permeability $\left(\dfrac{\text{g-mil}}{100 \text{ in}^2 \text{ 24 hr}}\right)$	Diffusivity $(\text{cm}^2/\text{secx}10^7)$
10PE33	.58	3000	1.49
5PE33	.25	500	.73
3PE33	.08	235	1.24
1PE33	.03	152	2.23
0PE33	.02	108	2.41
5PE28	.40	1030	.84
5PE33	.25	500	.73
5PE40	.15	250	.66

a mutual diffusion coefficient. The results are summarized in
Table V and plotted in Figure 5. The measurements on 10PE33 show
large error limits due to the extremely high flux levels (490 gm/
100 in^2/24 hours for a 1.7 mil film) which approach the evaporation rate of water (660 gm/100 in^2/24 hours) under our conditions.
Values of permeability and diffusivity are included for complete-
ness.

   Two opposing trends are evident in the dependence of the diffusion constant on water concentration. In the five samples in
the upper section of Table V with fixed urethane concentration
the diffusion constant decreases progressively with increasing
water concentration. For the samples in the lower portion of the
table with fixed PEO/PPO ratio, the diffusion coefficient increases
with increasing water concentration. This opposing dependence of
the diffusion constant on water concentration is also displayed
in Figure 5. The first result is consistent with other observations
(1,5) of a concentration dependent diffusion coefficient in various
polyurethanes which decreases with increasing water concentration.
This behavior can be explained by the clustering of sorbed water
leading to a progressive decrease in monomeric water, the effective
diffusing species, with increasing water concentration. In the
second series of polymers, the increase in water content is due
to a reduction in the hard segment concentration. Although the
resulting increase in water uptake may be accompanied by some increase in clustering, apparently the reduced tortuosity of the
diffusion path accompanying the reduction in the concentration of
the hard segment phase dominates the diffusion behavior.

   In order to compare our results with those appearing in Table
6 of Tobolsky and coworkers (2) it is necessary to calculate diffusion coefficients from their measured values of water flux P_w
in reverse osmosis. If P_w is interpreted as the product of a diffusion and solubility coefficient, the apparent values of the diffusion coefficient are as high as 1×10^{-5} cm^2/sec, about one-third
the value for the self diffusion coefficient of water and nearly
two orders of magnitude higher than those obtained on our samples.
It has been shown by Paul that the correct procedure involves a
correction for a polymer fixed frame of reference and the conversion
of the pressure differential to a concentration differential. This
leads to the following equation

$$D = \frac{P_w}{V_1} [(1-v_1)^2 \, (1-2x_1 \, v_1)] \qquad\qquad (3)$$

where V_1 is the molar volume of water, x_1 is the Flory-Huggin's
interaction parameter and the other terms have been defined previously. The results obtained by this procedure are plotted in
Figure 5. Despite the scatter, it is apparent that the diffusion

constants calculated from the reverse osmosis experiment are in the same range as the values which we obtain by steady state transmission.

CONCLUSIONS

It is clear from the results of this study that a wide range of water sorption levels and associated transmission rate can be achieved conveniently by appropriate variation of the proportion of polyethylene oxide and polypropylene oxide in the block co-polymer which comprises the soft segment and by varying the concentration of the hard segment phase. The absence of a direct proportionality between water uptake and PEO content expressed as the ratio of molecules of water to ethylene oxide units .indicates that the sorption process under immersion conditions is not simple but is affected by the block copolymer nature of the polyether soft segment. The characteristic decrease in diffusion constant with increasing water uptake in polymers with increasing ethylene oxide content dictates that high water transmission rates are to be achieved only at high swelling ratios. However, some comfort can be taken in the moderate transmission rate obtained in 3PE33 at only 8% water uptake, which is comparable to the water uptake in nylon 6.6. Since a reduction in hard segment concentration has been shown to increase both the water uptake and the diffusion constant as sample with 28% rather than 33% MDI might very well display a further improvement in transmission at an acceptable water sorption level. The approach used by Tobolsky and coworkers involving the incorporation of separate polyethyleneoxide and polypropyleneoxide segments in varying proportions appears to provide polymers with properties similar to those investigated in the present study, in so far as the results obtained under very different conditions can be compared. It would be instructive to analyze the behavior of these polymers by the techniques used in the present study to afford a direct comparison of overall sorption and transmission behavior.

REFERENCES

1. Schneider, N.S., Dusablon, L.V., Snell, E. W. and Prosser, R.A., J Macromol Sci-Phys B3 623-644 (1969).
2. Chen, C.T., Eaton, R.F., Chang, Y.S. and A. V. Tobolsky J. Appl Polym. Sci. 16 2105 (1972).
3. Scholten, H.G., Schuhmann, J.G and TenHoor, R.E., J. Chem & Eng. Data, 5 395-400 (1960).
4. Schneider, N.S., Allen, A.L., and Dusablon, L.V., J. Macromol Sci-Phys B3 767-776 (1969).
5. Paul, D.R., J. Polym Sci., Polymer Phys Ed., 11, 289 (1973).
6. Barrie, J.A. and Nunn A., This volume & Coatings and Plastics Preprints, 34 489 (1974).

EFFECT OF THICKNESS ON PERMEABILITY

Sun-Tak Hwang
Karl Kammermeyer
Department of Chemical and Materials Engineering
The University of Iowa
Iowa City, Iowa 52242

ABSTRACT

When the observed permeability changes as a function of membrane thickness, there are two reasons. One is the presence of boundary layers or interfacial resistances. The other is variation of permeability throughout the membrane, as in the case of laminated film composites of different permeability layers. Examples of both cases are presented and analyzed to elucidate internal and external phenomena of the membrane processes. This type of thickness analysis is completely general and can be applied to any membrane process.

INTRODUCTION

Permeability is a phenomenological coefficient, not a property of the membrane in general. Although it is well known that the value of permeability is a function of temperature and pressure, it is less understood that permeability can also vary with the thickness of the membrane. Permeability is an overall flow coefficient with no reference to the actual permeation mechanisms. Therefore, the numerical values of permeability alone have little meaning without specifying the experimental conditions, under which the permeability is measured (1).

If permeability is a property of the membrane and is homogeneous throughout the membrane, the value will be independent of the membrane thickness and becomes the intrinsic permeability. This is the case for most gas phase permeations through various

membranes. However, when either or both sides of the membrane are
in contact with liquid, the boundary resistance enters and the
permeability becomes dependent upon the membrane thickness (2-4).
There also is another way in which permeability can change as a
function of membrane thickness when the membrane acts as a heterogeneous medium.

This other mechanism of permeability dependency on membrane
thickness will be discussed together with additional examples of
boundary resistances from other investigator's studies.

THEORY

The simplest case of membrane permeation is that where the
membrane is homogeneous and there are no boundary resistances.
Then, the permeability can be taken as a product of diffusivity and
solubility, and it is independent of membrane thickness (1,2). Of
course the diffusivity is also uniform throughout the entire membrane. However, when the membrane exhibits two or more different
layers of structure with different permeabilities, and there are
boundary resistances, the observed overall permeability becomes a
function of membrane thickness. The interrelationship between permeability and diffusivity also becomes complicated and involves
membrane thickness. Therefore, the diffusivity is not a constant
value throughout the membrane.

The analysis of the thickness effect on permeability for the
general case can be made on the basis of electric circuit analogy.
Each phase in the membrane and the boundary layers presents its own
resistance to the mass transport. At steady state, the overall
permeability is defined by a permeation flux equation:

$$F = Q(C_1 - C_2)/L \tag{1}$$

The total resistance to permeation consists of resistances of different laminates and resistances of boundary layers or interfaces.

$$L/Q = (C_1 - C_2)/F = r_1 + r_2 + \sum_i L_i/Q_i \tag{2}$$

where $L = \sum_i L_i$.

When a membrane consists of a single homogeneous phase, Eq. (2)
becomes

$$1/Q = 1/DS + (r_1 + r_2)/L \tag{3}$$

Here, the membrane phase permeability is replaced by the product term of diffusivity and solubility. The analysis of boundary resistance can be successfully carried out by employing Eq. (3). When plotting the reciprocals of observed overall permeability against the reciprocals of membrane thickness, the product of diffusivity and solubility is calculated from the intercept and the sum of boundary resistances is obtained from the slope (1-3).

This simple approach would not work in the most general case, where a membrane shows two or more different zones of permeability. In that case, the plot of $\Delta C/F$ against membrane thickness gives a better result. Suppose there are two regions of different permeabilities in a membrane, for example, a dry layer and a swollen layer, as shown in Fig. 1. For this case, Eq. (2) can be written as

$$L/Q = \Delta C/F = r_1 + r_2 + L_a/D_a\, S_a + L_b/D_b\, S_b$$

$$= r_1 + r_2 + (1/D_a\, S_a - 1/D_b\, S_b)L_a + L/D_b\, S_b \qquad (4)$$

Therefore, when $\Delta C/F$ is plotted against membrane thickness, two segments of straight lines will result as shown in Fig. 2. From the slope of each straight line, the intrinsic permeability of that particular layer of the membrane can be obtained and the intercept gives the boundary resistances. Also, the thickness of each layer can be read from the break point. If the solubility of each region is known, the diffusivity of each layer can be calculated from the usual relationship:

$$Q_i = D_i\, S_i$$

Similarly, when there are many layers of different permeabilities in a membrane, the observed total permeability can be analyzed according to Fig. 3. The slopes of the straight lines are the reciprocals of the intrinsic permeabilities of different layers and the intercept is the sum of boundary resistances.

It is clear that the observed total permeability, Q, is a phenomenological quantity, that is, it is not the invariant property of the membrane. However, the intrinsic permeabilities, Q_i, are the material properties. These intrinsic permeabilities which are not measured directly can be obtained from the thickness analysis as outlined above.

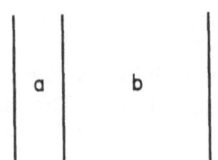

Figure 1   Two Layer Membrane

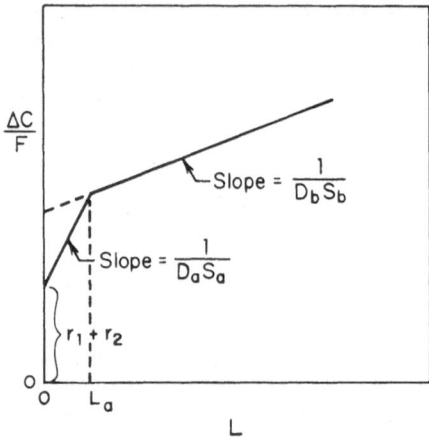

Figure 2   Thickness Analysis for Two Layers

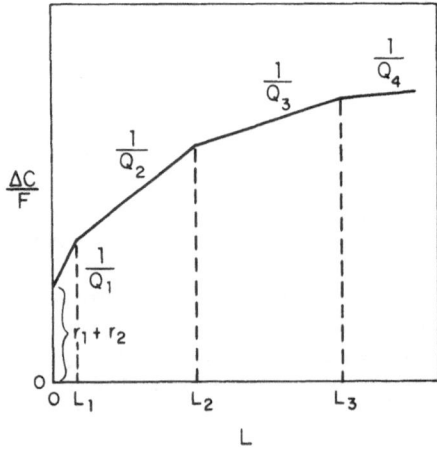

Figure 3   Thickness Analysis for Multilayer Membrane

RESULTS AND DISCUSSIONS

Examples of Boundary Resistances

The permeation of dissolved oxygen (2) and carbon dioxide (3) in water through a silicone rubber membrane showed definite boundary resistances by means of the thickness analysis. For an entirely different system, the same phenomenon was observed. In their study of a thin film composite membrane for single stage sea water desalination by reverse osmosis, Riley and his co-workers (5) found that the permeability changes with membrane thickness. The reciprocals of observed permeability of water through the acetyl cellulose acetate thin-film are plotted against the reciprocals of membrane thickness in Fig. 4. From the intercept, the limiting value of permeability is obtained as 8.31×10^{-7} (g)(cm)/(cm^2)(sec)(atm). The boundary resistance from the slope is 1.35×10^4 in corresponding units.

Another example of boundary resistances can be found in the study of gaseous hydrogen permeation through stainless steel by Phillips and Dodge (6,7). They found that the permeation rate is not proportional to the inverse of the membrane thickness. They attributed this deviation from the inverse-thickness relation to the interfacial mass transfer resistance. Their experimental data are plotted in Fig. 5. From the intercept of the least square fit, the limiting value of permeability at infinite thickness is obtained as 4.36×10^{-7} (std cc)(cm)/(cm^2)(sec)$\sqrt{\text{cmHg}}$, which is the product of DS. This compares well with the value 4.46×10^{-7} reported by Phillips and Dodge. From the slope of the straight line, the interfacial resistance is obtained as 1.29×10^5 in corresponding units.

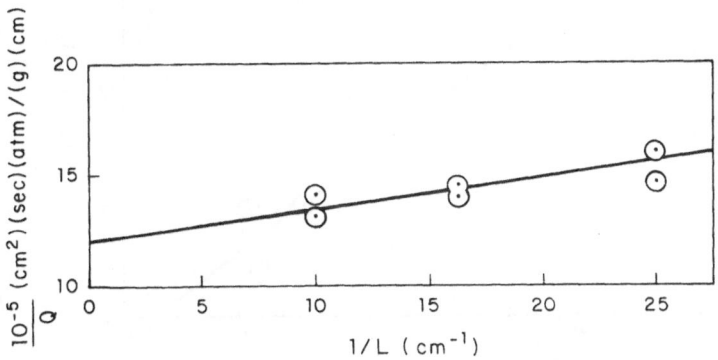

Figure 4   Water through Acetyl Cellulose Acetate

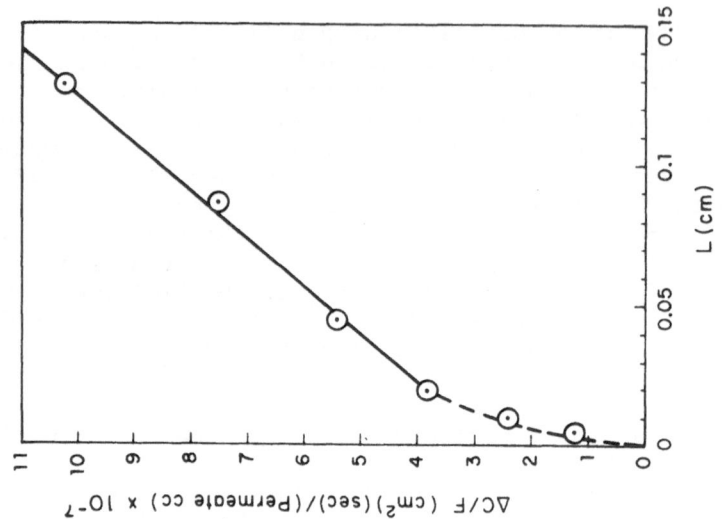

Figure 6   Water through Nylon 6

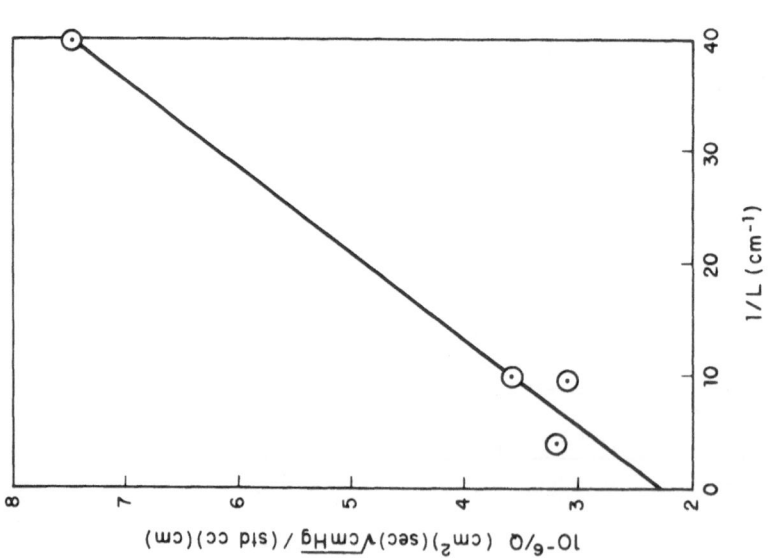

Figure 5   Hydrogen through Stainless Steel

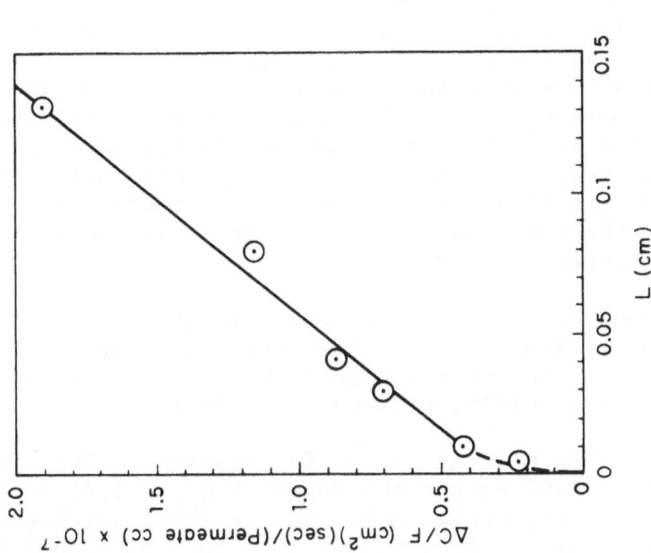

Figure 8   p-Dioxane through Polyethylene

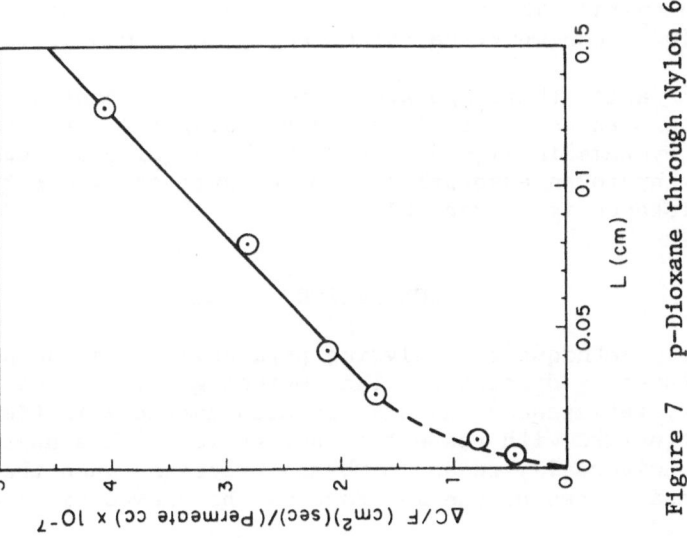

Figure 7   p-Dioxane through Nylon 6

Examples of Heterogeneous Membranes

Multiple laminates of thin films were prepared by Kim (8) in his study of actual concentration profiles across these membranes while pervaporation of various liquids took place. His experimental results strongly indicate that the resistance (or permeability) changes considerably within the membrane. Plots are made according to Eq. (2) for three systems in Figs. 6, 7, and 8. All three systems show the same straight line relationship between $\Delta C/F$ vs. membrane thickness except the initial regions. This fact indicates that there is a thin layer of high resistance region in the membrane and as membrane thickness increases a homogeneous low resistance region becomes thicker.

The reciprocal of the slope of the straight line will be called Q_h, which is the intrinsic permeability of the homogeneous phase in the membrane. The break point, beyond which the curve becomes a straight line, may be called critical thickness.

In Fig. 6, Nylon 6-water system is presented. The critical thickness is about 0.02 cm and the value of Q_h is 1.42×10^{-9} (cc permeate)(cm)/(cm^2)(sec). Below the critical thickness, the permeability value varies from 0.41×10^{-9} to 0.51×10^{-9} in the same units. In Fig. 7, the Nylon-p-dioxane system is shown. The critical thickness in this case is around 0.025 cm and the value of Q_h is 4.33×10^{-9} in the above units. Below the critical thickness, the permeability is less than 1.55×10^{-9}. Finally, in Fig. 8, pervaporation of p-dioxane through polyethylene is presented. The critical thickness is about 0.01 cm and the Q_h value is 8.25×10^{-9} in the same units. The permeability values range from 2.17×10^{-9} to 2.40×10^{-9} when membrane thickness is below 0.01 cm.

In all cases, there appears to be no significant amount of boundary resistances. This is not surprising for pure component permeation systems in liquid phase, which is quite different from the gaseous hydrogen adsorption on metal surface, where the interfacial resistance is appreciable.

CONCLUSION

A novel technique of analyzing permeability dependency on membrane thickness is discussed. This method gives a means of isolating the boundary resistances and also permits characterization of the membrane structure with respect to permeation. This approach can be applied universally to any membrane processes when there are boundary resistances or the membrane may be viewed as a composite laminate film.

NOMENCLATURE

C_1 permeate concentration in upstream

C_2 permeate concentration in downstream

D   diffusivity

F   permeation flux

L   membrane thickness

Q   permeability

Q_h intrinsic permeability of homogeneous phase

r_1, r_2 resistances of boundary layer

S   solubility of permeate in membrane

REFERENCES

1. S. T. Hwang and K. Kammermeyer, "Permeability as a Phenomenolog-
    ical Coefficient," a chapter in "Progress in Separation and
    Purification - Volume IV," John Wiley & Sons, Inc.,
    New York (1971).

2. S. T. Hwang, T. E. Tang, and K. Kammermeyer, J. Macromol. Sci.
    Phys., B5, 1 (1971).

3. S. T. Hwang and G. D. Strong, J. Polymer Sci., Symposium No. 41,
    17 (1973).

4. H. Yasuda and C. E. Lamaze, J. App. Polymer Sci., 16, 595
    (1972).

5. R. L. Riley, G. R. Hightower, and C. R. Lyons, "Thin film
    composite membrane for single-stage sea-water desalination by
    reverse osmosis," a paper presented at the joint NASA Ames
    Research Center/University of California at San Diego Confer-
    ence, Nov. 29-Dec. 1, 1972, Moffett Field, California.

6. J. R. Phillips and B. F. Dodge, A.I.Ch.E. Journal, 8, 93 (1963).

7. J. R. Phillips and B. F. Dodge, A.I.Ch.E. Journal, 14, 392
    (1968).

8. S. N. Kim and K. Kammermeyer, Separation Sci., 5, 679 (1970).

EFFECTS OF STEREOREGULARITY AND POLYMER SIZE ON THE DIFFUSION OF BENZENE IN POLYVINYL ACETATE

W.R.Brown and G.S.Park

University of Wales Institute of Science and Technology

Cardiff CF1 3NU, Wales, U.K.

It is known that changes in polymer tacticity can have a profound effect on physical properties. This mainly occurs through the onset of crystallinity in the regular polymers and previous investigations of diffusion in stereoregular polymers have mainly been concerned with crystalline or glassy materials.

It can be inferred, however, that since tacticity changes can affect the glass transition temperature even in non-crystalline polymers, these changes alter the proportion of free volume in the polymer. The free volume theory of diffusion has been very successful in explaining the concentration dependence of the diffusion of solvents in polymers (1)(2). For a range of polymers varying from polydimethylsiloxane with a glass transition temperature of -123°C to polyvinyl acetate with a glass transition temperature of 30°C a reasonable correlation between the diffusion coefficient and the concentration dependence of the diffusion coefficient with the glass transition temperature has been possible using the free volume theory (3). It appears probable that free volume changes in non crystalline elastomeric polymers should affect the diffusion coefficient when polymer stereoregularity is altered. The influence of free volume changes on diffusion coefficients are most noticeable in the region not far removed from the glass transition temperature and so polyvinyl acetate with a glass transition temperature of about 30°C is a useful material for investigating the effect of tacticity changes on the diffusion coefficient.

With very low molecular weight polymers the chain ends make a contribution to the free volume (4). Previous investigations of

the effect of polymeric weight on diffusion properties have not
shown any effect because an insufficient range of molecular
weights was covered. This limitation arose because the physical
characteristics of low molecular weight polymers made them
unsuitable for use in diffusion measurements. Blended polymers
in which the weight average molecular weight is about 30 times
greater than the number average molecular weight enable
dimensionally stable materials to be made even though large
numbers of polymer ends are present. This technique has been
used in the present investigation.

EXPERIMENTAL

Polyvinyl acetate Samples

Free radical polymerization of vinyl esters gives mainly
atactic polymer and the tacticity is affected to only a small
extent by the polymerization temperature. Isotactic polyvinyl
alkyl ethers can easily be obtained, however, by ionic
polymerization (5) and some of these ethers (6) can be cleaved to
the alcohol which on acylation gives essentially isotactic
polyvinyl esters. Preliminary work with the polyvinyl formates
showed that the isotactic material was not only insoluble in most
solvents but was an indifferent sorbent for inorganic vapours and
so the present studies have been restricted to the acetates.
Control of the molecular weight of the atactic polymers was
achieved by using transfer agents in free radical polymerizations.
Three polyvinyl acetates were prepared as follows.

<u>Isotactic polymer A</u> $M_n = 146,000$. Vinyl t-butyl ether was
prepared by refluxing t-butanol with vinyl 2-chloroethyl ether in
the presence of mercuric acetate catalyst (7). After purification
a 10% solution of the monomer in toluene was polymerized for an
hour to a 63% conversion in a serum cap vessel at -78°C using
0.6% of boron trifluoride etherate as a catalyst. After washing
well with methanol and drying under vacuum a 1% solution of the
isotactic polyvinyl ether in toluene was cleaved to the alcohol
with hydrogen bromide at 0°C (6). The precipitated alcohol was
washed well with methanolic sodium hydroxide and with methanol.
After drying under vacuum the alcohol was acetylated by heating at
80°C for 20 hours at 7% concentration in an equivolume mixture of
acetic acid and acetic anhydride containing 5% of pyridine (8).
The isotactic character of the resulting acetate was confirmed by
the absence of the 916 cm^{-1} absorption band (9) in the alcohol
formed by hydrolysis of the polyvinyl acetate. The molecular
weight was obtained from viscosity determinations on the polyvinyl
alcohol intermediate (10).

Atactic polymer B $M_n = 166,000$. Vinyl acetate was purified by
the partial polymerization method of Morton and Pirma (11). After
several outgassing cycles the monomer was polymerized to 15%
conversion using 2.4% of triethylamine as a transfer agent and 0.85%
of azoisobutyronitrile as initiator at 27°C for 18 hours. The polymer
was precipitated into petroleum ether and after purification by
several precipitations into petroleum ether from dichloromethane
solution the polymer was dried by heating at 90°C for several days
under vacuum. The molecular weight was obtained from the intrinsic
viscosity in acetone (12).

Low molecular weight atactic polymer C $M_m = 1,894$. Some of
the carefully purified vinyl acetate was polymerized with 0.75%
azoisobutyronitrile and 13% by volume of carbon tetrachloride at
27°C for 19 hours. A 15.5% yield of polymer was obtained which
after precipitation several times into petroleum ether from
dichloromethane solution and solvent removal by freeze drying on
the vacuum line gave a product which had a molecular weight of
1,894 using a Perkin-Elmer Hitachi vapour-phase osmometer.

Films of polymers A and B were cast on a plate glass surface
from filtered dichloromethane solutions. After detaching the films
the final traces of solvent were removed by heating the clamped
films at 70°C for several days on the vacuum line. Polymer C
behaved like a very viscous liquid and so could not be used
directly for sorption experiments. Since the mechanical
properties of a polymer depend essentially on the weight average
molecular weight it was possible to make a mechanically stable film
by mixing equal amounts of polymers B and C. The two polymers were
mixed by dissolving together in dichloromethane and then by
precipitation in petroleum ether. The mixed polymers were purified
by freeze drying and vapour phase osmometry gave a molecular
weight of 3922 for the mixture. Films of this polymers were made
in the same way as films of polymers A and B.

Expansion Coefficient and the Glass Transition Temperature

Since relatively small amounts of polymer were available the
specific volume against temperature relationship was determined by
measuring the refractive index with an Abbe refractometer (13).
The polymer specimens were draped on to the prism of the
refractometer which was subject to a programmed temperature change
of between 12 and 30°C hours^{-1}. The specimen temperature was
measured with a minute chromel alumel thermocouple and the whole
apparatus was kept in a dry box to avoid the condensation of
moisture. Films approximately 50 μm thick were found to be best
for these determinations.

Sorption Experiments

Rates of sorption of benzene vapour and sorption equilibria were obtained using the quartz spring balance and apparatus described elsewhere (14). Interval sorption measurements were used. In this technique sorption to equilibrium at one vapour pressure was followed by measurements of a further sorption resulting from an increase of vapour pressure. This was carried out for several sorption intervals (15). Since specimens as thin as 30 μm had to be used to enable the experiments to be completed in reasonable time the polymer specimens were supported on specially constructed, very light glass cradles. With the isotactic polymer A, however, it was found that even at 35°C 6% of benzene plasticized the polymer so much that rapid distortion of the film occurred. In experiments at 45°C this distortion was prevented by completely supporting the film on one face by allowing it to mould itself to a thin aluminium sheet. This resulted in sorption from one face only and allowance for this had to be made in calculating the diffusion coefficient.

RESULTS

The Glass Transition

At heating or cooling rates of 0.2–0.5°C min^{-1} all the polymers gave very clear breaks in the refractive index/temperature plots. A typical example is given in Figure 1 and reproducible results were found at different cooling rates or even when a programmed increase in temperature was used. Figure 1 shows a typical refractive index against temperature plot. The breaks in the plots give a direct measure of the glass transition temperature, T_g.

Table 1. Expansion Coefficient, α, Glass Transition Temperature, T_g, and Fractional Free Volume, f_v

Polymer	T_g °C	$\alpha_1 \times 10^4$	$\alpha_2 \times 10^4$	$\Delta\alpha \times 10^4$	$f_v \times 10^2$ 35°C	$f_v \times 10^2$ 45°C
A	25.8	5.89	1.68	4.21	2.887	3.308
B	31.4	6.26	1.62	4.64	2.639	3.103
C	23.6	6.31	1.43	4.88	3.056	3.544

A) Isotactic $M_n = 146,000$, B) Atactic $M_n = 166,000$
C) Atactic $M_n = 3,922$, $\alpha_1 =$ above T_g, $\alpha_2 =$ below T_g
$\Delta\alpha = \alpha_1 - \alpha_2$.

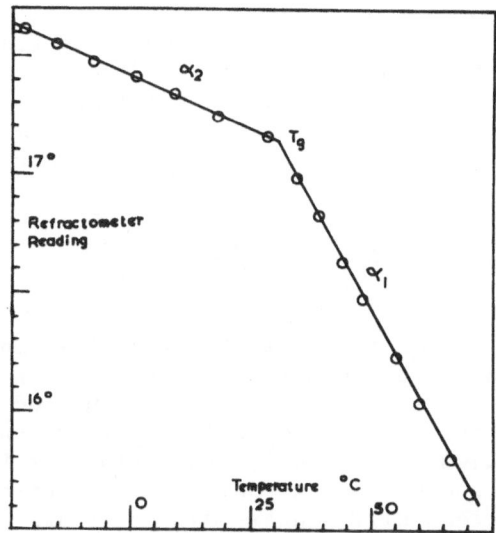

Fig.1   Refractive index (refractometer reading in degrees) as a
function of temperature (heating rate + 0.2°C min^{-1}) for an
atactic polyvinyl acetate polymerized at 115°C.

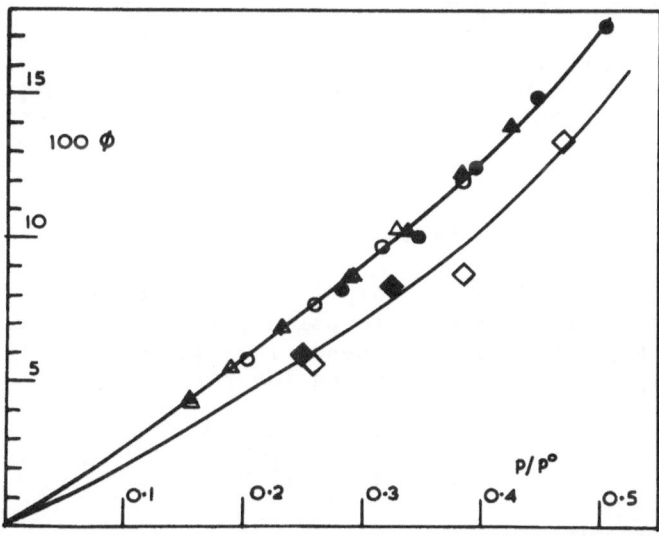

Fig.2   Equilibrium volume fraction, $\emptyset$, of sorbed benzene at
thermodynamic activity p/p° for various polyvinyl acetates.
p is the vapour pressure,   p° is the saturation vapour pressure.
●▲■ at 35°C, ○△□ at 45°C. ● ○ M_n = 166,000.
▲△ Mn = 3,922. ■□ Isotactic polymer.

W. R. BROWN AND G. S. PARK

The expansion coefficients, α, can be obtained from the slope of the refractive index/temperature plots. The Lorenz and Lorentz relationship enables the specific volume to be obtained from the refractive index in a constant composition system. It follows that for relatively small changes the expansion coefficient, α, is given by the relationship

$$\alpha = - S \frac{6n}{(n^2 - 1)(n^2 + 2)} \qquad (1)$$

where S is the slope of the plot of refractive index against temperature and n is the mean refractive index in the temperature range of interest. Use of equation 1 has enabled the expansion coefficient, α_1, above T_g and α_2 below T_g to be calculated from plots the type shown in Figure 1. The values of the glass transition temperature, the expansion coefficients and the expansion coefficient differences, $\Delta\alpha$, are shown in Table 1.

Equilibrium Sorptions

The extent of vapour sorption at equilibrium is shown as a function of the relative vapour pressure for all three polymers at 35°C and 45°C in Figure 1. The points for the two atactic polymers at both 35°C and 45°C fall on the same line but the much lower sorption by the isotactic polymer is very noticeable. The data give a reasonable fit to the Flory-Huggins relationship with a χ value of 0.36 for the atactic materials and 0.59 for the isotactic polymer.

The Diffusion Coefficient

Linear plots of vapour sorption were obtained against the square root of time for each sorption interval with all the polymers. From the slope of these plots the mean diffusion coefficient, $\overline{D}$, was obtained using the relationship

$$\overline{D} = \frac{\pi l^2}{16} \left(\frac{d(M_t/M_\infty)}{dt^{\frac{1}{2}}} \right)^2 . \qquad (2)$$

Here, l is the thickness of the film ($\frac{1}{2}l$ in the case of the isotactic material supported on an aluminium backing) and M_t and M_∞ are the quantities of vapour sorbed at time t and at infinite time. The mean diffusion coefficient, $\overline{D}$, is closely related to the average diffusion coefficient given by the integral in equation 3.

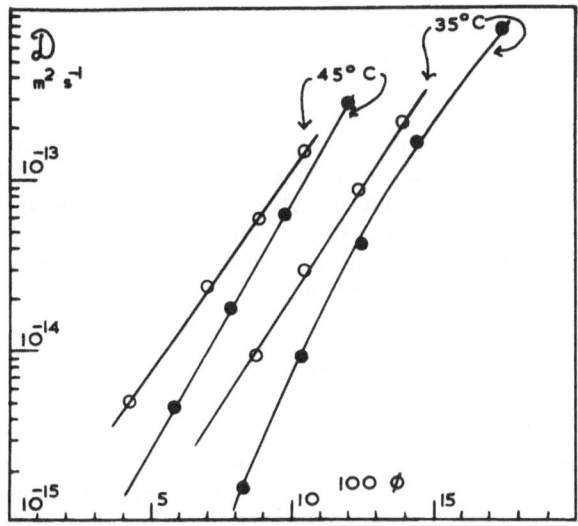

Fig.3 Effect of polymer molecular weight on the intrinsic
diffusion coefficient, $\mathcal{D}$, at volume fraction, $\emptyset$, of benzene.
● $M_n = 166,000$; ○ $M_n = 3,922$.

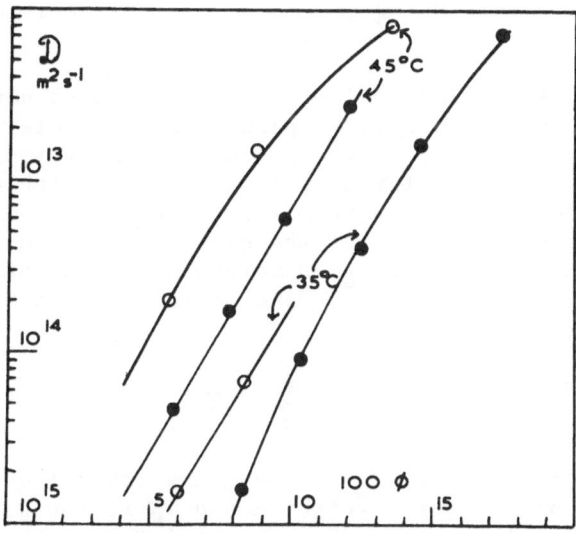

Fig.4 Effect of tacticity on the intrinsic diffusion
coefficient, $\mathcal{D}$, at volume fraction, $\emptyset$, of benzene.
● atactic polymer; ○   isotactic polymer.

$$\overline{D} = \int_{C_1}^{C_2} DdC/(C_2 - C_1) \tag{3}$$

In this expression D is the diffusion coefficient at concentration C and C_1 and C_2 are the steady concentrations throughout the polymer at the beginning and at the end of a sorption interval. Values of D as a function of C were obtained using a method described previously (15) and the intrinsic diffusion coefficient, $\mathfrak{D}$, was calculated from equation 4.

$$\mathfrak{D} = D/(1 - \emptyset)^{\frac{7}{3}} \tag{4}$$

Here $\emptyset$ is the volume fraction of solvent in the polymer solvent mixture at concentration C. The dependence of the intrinsic diffusion coefficient on concentration is shown in Figure 3 for the two atactic polymers B and D. The change on going from the high molecular weight atactic polymer to the isotactic one is shown in Figure 4.

DISCUSSION

Free Volume Changes and the Glass Transition Temperature

Low molecular weight polymers contain a high concentration of chain ends which disrupt the packing of the molecules and so introduce extra free volume which leads to a lowering of the glass transition temperature. This is shown in Table 1 where a decrease of the number average molecular weight from 166,000 to 3,922 results in a lowering of the glass transition temperature by 8°C. The dependence of T_g on molecular weight has been most investigated for polystyrene. Ueberreiter and Känig (16) obtained the relationship

$$\frac{1}{T_g} = \frac{1}{T_g^{\infty}} + \frac{A}{M} , \tag{5}$$

in which A is a constant, T_g is the glass transition temperature for a polymer of molecular weight M and T_g^{∞} is the glass transition temperature for a polymer of infinite molecular weight. Fox and Flory (17) proposed the relationship

$$T_g = T_g^{\infty} - \frac{K}{M} . \tag{6}$$

These two relationships are equivalent except at low values of
M and the constants A and K have the values of 0.515 and 1.2 x
10^5 for polystyrene. Our glass transition data for polyvinyl
acetates lead to A and K values of 0.35 mol(degree)$^{-1}$ and 3.16
x 10^4 mol degree   respectively. The value of K can be taken as
a measure of the average free volume associated with each chain
end and so it would appear that the chain ends in polystyrene make
a much larger contribution to the free volume in the polymer than
is found with the chain ends in polyvinyl acetate.

The glass transition temperature of isotactic polymethyl
methacrylate is as much as 60° lower than that for the atactic
material while no change in T_g is observed on going from atactic
to isotactic polymethyl acrylate (18). This difference can be
related to the much lower hindrance to rotation offered by the
side groups in the polymethyl acrylate. Meares (19) has proposed
that the higher glass transition temperature of polyvinyl acetate
compared to the isomeric polymethylacrylate probably is a result
of the greater hindrance to rotation afforded by the acetate side
groups. This would lead to the acetate polymer having a glass
transition temperature that is rather more sensitive to the
tacticity than is found for the acrylate. The 5°C lowering for
the glass transition temperature of the isotactic polyvinyl
acetate shown in Figure 1 supports this idea.

Lower glass transition temperatures are associated with
polymers having large proportions of free volume in the rubbery
state. The WLF relationship

$$f_v = f_0 + (T - T_g)\Delta\alpha , \qquad (7)$$

relates the fractional free volume, f_v, in a polymer at
temperature T to that at the glass transition temperature, f_0,
and to the difference, $\Delta\alpha$, of the cubical expansion coefficient
above and below T_g. This relationship enables values of the
fractional free volume to be calculated at the two experimental
temperatures of 35°C and 45°C provided that the 'universal value'
of 0.025 is assumed for f_0. The values of f_v have been included
in Table 1.

Free Volume Change and Diffusion Coefficient

Fujita (2) proposed the equation

$$D = A e^{-B/f_v} , \qquad (8)$$

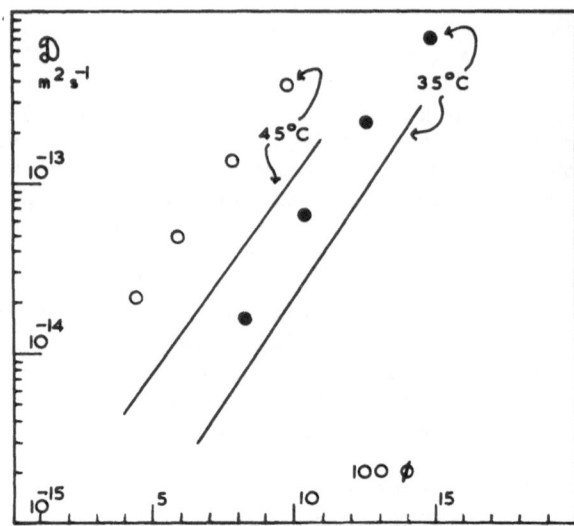

Fig.5 Comparison of the measured diffusion coefficient, $\mathcal{D}$,
for the low molecular weight polymer with values calculated from
the free volumes. 〜 measured diffusion coefficient,
○ ● calculated values, Ø is the volume fraction.

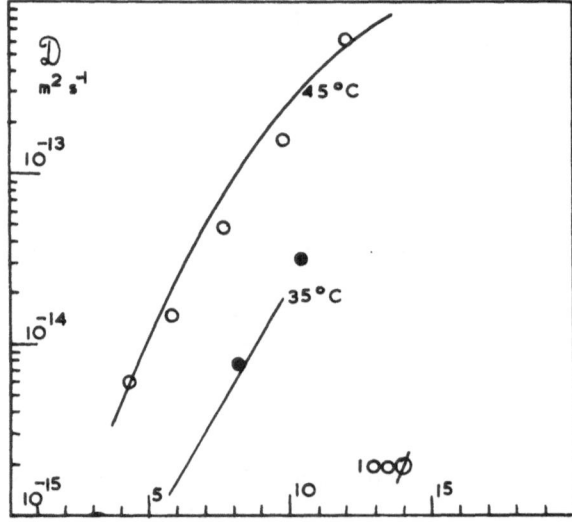

Fig.6 Comparison of the measured diffusion coefficient, $\mathcal{D}$,
for the isotactic polymer with values calculated from the free
volumes. 〜 measured diffusion coefficient, ○ ● calculated
values, Ø is the volume fraction.

to relate the diffusion coefficient, D, to the fractional free volume, f_V. In this relationship A and B are parameters depending on the polymer and on the diffusant. It is interesting to see to what extent this equation can predict changes in the diffusion coefficient from the calculated changes of f_V that occur on decreasing the molecular weight or increasing the isotacticity. Comparison of the diffusion coefficient, D_A, for the atactic polymer with that D_I for the isotactic or low molecular weight polymer using equation 8 gives

$$D_I = D_A \exp[B(f_I - f_A)/f_I f_A] \quad , \quad (9)$$

where f_A and f_I are the fractional free volumes for the atactic polymer and for either the isotactic or the low molecular weight one. Equation 9 holds at zero concentration of penetration. At higher concentrations the penetrant makes a contribution to the free volume of the system and then f_V can be replaced by $f_V + \phi(f_B - f_A)$ where ϕ is the volume fraction of penetrant and f_B is the effective free volume of the penetrant in the mixture. Applying this replacement to equation 9 then gives

$$D_I = D_A \exp\left[\frac{B(1 - \phi)(f_I - f_A)}{[f_I + \phi(f_B - f_I)][f_A +] \phi(f_B - f_A)]} \right] (10) .$$

Previous work on the diffusion of benzene (20) has suggested a mean value of 0.89 for B and 0.15 for f_B. From these values and taking the measured values of D_A in Figures 2 and 3 it is possible to substitute the free volume figures from Table 1 into equation 10 and to calculate the diffusion coefficients in the low molecular weight polymer and in the isotactic polymer. The calculated values at 35°C are much higher than the experimental ones for $\phi < 0.07$ and at 45°C for $\phi = 0$. This may arise from the errors caused by slow relaxation processes at temperatures just above T_g that have been discussed by Kishimoto and Matsumoto (21).

The calculated values at higher ϕ values fall on lines roughly parallel to the measured values as is shown for the low molecular weight polymer in figure 5 and the isotactic one in figure 6. The measured values for the isotactic polymer are not very far removed from the calculated ones. The low T_g of the low molecular weight polymer, however, leads to predicted diffusion coefficients in Figure 5 which are much larger than those actually obtained.

REFERENCES

1.   P.Meares, J.Polymer Sci., 1958, 27, 391.
2.   H.Fujita, Fortshr.Hochpolym.-Forsch., 1961, 3, 1.
3.   A.C.Newns, G.S.Park, J.Polymer Sci.,C, 1969, 22, 927.
4.   P.Meares, Trans.Faraday Soc., 1957, 53, 31.
5.   G.E.Schildknecht, S.T.Gross, A.O.Zoss, Ind.Eng.Chem., 1949,
     41, 1998.
     G.Natta, I.Bassi, P.Corradini, Makromolek Chem., 1956,
     18/19, 455.
6.   S.Okamura, T.Kodama, T.Higashimura, Makromolek Chem., 1962,
     53, 180.
     J.G.Pritchard, R.L.Vollmer, W.C.Lawrence, W.B.Black,
     J.Polymer Sci., A1, 1966, 4, 707.
7.   W.H.Watanabe, L.E.Conlon, J.Am.Chem.Soc., 1957, 79, 2828.
8.   O.L.Wheeler, S.L.Ernst, R.N.Crozier, J.Polymer Sci., 1958,
     8, 409.
9.   K.Fujii, T.Mochizuki, S.Imoto, J.Ukida, M.Matsumoto,
     J.Polymer Sci., A2, 1964, 2, 2327.
10. A.Beresniewicz, J.Polymer Sci., 1959, 39, 63.
11. M. Morton, I. Pirma, J.Polymer Sci., A, 1963, 1, 3043.
12. E.F.T.White, Ph.D. Thesis, University of London 1958.
13. R.H.Wiley, G.M.Brauer, J.Polymer Sci., 1948, 3, 455, 647, 704.
     R.B. Beevers, E.F.T.White, Trans Faraday Soc., 1960, 56, 744.
14. J.Crank and G.S.Park, "Diffusion in Polymers", Academic
     Press, London and N. York, 1968, 21.
15. W.R.Brown, R.B.Jenkins, G.S.Park, J.Polymer Sci.,C, 1973,
     41, 45.
16. K.Ueberreiter, G.Känig, J.Colloid Sci., 1952, 7, 569.
17. T.G. Fox, P.J.Flory, J.appl.Phys., 1950, 21, 581.
18. J.A.Shetter, J.Polymer Sci.B, 1963, 1, 209.
19. P.Meares, "Polymers, Structures and Bulk Properties",
     D. van Nostrand, London, 1965.
20. T.A.Garrett, G.S.Park, J.Polymer Sci.,C, 1967, 16, 601.
21. A.Kishimoto, K.Matsumoto, J.Polymer Sci.,A, 1964, 2, 679.

ACKNOWLEDGEMENTS

We thank the Paint Research Institute for financial support and Dr. K. Fujii for advice about the preparation of stereoregular polyvinyl esters.

A DISCUSSION OF THEORETICAL MODELS OF ANOMALOUS DIFFUSION OF VAPORS IN POLYMERS

J.H. Petropoulos and P.P. Roussis

Democritos Nuclear Research Center, Aghia Paraskevi

Athens, Greece

INTRODUCTION

The "anomalous" or "non-Fickian" diffusion of micromolecular vapors in homogeneous macromolecular substrates, a subject to which V. Stannett and his collaborators have eminently contributed, continues to develop both experimentally and theoretically. New experimental data have led to novel interesting theoretical concepts and insights. Nevertheless, orderly progress in this, like in any other, field also requires reassessment of the older models in the light of the more recent data, just as new theoretical treatments must pay due regard to older data. This task has been neglected to some extent, although it is admittedly hampered by the fact that most published data are not usually available in sufficient detail. However, present models of "non-Fickian" diffusion have not progressed much beyond qualitative or semiquantitative applications, so that their comparison at this level is quite useful.

Three main theoretical models can be distinguished, often existing in more than one version. These will be discussed in turn, with particular regard to the extent of their applicability to the data, their dependence on arbitrary assumptions and their physical meaningfulness.

The experimental evidence to be considered is largely in the form of sorption kinetic curves (M_t or M_t/M_∞ vs t or $\sqrt{t}$, where M_t, M_∞ represent the amount of penetrant absorbed or desorbed at time t and at final equilibrium respectively) or permeation curves (which represent the amount of penetrant permeated as a function of t and yield the time lag L^a as the intercept on the t axis). The anomalous or "non-Fickian" features found in such data have been discussed in detail e.g. by Fujita[1]

MOLECULAR RELAXATION MODEL

Physically, this model takes account of the molecular rearrangement in the polymer necessary to accomodate a change in penetrant content. Near or below the glass transition temperature, such molecular relaxations will tend to be quite slow on the time scale of the diffusion process. Putting it in simple language, during absorption (at constant external penetrant activity) the polymer network can "open up" immediately only to a limited extent (since only very small polymer segments can move practically instantaneously). Further "opening up" of the polymer structure will occur gradually (as larger polymer segments can rearrange) until the final swelling equilibrium is attained. Thus, one can distinguish an "initial state" (corresponding to whatever molecular rearrangements can occur practically instantaneously) in addition to that of final (true) equilibrium. In absorption the latter state will correspond to the more "open" structure, the converse being true for desorption.

Different versions of this model have been applied to systems where the vapor is a good swelling agent of the polymer. A more "open" structure may be assumed to enhance the mobility of penetrant, just as it will accomodate more penetrant (i.e. enhance the distribution or solubility coefficient). The treatments of Crank[2] and of Long and Richman[3] are based on one of the above effects respectively. They employ the classical formulation for unidimensional "Fickian" diffusion, namely

$$\frac{\partial C}{\partial t} = \frac{\partial}{\partial X} \left(D \, \frac{\partial C}{\partial X} \right) \tag{1}$$

where t is the time, X the distance coordinate ($0 \leqslant X \leqslant \ell$, ℓ=thickness of the membrane), C(X,t) the concentration of penetrant in the polymer and D(C) its diffusion coefficient. The following boundary conditions hold for C(X,t)

$$C(0,t) = C_o, \quad C(\ell,t) = C_\ell, \quad C(X,0) = C_1 \tag{2}$$

where C_o, C_ℓ, C_1 are constants; $C_o > C_\ell = C_1$ in the usual permeation experiments, and $C_o = C_\ell$ in sorption experiments (with $C_o > C_1$ for absorption and $C_o < C_1$ for desorption).

The above formulation is modified by making either D (Crank) or C_o (Long and Richman) a function of t. The values of D(C,t) and of $C_o(t)$ which refer to the "initial state" mentioned above are denoted by D(C,0) and $C_o(0)$ respectively and the rate of approach to the respective final equilibrium values D(C,∞), $C_o(\infty)$ is assumed to follow first order kinetics (although this must be qualified because the respective rate constants or relaxation frequencies β_2, β_1 may depend on C). Accordingly, Equations (1) and (2) have to be solved simultaneously with either (Crank)

$$\frac{\partial D(C,t)}{\partial t} = \frac{\partial D(C,0)}{\partial C} \frac{\partial C}{\partial t} + \beta_2 \{D(C,\infty) - D(C,t)\} \tag{3}$$

or (Long and Richman)

$$C_o(t) = C_o(\infty) - \{C_o(\infty) - C_o(0)\} \exp(-\beta_1 t) \qquad (4)$$

In Equation (3), $D(C,\infty)$ for the systems under consideration is an increasing function of C, usually of nearly exponential form, and $D(C,0)$ is assumed to be likewise. The physical considerations set out above imply that $D(C,\infty) > D(C,0)$ for absorption and conversely for desorption. These functions coincide at $C=C_1$ and diverge increasingly as $C \to C_o$. Fortunately, the results obtained for sorption kinetics and the permeation time lag are not very sensitive to the precise functional form of $D(C,0)$ and $\beta_2(C)$[2], but assumptions are necessary as to the magnitude of these quantities. Corresponding assumptions are necessary with respect to Equation (4). Here, however, the physical basis is more solid. The reality of the phenomenon has been demonstrated experimentally and even the kinetics of the variation of $C_o(t)$ can be followed[3]. Furthermore, for sorption experiments covering small concentration intervals β_1 and D can be assigned constant values as an approximation. In this way Long and Richman[3] obtained and applied an analytic expression for M_t.

Both approaches explain properly the most prominent "non-Fickian" features; namely absorption M_t vs $\sqrt{t}/\ell$ curves, which are usually S shaped and not superposable when ℓ is varied, and time lags which exceed the calculated "Fickian" values and are also not subject to the ℓ^2 scaling law[4,5]. Crank's approach can explain reasonably well (on a semiquantitative level) the often very marked difference in shape between absorption and desorption curves; but can not really deal with "two-stage" absorption M_t vs $\sqrt{t}$ curves[1,6]. The Long and Richman approach, on the other hand, leads naturally to such curves, when the relaxation process is sufficiently slow as compared with diffusion. Here too, reasonable semiquantitative success has been claimed.[3] By some elaboration of this treatment (involving a second relaxation process), Fujita and Kishimoto[1] were able to reproduce remarkably well certain general and characteristic changes in shape of experimental absorption M_t vs $\sqrt{t}$ curves over successive small concentration intervals. The diffusion coefficient and relaxation frequencies were strongly increasing functions of C, whereas the quantity $\{C_o(0)-C_1\}/\{C_o(\infty)-C_1\}$ varied inversely with C. The former quantities were most probably assigned constant values within each sorption interval as an approximation, although this point has not been clarified. It is also noteworthy that consistent semiquantitative interpretation of "non-Fickian" sorption curves and permeation time lags over successive small concentration intervals was achieved in the system water-polyacrylamide[7] by means of the Long and Richman model.

In view of such success, it is regrettable that this approach has not been tested further. Its application to desorption is of particular interest, since predictions differing from those of Crank's treatment may be expected. In fact, simple variable transformation

shows that if the magnitude of $C_0(\infty)-C_0(0)$ is the same for the absorption-desorption pair, then the two curves superpose on the Long
and Richman model (with β_1,D constant). They separate if D and/or
β_1 are functions of C. This is illustrated in Figure 1, which shows
examples calculated by us using a numerical explicit method to solve
Eqs.(1),(2),(3) and (4) simultaneously. $D(C,0)$ and $D(C,\infty)$ were represented by polynomial functions (resulting in D-C relations of
smaller curvature than exponential ones as is often the case in practice) and so was $\beta(=\beta_1=\beta_2)$ when it was not constant. The results
show that the S shape becomes less noticeable and is ultimately eliminated as the dependence of D on C (or t) becomes more marked (cf.
curves D,E,F). The reverse is true in absorption (cf. curves B,C).
In polymer-penetrant systems of the type under discussion here, S
shaped desorption curves, though not impossible to find[8], are certainly rare. Hence, in order to deal with desorption, the Long and
Richman treatment must take into account at least the variation of
D with C within the sorption interval. Whether inclusion of any time dependence of D is also necessary can only be decided in each specific case by means of more detailed quantitative analysis.

One may also examine the possible dependence of β_1 on C during
a sorption experiment through kinetic analysis of $C_0(t)$. It is a
matter of some surprise that no such detailed kinetic analyses have
appeared. All that can be said about published[3] direct quantita-

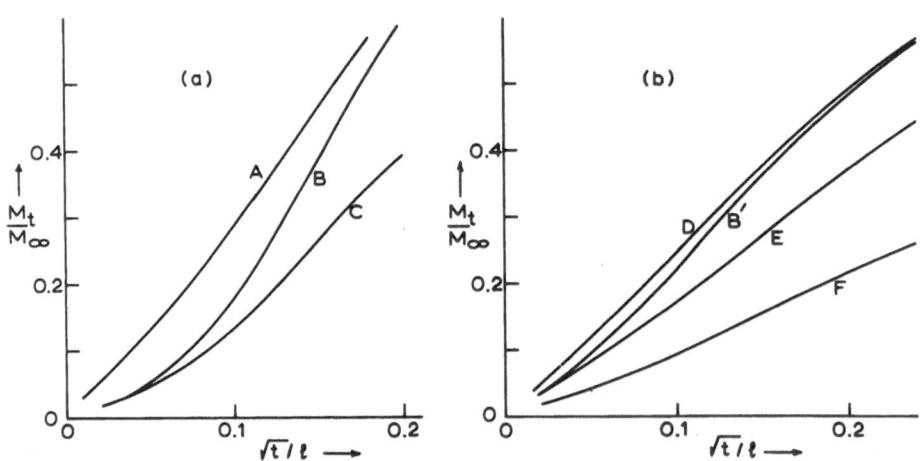

Figure 1.   Absorption (Fig. 1a) and desorption (Fig. 1b) curves for
$C_0(t)$, $D(C,t)$ with $\beta=\beta_1=\beta_2$, $\ell^2\beta=100$; $C_0(\infty)=1$, $C_1=0$ (curves
A,B,C), $C_0(\infty)=0$, $C_1=1$ (B$'$,D,E,F); $C_0(0,0)=1$ (A), 0.4 (B,C),
0.3 (D,E,F), 0.6 (B$'$); $D(C,\infty)=1$ (F), $D(C,\infty)=1+5C$ (C,E),
$D(C,\infty)=1+4C+15C^2$(A,B,B$'$,D); $D(C,0)=D(C,\infty)$(B,B$'$,C,D,E,F),
$D(C,0)=1+2C+7C^2$ (A).

tive observations of $C_o(t)$ (methyl iodide in cellulose acetate), which cover early times only, is that they are not incompatible with first order kinetics (i.e. constant β_1). A conclusion not much firmer than this (this time because of experimental scatter) applies to the second stage (where C is supposed to be very nearly uniform throughout the membrane) of cellulose-water M_t vs $\sqrt{t}$ curves[9] replotted by us. On the other hand, second stage M_t vs $\sqrt{t}$ curves for cellulose acetate-acetone[6] replotted by Kwei[10] indicate zero order kinetics. It may be noted that zero order kinetics is not contradicted by the cellulose acetate-methyl iodide data as far as they go, but certainly does not fit those for cellulose-water. The detailed study of two-stage M_t vs $\sqrt{t}$ curves in various systems by Odani et al[11,12] was confined to observation of the inflection point of the second stage from which β_1 could be deduced[1] on the basis of the Long and Richman model. The fact that in one system values of β_1 so obtained agreed with parallel results on relaxation frequency of creep[13] is a noteworthy success. However, the exponential dependence of β_1 on C_1, which was found to hold fairly generally, may be due entirely to a dependence of β_1 on penetrant activity a. Thus, it need not imply variation of β_1 during a sorption experiment, in which the membrane surface is exposed to penetrant at constant $a(=a_o)$.

   We have examined briefly the effect on the $C_o(t)$ kinetics of three different functional dependences of β_1 on C. Some examples are shown in Figure 2, together with one of the effect of replacing

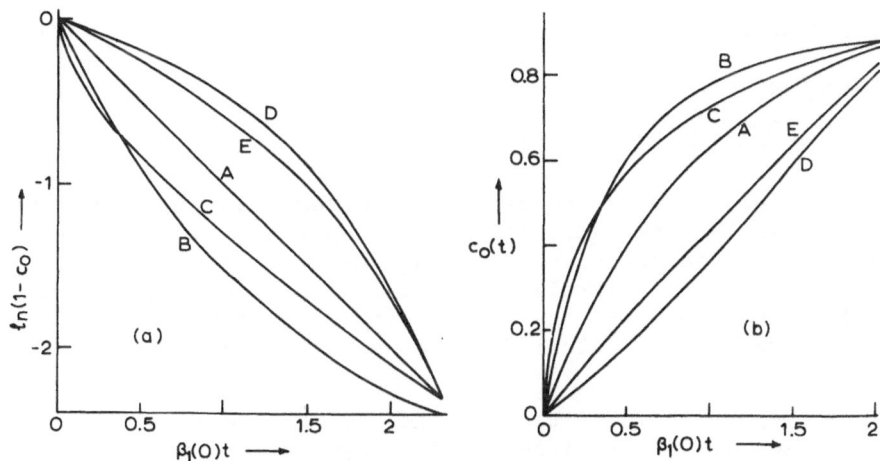

Figure 2.   Kinetics of $c_o(t)=\{C_o(t)-C_1\}/\{C_o(\infty)-C_1\}$. Curve A: $\beta_1=1$; B: $\beta_1=0.230$ to 23.2 (rectangular spectrum); C: $\beta_1(c_o)=0.15$ $(1+b_1c_o)$; D: $\beta_1(c_o)=3.38\exp(b_2c_o)$; E: $\beta_1(c_o)=2.21/(1-b_3c_o)$; $b_1,b_2,b_3=$ constants; $\beta_1(1)/\beta_1(0)=10$.

β_1 by a continuous spectrum of relaxation frequencies (rectangular in shape and extending over two decades). Significant deviations from first order kinetics are found in all cases (Figure 1a); but good approximations to zero order kinetics are given by the inverse linear and exponential $\beta_1(C)$ functions (Figure 1b). Such an explanation of the aforementioned second stage data on cellulose acetate-acetone, however, seems unlikely because in this system the position of the second stage curve on the t axis appears to be particularly insensitive to C[6,11]. Resolution of difficulties of this kind is, in our view, of considerable importance for the further validation of the Long and Richman approach.

Another criticism of the said treatment concerns its mathematical formulation. The reason is that the concentration gradient is not the proper driving force for diffusion if C is also affected directly by the relaxation process(this emerges more clearly if one replaces the relaxation process by a chemical reaction; the proper driving force of diffusion is not the total concentration gradient, but only that of unreacted penetrant). This criticism is met by using the chemical potential gradient of the penetrant as the driving force to obtain[5]

$$\frac{\partial C}{\partial t} = \frac{\partial}{\partial X} \left(D_T S \frac{\partial a}{\partial X} \right) = \frac{\partial}{\partial X} \left(P_T \frac{\partial a}{\partial X} \right) \tag{5}$$

where $D_T, P_T, S = C/a$ are the "thermodynamic" diffusion, permeability and solubility coefficients respectively. A strict irreversible thermodynamics approach has been given[14] but not applied in practice. The boundary conditions for $a(X,t)$ are completely analogous to Eq.(2), namely ($a_o, a_\ell, a_1 = $const.),

$$a(0,t) = a_o \ , \quad a(\ell,t) = a_\ell \ , \quad a(X,0) = a_1 \tag{6}$$

and Equation (4) is replaced by[15]

$$\frac{\partial C(a,t)}{\partial t} = \frac{\partial C(a,0)}{\partial a} \frac{\partial a}{\partial t} + \beta_1 \{C(a,\infty) - C(a,t)\} \tag{7}$$

Equation (7) is analogous to Equation (3) rather than Equation (4); thus making clear that $C = C(a,t)$ is affected by the relaxation process throughout the membrane, not just at the surface (i.e. the variation of C_o with t is merely a manifestation of the time dependence of S). One, therefore, has to specify a complete "initial sorption isotherm" $C(a,0)$ exactly analogous to $D(C,0)$ in Equation (3).

The Crank approach is little affected by the new formulation since D, D_T do not as a rule differ much in systems of the kind under consideration. The main point of some significance is whether $D_T(a,t)$ or $D_T(C,t)$ is more appropriate (or useful). It will be noted that in the latter case additional time dependence is built into D_T if $C = C(a,t)$ in a combined Crank-Long and Richman model (cf. curves A, B, C in Figure 1a). This issue can be decided only by detailed quantitative applications to experimental data. Otherwise, the results obtained from the new formulation (cf. curves C,D,E in Figure 3) are not generally very different from those of the previous one (so that the practical utility of the latter is preserved). In our calculations we used $P_T(a,t)$ in place of $D_T(a,t)$, replacing Equation (3) by an exactly analogous one for $P_T(a,t)$[15]. This is quite acceptable theoretically and leads to a particularly simple and useful expression for the permeation time lag L^a, namely

$$L^a - L_S^a = \int_0^\infty dt \int_{a_\ell}^{a_o} \{P_T(a,\infty) - P_T(a,t)\}\, da \Big/ \int_{a_\ell}^{a_o} P_T(a,\infty)\, da \qquad (8)$$

where L_S^a refers to a "Fickian" system with $S=S(a,\infty)$, $P_T=P_T(a,\infty)$. With the aid of Eq.(8) the behavior of the quantity $L_T^a(a_o,a_\ell=0)$ for polyvinyl acetate-methanol could be reproduced reasonably well[16].

A full treatment becomes imperative, and the individual Crank and Long and Richman approaches correspondingly break down, when the penetrant is a poor swelling agent of the polymer. The limited but extremely interesting experimental information about such systems relates mainly to water in hydrophobic polymers and is due in large measure to Stannett et al[17-20]. In the cellulose acetate-water system, for example, permeation experiments[19] indicate "Fickian" behavior ($L^a=L_S^a$), whereas the sorption M_t vs $\sqrt{t}/\ell$ curves[20] are

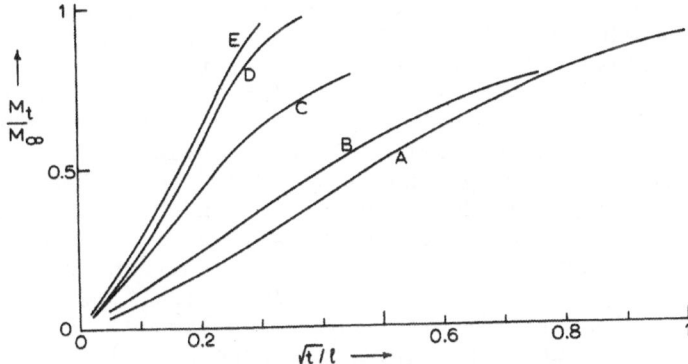

Figure 3. Absorption curves for $S(a,t)$, $P_T(a,t)$ with $a_o=1$, $a_1=0$, $\ell=1$; $P_T(a,\infty)=1$ (curves A,B), $P_T(a,\infty)=1+5a+19a^2$ (C,D,E); $C(a,\infty)=0.8a+0.2a^2$; $P_T(a,0)=1$ (A), $P_T(a,0)=1+0.5a+a^2$ (B), $P_T(a,0)=1+2a+7a^2$ (C,D,E); $C(a,0)=0.55a+0.05a^2$ (A,B,C,D), $C(a,0)=C(a,\infty)$ (E); $\ell^2\beta=10$ (A,B), $\beta=\beta(0)(1+5a+19a^2)$, $\ell^2\beta(0)=0.2$ (C), 2 (D,E), $\beta=\beta_1=\beta_2$.

clearly "non-Fickian" (S shape and dependence on ℓ). These findings can only be reconciled by assuming that S and D_T respectively increase and decrease with time; the corresponding "non-Fickian" effects neutralising one another in the time lag, but only partly so in sorption kinetics. This is shown particularly well in terms of the parameters S and P_T. Equation (8) shows that $L^a = L^a_s$ if P_T is independent of t. At the same time, Equation (5) shows that in sorption kinetics there will be a residual "non-Fickian" influence from S(a,t) (cf. also curve A in Figure 3). This, together with the fact that D_T is practically independent of a or C (as deduced from permeation measurements[19] and careful analysis[21] of sorption kinetics), explains further why the desorption M_t vs $\sqrt{t}$ curves[20] are also S shaped. The data for water diffusion in the more hydrophobic ethyl cellulose[18] can be interpreted similarly, by assuming a stronger tendency of D_T to decrease with time, so that now P_T will also tend to diminish concurrently[15,16]. This leads to $L^a < L^a_s$ and a tendency for the S shape of the absorption M_t vs $\sqrt{t}$ curves to become less pronounced (cf. curve B in Figure 3), in conformity with experiment[18,16] (although desorption curves are more difficult to interpret).

Physically, this behavior has been explained[15-16] by the increasing tendency of penetrant molecules to cluster (and hence to become immobilized) as the penetrant becomes a poorer swelling agent of the polymer (i.e. as penetrant-penetrant and polymer-polymer interactions become more favored energetically over polymer-penetrant ones). In normal "Fickian" diffusion this causes the D(C) function to change from strongly increasing to markedly decreasing (as in ethyl cellulose-water) with a concentration-independent D as an intermediate stage (e.g. cellulose acetate-water). In "non-Fickian" diffusion, the more "open" structure, referred to in the description of molecular relaxation at the beginning of this Section, should enhance clustering, if such tendency exists. The effect on S will be as before (since more penetrant can be accomodated whether clustering occurs or not), but that on D_T will obviously be reversed (since increased clustering means lower D_T). The relative magnitude of the opposing time dependences of S and D_T will determine the behavior of P_T. More specifically, P_T will be independent of t (cf. cellulose acetate-water) if the additional penetrant entering the substrate (during absorption) as a result of the time dependence of S goes into clusters (or is otherwise immobilized). A P_T decreasing with t will be obtained (cf. ethyl cellulose-water), if macromolecular rearrangement resulting in a growth of clusters (possibly around suitable polymer groups) at the expense of monomeric penetrant can occur.

We conclude from the above that the molecular relaxation model can explain in a physically meaningful way a wide diversity of "non-Fickian" behavior and can correlate this with "Fickian" diffusion behavior. Further data are certainly needed particularly in con-

nection with poor swelling agent systems and the kinetics of the
relaxation process. Perhaps the chief difficulty of this model lies
in the use of the "initial state", which is poorly characterized physically, and introduces in effect a significant number of adjustable
or semi-adjustable parameters. Independent information on this point
would undoubtedly be very useful. Furthermore, there are phenomena
which apparently lie beyond the scope of this model. Anomalies (inflection points or maxima) in permeation curves[1,22] and diffusion-
induced molecular orientation effects[23] are cases in point.

DIFFUSION WITH CONVECTION MODEL

When a glassy polymer is very strongly swollen by the penetrant,
zero order absorption kinetics are often obtained[24,25]. More detailed observation reveals a sharp (discontinuous) penetrant front
which separates the highly swollen outer region from the inner glassy core and advances at constant velocity v. Such so called Case II
diffusion processes are considered to be an extreme form of "non-
Fickian" diffusion, which is completely rate-controlled by the swelling stress and the concomitant rather far-reaching structural changes. A more precise and detailed physical picture of the phenomenon
is still lacking.

A phenomenological model of "non-Fickian" diffusion has been
proposed, however, in which the "Fickian" (Case I) and Case II mechanisms are combined additively through the equation

$$\frac{\partial C}{\partial t} = \frac{\partial}{\partial X} \left(D \frac{\partial C}{\partial X} - vC \right) \tag{9}$$

Equation (9) has been solved, assuming D also to be constant, by
Peterlin[26,27] and by Frisch et al[28] from somewhat different physical view points. The former author considers D to be associated
with a Fickian diffusion front preceding the Case II boundary and
hence to refer to glassy polymer. According to the latter treatment,
on the other hand, D refers to the swollen region (there being no
significant penetration into the glassy core beyond the sharp boundary). The physical implications of these points of view will not be
discussed in any detail, since we are here concerned chiefly with
the kinetics (which is of the same form in both cases). The general
absorption formula is complicated, but at sufficiently short and long
times, it reduces to the following simple kinetic expressions respectively (assuming a semi-infinite medium)

$$M_t = k_1\sqrt{t} + k_2 t \tag{10}$$

$$M_t = k_3 + k_4 t \tag{11}$$

where k_1, k_2, k_3, k_4 are constants deducible from the parameters of the
appropriate model. Observations of the position of the penetrant
front can be analyzed kinetically also on the basis of Equations (10),

(11). An expression for the permeation time lag has been deduced[28], but not applied experimentally so far.

This treatment has proved particularly useful for the description of the penetration kinetics of solvents or swelling agents (including binary mixtures) from the liquid phase. The work of Kwei et al[29] is particularly noteworthy for covering the full range from Case I to Case II kinetics by changing either the penetrant or the cross-link density of the epoxy polymer used. Conformity to Equation (10) has been claimed[24,28], although this claim is not sufficiently well documented.

Evidence for this model in the case of penetrant vapors is limited. A tendency to pass from Case II to Case I sorption kinetics with decrease in a_o was demonstrated by Hopfenberg et al[30] for n-pentane in biaxially oriented polystyrene. The full range between Cases I and II was far from covered, however, (because sorption became too slow at low a_o) and Equation (10) was not applied. Kwei[10] demonstrated Case II kinetics in some older absorption curves by replotting them on a t scale. Other absorption M_t vs $\sqrt{t}$ curves, including two-stage ones, exhibit a substantial linear or quasilinear part when replotted on a t scale, after an initial convex upwards portion as required by Equations (10) and (11). Conformity to Equation (10) is claimed in some cases at least, but no details are given[10]. In a further more quantitative test of the approach of Frisch et al, sorption and penetration kinetics of acetone in polyvinyl chloride[31] were reproduced quite well by means of the parameters D, v, C_X and Equation (11). However, the value of the latter parameter (denoting the concentration below which the penetrant profile becomes discontinuous) turns out to be somewhat unreasonably low in our view (less than 10% of C_o).

The interpretation of two-stage absorption M_t vs $\sqrt{t}$ curves by the diffusion with convection model is noteworthy, because it differs radically from that of Long et al. The latter view has, of course, been confirmed by direct observation of penetrant distribution in some systems[3], but need not apply in all cases (cf. the differences in second-stage sorption kinetics discussed previously). The present model, however, runs into considerable difficulties in connection with absorption curves possessing concave upwards sections on a t scale, which are not particularly scarce in either recent or older literature[8,32]. Also considerable difficulty can be foreseen where desorption is concerned, since nearly "Fickian" desorption curves are found to correspond to absorption curves with marked Case II characteristics[30,32]. Even in the case of absorption, considerable further and varied testing will be required to assess the extent of the applicability of the model (e.g. to successive narrow concentration intervals) and the physical meaningfulness of the parameter v.

DIFFERENTIAL SWELLING STRESS MODEL

This model is based on the consideration that uneven distribution of penetrant across the polymer film during diffusion causes a correspondingly uneven swelling tendency along the plane of the membrane[2]. At any t, regions of low C are forced to expand, whereas regions of high C are correspondingly prevented from extending fully by the mutually exerted differential swelling stresses f(X,t). The membrane as a whole will expand to an area $\overline{A}(t)$ determined by the condition of mechanical equilibrium, namely

$$\int_0^\ell f(X,t)dX = 0 \tag{12}$$

where f is positive when tensile or negative when compressive and is determined by the corresponding local elastic modulus G and strain $s=(\overline{A}-A)/A$; A represents the value of the membrane area corresponding to the local concentration of penetrant in the absence of constraint (i.e. at swelling equilibrium). A linear dependence of A on C can be reasonably expected, i.e. (α=constant)

$$\overline{A}(C) = A(0)(1 + \alpha C) \tag{13}$$

Crank[2] assumed the diffusion coefficient to depend on strain linearly (increasing or decreasing according as s is positive or negative). He then imposed some drastic simplifications including: (i) dependence of G on C only (pure elasticity);(ii) discontinuous increase of D with C from one value to another, giving rise to a sharp division of the membrane into only fully swollen and dry parts; and (iii) equality of the ratio of the values of D at maximum and zero strain in swollen and dry parts.

The resulting model can be handled analytically. Its practical usefulness is quite obviously rather restricted, although it does yield some useful predictions[2]. Its main virtues include a solid physical basis (note the close analogy with thermally induced stresses) providing a link with the mechanical properties of the polymer-penetrant system, and the possibility of correlating sorption and transverse swelling kinetics.

For these reasons, it appeared worthwhile to refine this model by assuming the polymer to exhibit linear viscoelasticity, characterized by instantaneous and long time moduli G_0, G_∞ respectively and stress relaxation frequency β, all exponential functions of C. The time dependence of f (based on the simple mechanical equivalent of a spring in parallel with a Maxwell element) was given by

$$\frac{\partial f}{\partial t} = G_0 \frac{\partial s}{\partial C} \frac{\partial C}{\partial t} + s \frac{\partial G_\infty}{\partial C} \frac{\partial C}{\partial t} + \{\frac{1}{G_0-G_\infty} \frac{\partial(G_0-G_\infty)}{\partial C} \frac{\partial C}{\partial t} - \beta\} (f-G_\infty s) \tag{14}$$

Furthermore, D was assumed to depend on f rather than on s. D(C,f) was exponential in both C and f (increasing with C and either increasing or decreasing with f according to whether f is tensile or compressive), in the absence of a precise theory linking D to f.

Our computations (see Figure 4) have so far been concerned with sorption kinetics of polymer-good swelling agent systems and the influence on this of (i) the rate of stress relaxation relative to diffusion, as measured by $\ell^2 \beta(C_1)/D(C_1,0)$ (cf. curves A,C,D); (ii) the extent of "softening", i.e. lowering of G_o, G_∞ with increasing C, (cf. curves A,B,D) and (iii) the magnitude of the stress dependence of D (cf. curves A,D,E). An explicit numerical method was used to solve Equations (1) and (2) simultaneously with Equations (13) and (14). It will be noted that absorption curves varying from "Fickian" (curve A) to "non-Fickian" with sections concave upwards on a t scale (curves D,E), are obtained. In fact, the theoretical results of Figure 4a parallel quite closely the experimental plots for n-pentane in cast polystyrene given in Figure 3 of Reference (32) (bearing in mind that the main effect of increasing temperature in these experiments would be reduction in maximum stress and increase in $\ell^2\beta/D$). Figure 4a also shows quite clearly that one need not necessarily invoke Case II mechanisms to interpret absorption M_t vs t curves exhibiting substantial linear portions. Finally, Figure 4b indicates that the correct shapes of absorption and desorption M_t vs $\sqrt{t}$ plots can be predicted.

The above results are, of course, limited and mostly of qualitative significance. The model obviously needs to be applied more widely and also to be extended in a number of directions. Inclusion of the influence of f on the solubility coefficient seems necessary if there is to be any hope of interpreting "two-stage sorption". Sorption-induced semi-permanent molecular orientation effects can be explained if elastic limits are exceeded. Clearly, the model could become very complicated. It has the advantage, however, that the additional parameters should normally be deducible from the mechanical properties, although suitable information of this kind is scarce at present. However, there remain difficulties in relating bulk properties to diffusion parameters which are essentially dependent on local properties. For example, tensile stress can generate microcracks in glassy polymers. The resulting large increase in D explains the permeation curve anomalies referred to earlier[22,33]. However, D is so sensitive to the presence and mode of interconnection of such microchannels, that there is little hope of establishing a general D-f relation under such conditions.

CONCLUSION

Each of the theoretical models considered above has special features which enhance its utility for certain applications. They differ, however, when their present status is assessed on the basis of the general criteria mentioned in the Introduction. The molecular relaxation model has been applied more extensively that the other two,

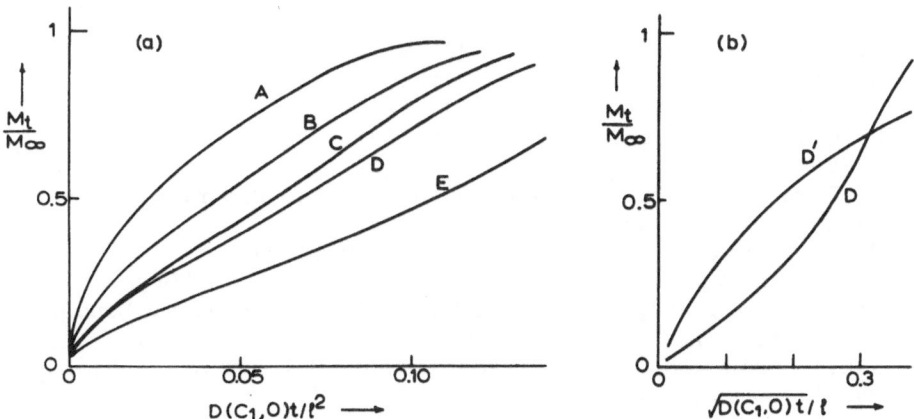

Figure 4.   Absorption curves for $D=D(C,f)$ with $G_o(C)/G_\infty(C)=5$; $G_o(C_o)/$
$G_o(C_1)=1$; $\alpha C_o=0.1$; $D(f=f_{max})/D(f=0)=20$; $f_{max}=G_o(C_1)\alpha C_o$;
$D(C=C_o)/D(C=C_1)=20$; $\ell^2\beta(0)/D(C_1,0)=0.1$; $\beta(C_o)/\beta(C_1)=20$;
except that: $D=D(C)$ (curve A, "Fickian"); $G_o(C_o)/G_o(C_1)=$
0.4 (B); $\ell^2\beta(C_1)/D(C_1,0)=1$ (C); $D(f=f_{max})/D(f=0)=100$ (E);
desorption curve(D').

although significant gaps remain. In consequence, its deficiencies
are more likely to be apparent. However, limitations can be foreseen in the case of the other models, particularly the diffusion
with convection one. Against this, one must note that the latter
approach has as yet been elaborated relatively less than the others.
On the other hand, it lacks a really precise physical basis. From
this point of view, we would consider the differential swelling
stress model as the most satisfactory one and feel that a more detailed correlation between sorption and swelling kinetics on one
hand and mechanical properties on the other, should prove to be a
fruitful line of future research.

REFERENCES

(1)   H. Fujita, Fortschr. Hochpolymer Forsch., 3, 1 (1961).
(2)   J. Crank, J. Polymer Sci., 11, 151 (1953).
(3)   F.A. Long and D. Richman, J. Am. Chem. Soc., 82, 513 (1960).
(4)   H.L. Frisch, J. Chem. Phys., 37, 2408 (1962).
(5)   J.H. Petropoulos and P.P. Roussis , J. Chem. Phys., 47,
      1491 (1967).
(6)   E. Bagley and F.A. Long, J. Am. Chem. Soc., 77, 2172 (1955).
(7)   A. Kishimoto and T. Kitahara, J. Polymer Sci., A1, 5,
      2147 (1967).
(8)   F.A. Long and R.J. Kokes, J. Am. Chem. Soc., 75, 2232 (1953).

(9)   A.C. Newns, Trans. Faraday Soc., 52, 1534 (1956).
(10) T.K. Kwei, J. Polym. Sci., A2, 10, 1849 (1972).
(11) A. Kishimoto, H. Fujita, H. Odani, M. Kurata and M. Tamura,
      J. Phys. Chem., 64, 594 (1960).
(12) H. Odani, S. Kida and M. Tamura, Bull. Chem. Soc. Japan, 39,
      2378 (1966).
(13) M. Tamura, K. Yamada and H. Odani, Repts. Progr. Polymer Phys.
      Japan, 6, 163 (1963).
(14) H.L. Frisch, J. Chem. Phys., 41, 3679 (1964).
(15) J.H. Petropoulos and P.P. Roussis, Chap. 19 in "Organic Solid
      State Chemistry", Ed. G. Adler, Gordon and Breach, London,
      p. 343 (1969).
(16) J.H. Petropoulos and P.P. Roussis, J. Polymer Sci., C, No 28,
      243 (1969).
(17) V. Stannett and J.L. Williams, J. Polymer Sci., C, No 10, 45
      (1965).
(18) J.D. Wellons and V. Stannett, J. Polymer Sci., A1, 4, 593
      (1966).
(19) J.D. Wellons, J.D. Williams and V. Stannett, J. Polymer Sci.,
      A1, 5, 1341 (1967).
(20) H.B. Hopfenberg, F. Kimura, P.T. Rigney and V. Stannett, J.
      Polymer Sci., C, No 28, 243 (1969).
(21) P.P. Roussis, to be published.
(22) P. Meares, J. Polymer Sci., 27, 405 (1958).
(23) P. Dreschel, J.L. Hoard and F.A. Long, J. Polymer Sci., 10,
      241 (1953).
(24) T. Alfrey, Jr., E.F. Gurnee and W.O. Lloyd, J. Polymer Sci.,
      C, No 12, 249 (1966).
(25) A.S. Michaels, H.J. Bixler and H.B. Hopfenberg, J. Appl. Po-
      lymer Sci., 12, 991 (1968).
(26) A. Peterlin, J. Polymer Sci., B, 3, 1083 (1965).
(27) A. Peterlin, Makromol. Chem., 124, 136 (1969).
(28) T.T. Wang, T.K. Kwei and H.L. Frisch, J. Polymer Sci., A2, 7,
      2019(1969).
(29) T.K. Kwei and H.M. Zupko, J. Polymer Sci., A2, 7, 867 (1969).
(30) H.B. Hopfenberg, R.H. Holley and V. Stannett, Polymer Eng.
      Sci., 9, 242 (1969).
(31) T.K. Kwei, T.T. Wang and H.M. Zupko, Macromolecules, 5,
      645 (1972).
(32) R. Holley, H.B. Hopfenberg and V. Stannett, Polymer Eng. Sci.,
      10, 376 (1970).
(33) B. Rosen, J. Polymer Sci., 47, 19 (1960).

TRANSPORT BEHAVIOUR OF ASYMMETRIC CELLULOSE ACETATE MEMBRANES

IN DIALYSIS AND HYPERFILTRATION

W. Pusch

Max-Planck-Institut für Biophysik

6 Frankfurt am Main, Germany

ABSTRACT

The transport of salt solutions across asymmetric cellulose acetate membranes is discussed using the linear relationships of the thermodynamics of irreversible processes. The corresponding transport coefficients such as the mechanical permeability; l_p, the osmotic permeability, l_π, and the reflection coefficient, σ, are determined by dialysis experiments for different salt solutions. The experimental results manifest a strong dependence of l_p and l_π on the salt concentration. This is shown to be due to concentration polarization within the porous sublayer of the asymmetric membrane. The dependence of the mechanical permeability, l_p, on concentration is then estimated using a Nernst-Planck equation for the salt transport within the porous sublayer of the asymmetric membrane. The influence of the concentration profile within the porous sublayer, caused by concentration polarization effects, on measured membrane potentials is also discussed. Furthermore, a relationship between salt rejection, r, measured in hyperfiltration and the three transport coefficients, the salt concentration of the brine, as well as the pressure difference, ΔP, across the membrane is derived and compared to experimental findings.

INTRODUCTION

During the years 1952-1962 a new technique for the production of drinking water from brackish and sea water was developed within a research program in the United States. This new technique is attractive in its simplicity. In the early work, based on the

233

principles of osmosis with semipermeable membranes, salt solutions
were contacted with suitable membranes under pressure and selective
transport of water was observed. These membranes should be permeable to water but more or less impermeable to salt. Looking for
suitable membranes, Reid and Breton {1} demonstrated that membranes
cast from solutions of cellulose acetate in acetone reject salt in
"reverse osmosis" or hyperfiltration very well. They used cellulose
acetate membranes, termed homogeneous today, which had a thickness
of about 6 μm. Due to their very low water permeability (water
flux at 40 atm pressure drop: $q \simeq 3 \cdot 10^{-5}$ cm/sec) these membranes
could not be used for the economic production of drinking water
from brackish or sea water. An obvious development of much thinner
membranes was not pursued because Loeb and Sourirajan {2} developed
asymmetric membranes. They developed the following method of
preparation of their modified cellulose acetate membranes. From a
casting solution of cellulose acetate, magnesium perchlorate, water,
and acetone (ratio 22.5:1.1:10:66.4 weight-%) the membranes are
cast on a glass plate at -10°C. After an evaporation period for
the solvent of about 1 to 3 minutes at the same temperature, the
glass plate with the membrane is immersed into an ice water bath.
It is kept there for about one hour during which the ice water is
renewed several times. Thereafter, the membrane is removed from
the plate and heated by immersion in hot water (annealed) for about
5 minutes at 65°C to 95°C. The salt rejection of these membranes
increases with increasing annealing temperature while the water
permeability decreases with increasing annealing temperature.
Highly-annealed membranes exhibit the same salt rejection as the
membranes used by Reid and Breton, but the water permeability of
the highly-annealed membranes is much higher than that of the homogeneous membranes of the same thickness. Electron microscopic investigations by Merten et al. {3} proved these modified membranes
to consist of a fine-pored matrix with a very thin, dense layer of
cellulose acetate on the so-called air-dried surface. The dense
surface layer was estimated to be about 500-3000 Å thick, the total
membrane thickness being 100 μm. The porous substructure was
estimated to have a pore size in the order of 0.1 μm. It may be
concluded that the asymmetric membrane consists essentially of two
single membranes. The active layer at the air-dried surface is
equivalent to a homogeneous membrane of about 500-3000 Å and is
responsible for the salt rejection. The porous sublayer is a very
good natural support for the active layer. Since the water permeability is determined mainly by the thickness of the active
layer, which is nearly independent of the thickness of the whole
membrane, the volume flux across the membrane is nearly independent
of the overall thickness of the membrane. Since these were the
first asymmetric membranes, there was much interest in the evaluation of their transport properties by experimentally determined
transport coefficients.

TRANSPORT EQUATIONS

For a general description of transport through artificial membranes one may use the phenomenological relations of thermodynamics of irreversible processes. Thereby, one gets linear relations between fluxes, Φ_i, and conjugated general forces, X_i, like $\Phi_i = \sum_k \ell_{ik} \cdot X_k$. The ℓ_{ik} depend on the physico-chemical state of the system, i.e. on temperature, T, pressure, P, and mean salt concentration, c_s, for instance, in a binary system. With these linear relations it is possible to give an overall phenomenological description of transport through synthetic membranes, as has been shown by several investigators for ion exchange membranes {4,5,6}. In this section, the transport phenomena in asymmetric cellulose acetate membranes will be described by using the linear relationships of the thermodynamics of irreversible processes.

Given an isothermal system, with a membrane separating two salt solutions of different salt concentrations c_s' and c_s'' (Figure 1) which are kept under different hydrostatic pressures P' and P'' and different electrical potentials E' and E'', a water flux Φ_w {mol/cm^{2}sec}, a salt flux Φ_s {mol/cm^{2}sec}, and an electric current j {A/cm^2} will result. Since Φ_w and Φ_s cannot be determined directly by experiment, it is useful to introduce other fluxes. These are the volume flux q {cm/sec} and the so-called chemical flux χ {cm/sec}. Between the original fluxes, Φ_w and Φ_s, and the new fluxes, q and χ, the following relations exist:

$$q = V_s \Phi_s + V_w \Phi_w \quad (1) \qquad \chi = \Phi_s/c_s - \Phi_w/c_w \quad (2)$$

where V_w and V_s are the partial molar volumes of water and salt, respectively. As can be seen from Equation (2), the chemical

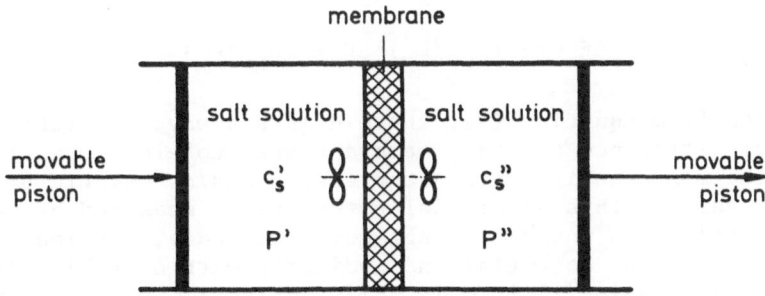

Figure 1: Diagrammatic presentation of an isothermal membrane system.

flux is equal to the difference in speed of salt and water. It is
a measure of the variation of concentration with time within the
two phases (') and (") and is defined by:

$$\chi \equiv \frac{1}{A}\frac{V'}{\bar{c}_s}\cdot\frac{dc'_s}{dt} = -\frac{1}{A}\frac{V''}{\bar{c}_s}\frac{dc''_s}{dt} \tag{3}$$

where V' and V'' are the volumes of phase (') and ("), respectively;
A is the effective membrane area (cm2); $\bar{c}_s = (1/2)(c'_s + c''_s)$ is the
mean salt concentration; and dc_s/dt is the variation of concentration with time within the two phases. As Schlögl {7} has shown,
the following linear relations exist between fluxes and conjugated
forces for a system near equilibrium:

$$j = L_E\Delta E + L_{EP}\Delta P + L_{E\Pi}\Delta\Pi \tag{4}$$

$$q = L_{EP}\Delta E + L_p\Delta P + L_{P\Pi}\Delta\Pi \tag{5}$$

$$\chi = L_{E\Pi}\Delta E + L_{P\Pi}\Delta P + L_\Pi\Delta\Pi \tag{6}$$

where $\Delta P = P' - P''$(at), $\Delta E = E' - E''$ (v), and $\Delta\Pi = \Pi' = \Pi''$ (at) are
the pressure difference, potential difference, and osmotic pres-
sure difference across the membrane, respectively. Thereby, ΔE has
to be determined by reversible electrodes like Ag/AgCl-electrodes
if chlorides are used as electrolyte. The number of independent
transport coefficients, L_{ik}, is reduced to six by the Onsager reciprocal relationships which require $L_{ik} = L_{ki}$. Now, if one is not
interested in electro-osmotic phenomena, measurements are mostly
performed under the special boundary condition of electrical current, j, equal to zero. Under this boundary condition Equation (4)
can be used to calculate the electrical potential difference, ΔE,
across the membrane for j = 0 yielding:

$$\Delta E = -(L_{EP}/L_E)\cdot\Delta P - (L_{E\Pi}/L_E)\cdot\Delta\Pi \tag{7}$$

As is seen from Equation (7), there exists always an electrical
potential difference across a membrane when no electrical current
passes the membrane if there are pressure and/or osmotic pressure
differences. If this electrical potential is measured by means of
calomel-electrodes instead of Ag/AgCl-electrodes, for instance, it
is termed membrane potential and indicated with $\Delta\Psi$ {mV}. The
linear relationship (7) is only valid for small trans-membrane
concentration differences. As was shown by Schlögl {7}, the membrane potential for larger concentration differences can be estimated by integrating the corresponding Nernst-Planck equations.

Introducing Equation (7) into Equations (5) and (6) and re-arranging algebraically one obtains the following two equations:

$$q = \ell_p \Delta P + \ell_{\pi p} \Delta \Pi \quad (8) \qquad \chi = \ell_{\pi p} \Delta P + \ell_\pi \Delta \Pi \quad (9)$$

where

$\ell_p = L_P - L_{EP}^2 / L_E$ = mechanical permeability of the membrane (cm/sec·at)

$\ell_\pi = L_\Pi - L_{E\Pi}^2 / L_E$ = osmotic permeability of membrane (cm/sec·at)

$\ell_{\pi p} = L_{P\Pi} - L_{EP} L_{E\Pi} / L_E$ = coupling coefficient (cm/sec·at)

As can be seen from these relations, the mechanical and osmotic permeability of a membrane measured at j = 0 is always smaller than the corresponding value measured at ΔE = 0. This effect is especially pronounced with membranes of large fixed charge concentrations (ion exchange membranes of high fixed charge capacity). With the definition of the reflection coefficient, $\sigma \equiv -\ell_{\pi p} / \ell_p$, given by Staverman {8}, Equations (8) and (9) yield:

$$q = \ell_p (\Delta P - \sigma \Delta \Pi) \quad (10) \qquad \chi = -\sigma \ell_p \Delta P + \ell_\pi \Delta \Pi \quad (11)$$

Thus, the transport behaviour of a synthetic membrane is characterized by the three independent transport coefficients ℓ_π, ℓ_p, and σ. The reflection coefficient, σ, is a measure of the permeability of the membrane to salt. If σ = 1, no salt can pass the membrane (semipermeable membrane); if σ = 0, solute and solvent cross the membrane in the same concentration ratio as they have in the adjacent bulk solution. Using the linear relations (10) and (11), the transport coefficients can be determined by measuring the dependence of volume flux, q, and chemical flux, χ, on pressure difference, ΔP, and osmotic pressure difference, $\Delta \Pi$.

Experimental Determination of Transport Coefficients

With the cell shown in Figure 2 the transport coefficients of a membrane can be measured. A detailed description of this cell was already given elsewhere {9,10}. The three transport coefficients can be determined as follows.

1. Determination of ℓ_p at $\Delta \Pi$ = 0

Both cell compartments - phase (') and (") - contain salt solutions of equal concentration. Measuring the volume flow Q = A·q, as a function of ΔP using about four to five different values of ΔP, one gets ℓ_p from the slope of the corresponding straight line in a Q-ΔP diagram if the effective membrane area, A, is known.

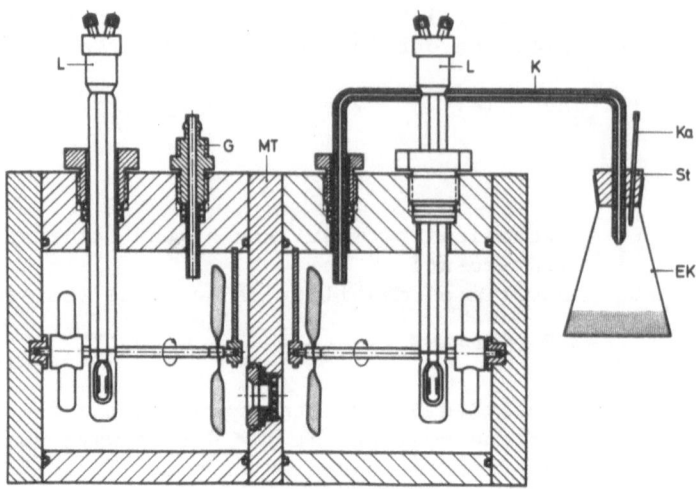

Figure 2:   Cross-sectional view of dialysis cell

2.   Determination of ℓ_p and σ at $\Delta\Pi$ = constant = 0

An osmotic difference $\Delta\Pi$ between the two cell compartments is established ($c_s' > c_s''$) and the volume flow Q is measured as a function of ΔP. From the slope of the corresponding straight line in a Q–ΔP diagram, one again gets ℓ_p. In addition, from the intersection of the straight line with the abscissa, one gets σ = $(\Delta P/\Delta\Pi)_{q=0}$. Thereby, a mean value $\overline{\Delta\Pi}$ is determined from $\Delta\Pi_a$ at the beginning and $\Delta\Pi_e$ at the end of the measurement (arithmetical mean). The osmotic pressure difference $\Delta\Pi_e$ is obtained by conductivity titration of the two salt solutions after the experiment is finished.

3.   Determination of ℓ_π and $\sigma\ell_p = \ell_{\pi p}$ at $\Delta P = 0$

At time t = 0 a concentration difference between the two cell compartments is set up ($c_s' > c_s''$). The variation of concentration with time is then followed by measuring the conductivity of both phases at a time interval of about 24 hours. After the measurements are completed (about ten days) the cell constants of the conductivity cells used are determined once more by conductivity titrations of the salt solutions from both compartments. Assuming a linear relation between the salt concentration and conductivity for small concentration differences, one gets the salt concentration from a calibration curve. The volume flow Q is obtained by determining the volume of solution which flowed into an Erlenmeyer flask during about 24 hours. From the curves of concentrations, c_s' and c_s'', and the volume increase, ΔV, with time, a mean value of the chemical flux, χ, and the volume flux, q, as well as a mean value $\overline{\Delta\Pi}$ is estimated by averaging over 24 hours. Plotting $\overline{\chi}$ and $\overline{q}$ as functions of $\overline{\Delta\Pi}$, one gets straight lines. The slopes of the calculated regression lines yield the coefficients ℓ_π and $-\sigma\ell_p = \ell_{\pi p}$, respectively. If ℓ_p is known from a different measurement, a value for σ is also obtained.

Using solutions of NaCl, Na_2SO_4, NaF, $CaCl_2$, and saccharose the mechanical permeability, ℓ_p, was measured as a function of solute concentration ($0 < c_s < 1$ mol/1) at 20°C. Using solutions of NaCl, Na_2SO_4, NaF, and $CaCl_2$ the reflection coefficient, σ, was also measured. The corresponding experimental results are shown graphically in Figures 3, 4, and 5. In addition, ℓ_π and $\ell_{\pi p}$ were determined as a function of concentration ($0 < c_s < 0.5$ mol/1) at 25°C using NaCl and Na_2SO_4 solutions. The corresponding results for ℓ_π are also graphically represented in Figure 6. The curve $\ell_{\pi p}$ as a function of c_s has the same shape as was shown recently {11}.

As one can see from these figures, there exists a strong dependence of ℓ_p, ℓ_π, and $\ell_{\pi p}$ on salt concentration. On the other hand, the reflection coefficient, σ, varies only slightly with salt concentration. If one calculates σ from the measured value $\sigma\ell_p$ and ℓ_p measured by methods (3) and (1) respectively, it is found that these values agree with the σ values measured directly using method (2). The experimental results show, above all, that

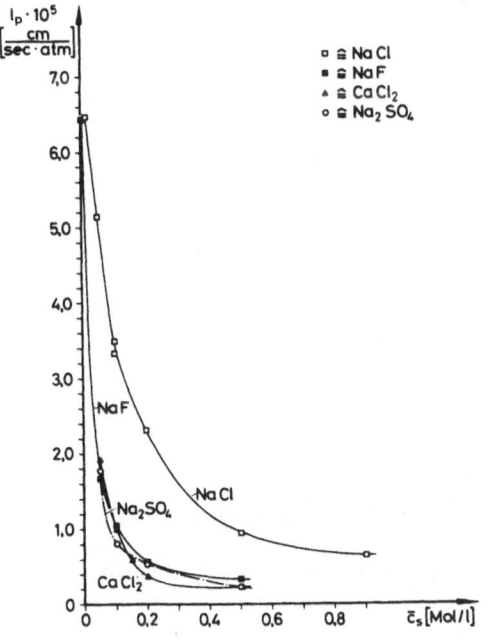

Figure 3: Mechanical permeability, ℓ_p, as a function of salt concentration $c_s = \overline{c}_s$ at 20°C using different salt solutions.

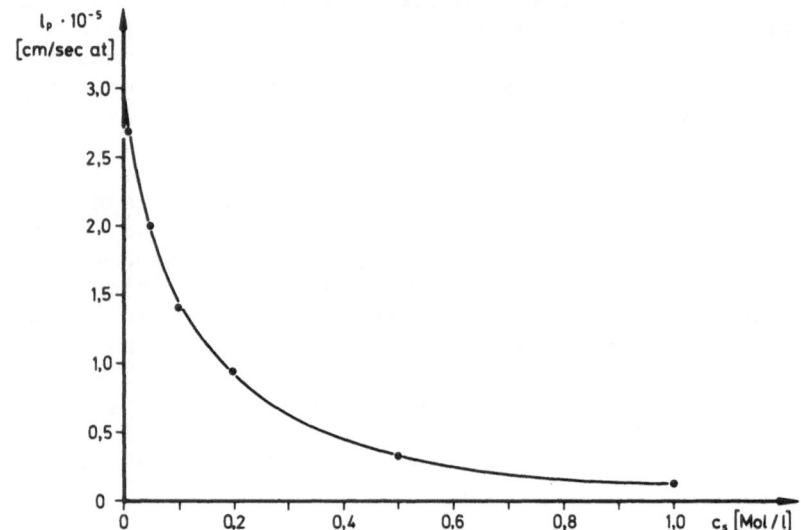

Figure 4: Mechanical permeability, ℓ_p, as a function of saccharose concentration at 25°C. With regard to the concentration dependence of ℓ_p in this case, the increase of viscosity of saccharose solutions with concentration has to be taken into account.

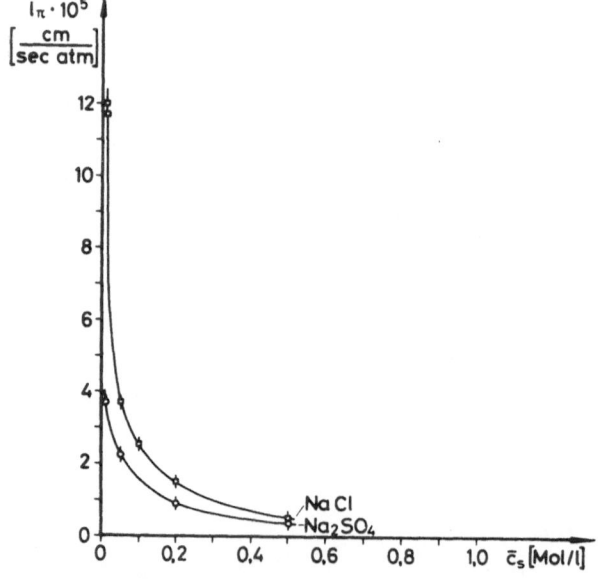

Figure 5: Osmotic permeability, ℓ_π, as a function of mean salt concentration $\bar{c}_s$ at 25°C for NaCl and Na_2SO_4 solutions.

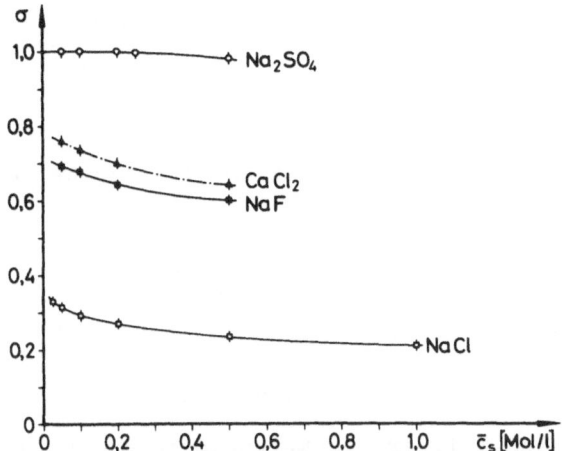

Figure 6:   Reflection coefficient, σ, as a function of salt concentration, $\bar{c}_s$, at 20°C using NaCl, NaF, Na_2SO_4, and $CaCl_2$ solutions
(NaCl - (I) curve $\hat{=}$ virgin membrane was used; NaCl - (II) curve $\hat{=}$
membrane pressurized at 50 atm before use).

a consistent description of the transport through asymmetric cellulose acetate membranes is possible for dialysis experiments. In
the next section, corresponding considerations are examined for
hyperfiltration experiments.

Relation Between Salt Rejection and Transport Coefficients

     In hyperfiltration experiments the volume flux, q, and the
salt rejection, r, are determined. Here, r is defined as follows:

$$r = (c_s' - c_s'')/c_s' \tag{12}$$

If one would like to compare hyperfiltration with dialysis experiments, one has to relate r to the transport coefficients and the
pressure differences across the membrane. This can be done as was
shown recently {12} if the following boundary condition is used:

$$c_s'' = \Phi_s/q \tag{13}$$

This boundary condition states that the composition of the product
is determined by the composition of the solution which crosses the
membrane. Introducing (13) into Equation (12) which can be rear-

ranged to yield $r = 1 - c_s''/c_s'$, one arrives at the following relation:

$$r = 1 - (\Phi_s/c_s' q) \tag{14}$$

Assuming $q \simeq V_w \Phi_w$ and $c_w' V_w' \simeq 1$, Equation (14) can be rearranged to yield:

$$r = (1/q)(\Phi_w/c_w' - \Phi_s/c_s') \tag{15}$$

Using Equation (2), one arrives at:

$$r \simeq -\chi/q \tag{16}$$

If Equations (10) and (11) are substituted in Equation (16) and if the approximation $r \simeq \Delta\Pi/\Pi'$ is used, one obtains a quadratic equation for r which can be solved to give:

$$r = (1/2\sigma)\{(\Delta P/\Pi' + \ell_\pi/\ell_p) - \sqrt{(\Delta P/\Pi' + \ell_\pi/\ell_p)^2 - 4\sigma^2(\Delta P/\Pi')}\,\} \tag{17}$$

Equation (17) contains the dependence of r on the pressure difference, ΔP, the concentration of the brine solution, c_s' (Π'), and the ratio ℓ_π/ℓ_p. The dependence of r on the ratio ℓ_π/ℓ_p is very interesting with regard to the different salt rejection of ion exchange membranes compared to asymmetric cellulose acetate membranes possessing the same reflection coefficient.

Figures 7, 8, and 9 present r as a function of some interesting variables using different parameters such as σ, Π', and ℓ_π/ℓ_p. From Figure 7 one can see that $r \to \sigma$ if $\Delta P \to \infty$. This theoretical result agrees with theoretical findings obtained by Kedem and Spiegler {13}. Figure 8 shows the dependence of r on pressure difference ΔP for different salt concentrations (different values of Π'). As can be seen from this figure, r approaches its limiting value, σ, much more slowly with increasing salt concentration. This theoretical prediction is in agreement with experimental results which are graphically shown in Figure 10. Another interesting conclusion from Equation (17) is the dependence of r on ℓ_π/ℓ_p if σ, Π', and ΔP are kept constant (Figure 9). As ℓ_π/ℓ_p increases, r decreases. This result seems to be important in connection with the different salt rejection of ion exchange membranes and asymmetric cellulose membranes which have nearly the same reflection coefficient. From experiments it is well known {14} that ion exchange membranes having $\sigma \simeq 0.9$ will exhibit a salt rejection of only 0.4 - 0.5 at 100 at whereas asymmetric cellulose acetate membranes with the same reflection coefficient will give about 90% salt rejection ($r = 0.9$). Equation (17) shows that this difference is due to the different values of ℓ_π/ℓ_p for ion exchange membranes and cellulose acetate membranes. For asymmetric cellulose acetate membranes $\ell_\pi/\ell_p \simeq 1$ whereas for ion exchange membranes $5 < \ell_\pi/\ell_p < 20$ because of the lower water permeability of these membranes.

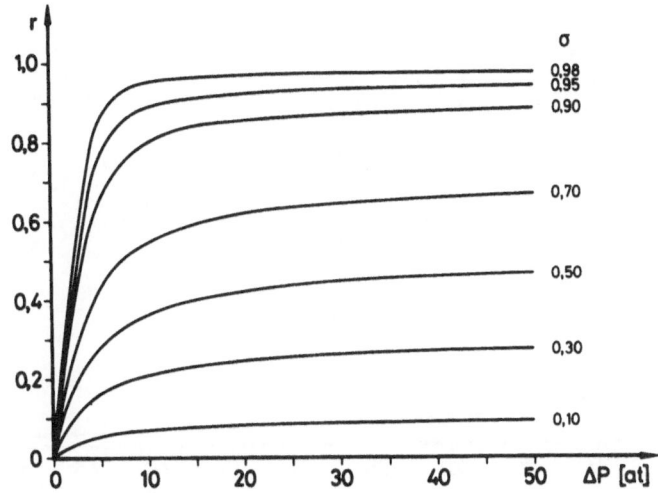

Figure 7:   Salt rejection, r, as a function of ΔP using different parameter values of σ ($c_s' = 0.1$ m NaCl; $\ell_\pi/\ell_p = 1$; $\Pi' = 4.56$ atm)

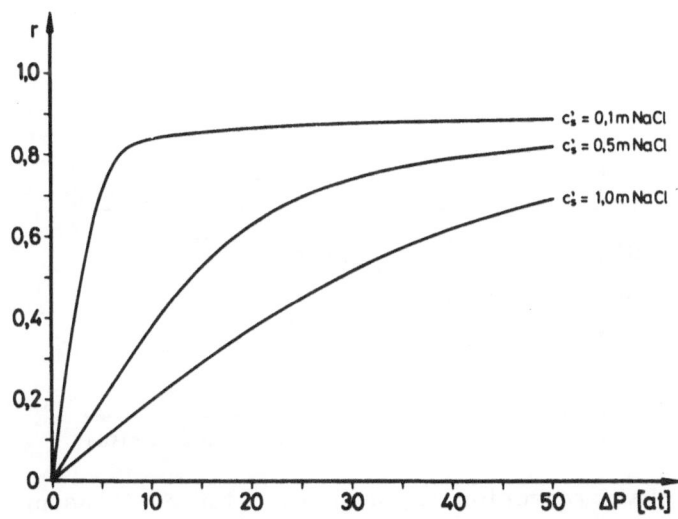

Figure 8:   Salt rejection, r, as a function of the pressure difference, ΔP, using different brine concentrations, c_s'. ($\sigma = 0.9$; $\ell_\pi/\ell_p = 0.95$)

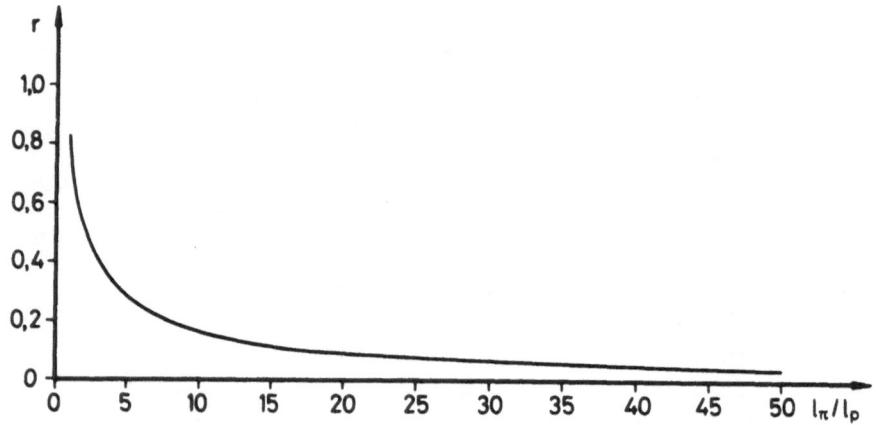

Figure 9: Salt rejection, r, as a function of ℓ_π/ℓ_p ($c_s' = 0.1$ m NaCl solution; $\Delta P = 10$ atm; $\sigma = 0.9$)

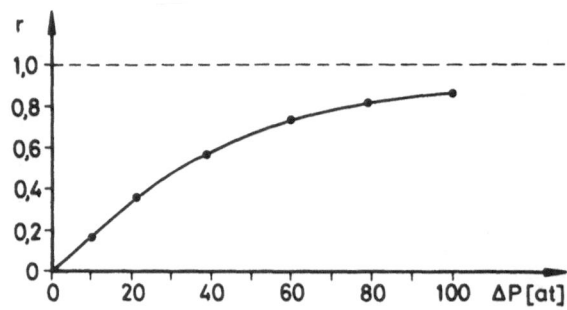

Figure 10: Salt rejection, r, as a function of ΔP using 1 m NaCl solution and an asymmetric cellulose acetate membrane annealed at 82.5°C

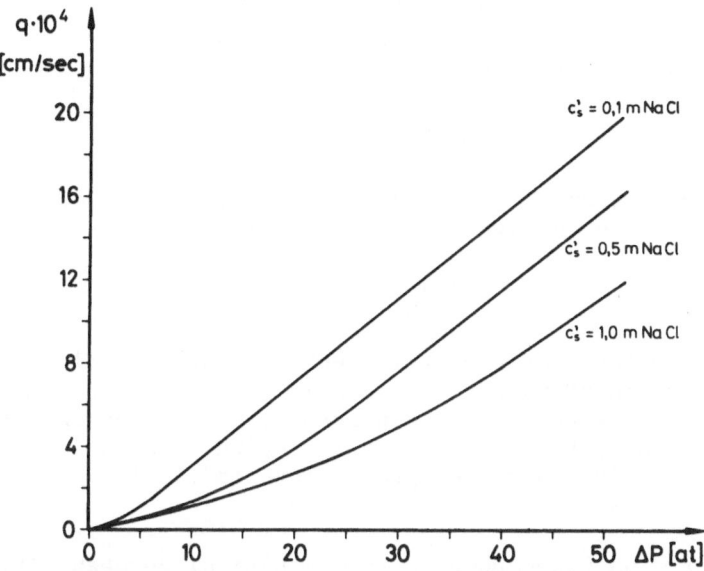

Figure 11: Volume flux, q, as a function of pressure difference, ΔP, in hyperfiltration using different brine concentrations, c_s'. Calculated from Equation (18) with σ = 0.9; ℓ_π/ℓ_p = 0.95.

One can also obtain a relation between volume flux, q, and pressure difference, ΔP {12}. Since in hyperfiltration ΔΠ is itself a function of ΔP, it is not an independent variable as it would be in dialysis measurements. Using the boundary condition (13), the linear relationship (11) for χ, and taking into account that r ≈ ΔΠ/Π', one arrives at:

$$q = \{(\ell_\pi/\ell_p - \sigma^2)/(\ell_\pi/\ell_p - r\cdot\sigma)\}\cdot\ell_p\cdot\Delta P \tag{18}$$

Figure 11 shows this relationship graphically. As can be seen, there exists no linear relationship between q and ΔP in hyperfiltration experiments at low pressures as has been frequently assumed in the past. Arriving at the asymptotic value of r, one gets (ΔP → ∞):

$$q \sim \ell_p\cdot\Delta P \tag{18a}$$

That means, at sufficiently high pressures there is a linear relationship between q and ΔP. From the slope of the corresponding asymptotic line one obtains the mechanical permeability, ℓ_p. On the other hand, one can again obtain the reflection coefficient, σ, from the intersection of that asymptotic line with the abscissa. Using r ≈ ΔΠ/Π' and Equation (10), one gets:

$$q = \ell_p(\Delta P - r \cdot \sigma \cdot \Pi') \qquad\qquad (19)$$

For the asymptotic case, that yields {15}:

$$q = \ell_p(\Delta P - \sigma^2 \Pi') \qquad\qquad (19a)$$

From the intersection of the asymptotic straight line of a plot of q versus ΔP one thus arrives at the following relation:

$$\sigma^2 = (\Delta P/\Pi')_{q=0} \qquad\qquad (20)$$

There is no complete experimental check presently available which permits comparison of theoretical predictions with experimental findings. The only relationship presently checked by experiment is the statement that, for the asymptotic region, r should approach the value of σ. This has been proved invalid {16,17}.

Concentration Dependence of the Mechanical Permeability

Measuring ℓ_p in hyperfiltration experiments by the use of Equation (18a), one finds that there is only a small dependence of ℓ_p on salt concentration (about 10% variation within the concentration range $0 < c_s' < 1$ mol/l NaCl). At first glance, that seems to contradict the experimental results obtained with dialysis experiments since a very strong dependence of ℓ_p on solute concentration was observed for dialysis. The reason for this difference can be found in the different boundary conditions. Whereas, in hyperfiltration the boundary condition of free outflow exists at the low pressure side, in dialysis experiments the salt concentration on both sides of the membrane is maintained constant. Therefore, one should check the influence of different boundary conditions in dialysis measurements on the value of the mechanical permeability, ℓ_p. To investigate that, measurements with the following boundary conditions have been made.

1. The salt concentration c_s' was varied between 0 and 0.2 m NaCl keeping $c_s'' = 0$ (pure water in the corresponding cell compartment). Thereby, the active layer of the membrane was juxtaposed with the salt solution. The volume flow was directed from phase (') to phase (") by adjusting the pressure difference, ΔP, across the membrane.

2. The same arrangement as in case 1 but the membrane was turned over so that the active layer was juxtaposed with pure water. Thereby, the volume flow, Q, was directed from phase (") to phase (').

3. The salt concentration c_s'' was varied between 0 and 0.2 m NaCl keeping $c_s' = 0.2$ m NaCl. The active layer was again juxtaposed with phase ('), the solution of constant salt concentration. The volume flow was directed from phase (') to phase (").

The corresponding experimental results of these measurements are represented graphically in Figure 12. As can be seen from this figure, the experimentally determined values of ℓ_p depend strongly on the salt concentration of that solution which is adjacent to the porous surface of the asymmetric cellulose acetate membrane. It should be already pointed out here that the fact that the measured value of ℓ_p depends on the direction of the volume flux and in an asymmetric way on the salt concentrations c_s' and c_s'' indicates that one is far off from the linear range of the raltions of thermo-dynamics of irreversible processes. Within the linear range there should be no asymmetric behaviour of any kind of membrane.

The experimental results demonstrate that the transport coefficients of an asymmetric cellulose acetate membrane measured far off the linear range depend strongly on the salt concentrations of both solutions, c_s' and c_s'', and on the direction of the volume flow. This statement agrees with conclusions drawn by Jagur-Grodzinski and Kedem {18} treating theoretically a double-layer membrane. Here, this is a consequence of the asymmetric structure of the modified cellulose acetate membrane. With a volume flow through the membrane, a concentration profile within the porous sublayer

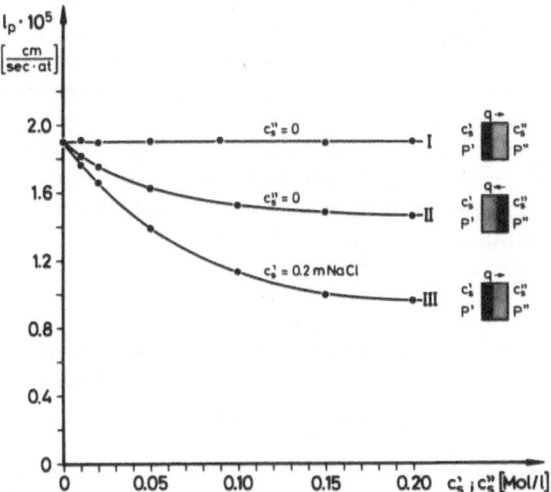

Figure 12: Volume flux, q, as a function of c_s' or c_s'', respectively, using different boundary conditions with curves I, II, and III.

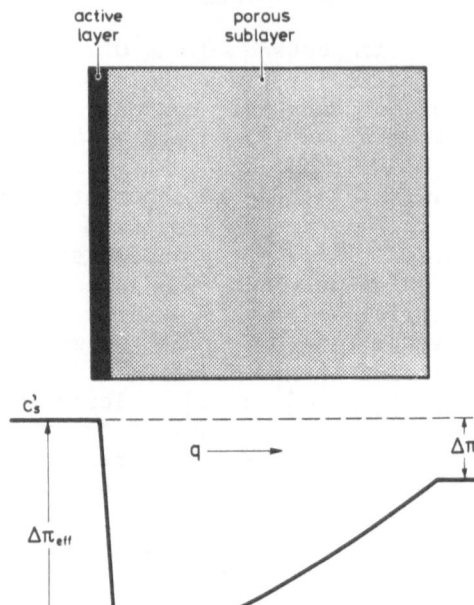

Figure 13; Concentration
profile within the porous
sublayer of **an asymmetric
cellulose acetate membrane.**

is developed by the interaction of salt rejection at the active
layer, volume flow, and diffusion of salt within the porous sublayer. Therefore, a concentration profile develops as is shown in
Figure 13. Thereby, an osmotic pressure difference $\Delta\Pi^m$ across the
active layer is developed. This osmotic pressure difference is
responsible for the fluxes through the active layer and thus, for
the fluxes through the entire membrane if one assumes that the
reflection coefficient of the porous sublayer is zero or nearly
zero. Furthermore, this concentration profile is also responsible
for the strong concentration dependence of the transport coefficients within the linear range where no asymmetric transport
behaviour should occur. This will be demonstrated in the following sections.

Theoretical Treatment of the Influence of the Porous Sublayer on
the Mechanical Permeability in Dialysis Experiments
near Equilibrium

Consider a membrane which is composed of two single membranes
(Figure 13). One of these is assumed to be a homogeneous cellulose
acetate membrane (active layer) of a thickness of about 2000 Å.
The other single membrane is assumed to be porous (sublayer) with
pore radii r > 100 Å and a thickness of $\delta \simeq 100$ μm. This double

layer model was proposed by Merten et al. {19} as a consequence of
electron microscopic investigations. With this model it should be
possible to estimate the influence of the concentration profile
on the transport properties of the membrane by solving the corresponding differential equation for the mass transport in the porous
sublayer. Now, the following assumptions are made {12}:

a) The transport properties of the active layer can be described
    using the linear relations of thermodynamics of irreversible
    processes. The following relationships are used.

$$q = \ell_p'(\Delta P^m - \sigma\Delta\Pi^m) \tag{21}$$

$$\chi = -\sigma\ell_p'\Delta P^m + \ell_\pi'\Delta\Pi^m \tag{22}$$

where ℓ_p' = mechanical permeability of the active layer
                       (cm/sec·at)
        ℓ_π' = osmotic permeability of the active layer
                       (cm/sec·at)
        ΔP^m = pressure difference across active layer (at)
        $\Delta\Pi^m$ = osmotic pressure difference across the active
                       layer (at)

b) The reflection coefficient of the porous sublayer is assumed to
    be zero ($\sigma_p = 0$). Therefore, the following linear relation-
    ship between volume flux, q, and pressure difference, $\Delta P''$,
    exists:

$$q = \ell_p''\Delta P'' \tag{23}$$

where ℓ_p'' = mechanical permeability of porous sublayer
                       (cm/sec·at)
        $\Delta P''$ = $P^m - P''$ = pressure difference across porous
                       sublayer (at)

The chemical flux, χ, across the porous sublayer can be determined by the salt and water flux using Equations (1) and (2).

c) The main influence of the porous sublayer is on the salt flux
    and thereby, on the concentration profile within the support-
    ing layer. Due to this concentration gradient the effective
    osmotic pressure difference $\Delta\Pi^m$ which determines the transport
    behaviour of the active layer deviates from the osmotic pres-
    sure difference $\Delta\Pi = \Pi' - \Pi''$ across the entire membrane. With
    the simplified Nernst-Planck Equation the following relation
    for the salt flux is obtained:

$$\Phi_s = -D_s(dc_s/dx) + c_s q \tag{24}$$

where Φ_s = salt flux across the entire membrane
(mol/cm^2sec)

c_s = local salt concentration in the porous sublayer
(mol/cm^3)

D_s = diffusion coefficient of salt within porous
sublayer (cm^2/sec)

x = coordinate perpendicular to membrane surface
(cm)

In the stationary state the salt flux, Φ_s, is constant through the
entire membrane. Therefore, one can integrate the differential
equation with the following boundary conditions: $c_s = c_s''$ at $x = \delta$
and $c_s = c_s^m$ at $x = 0$ (δ = thickness of porous sublayer). Here, the
origin ($x = 0$) is chosen to be at the interface between the active
layer and the porous sublayer. With these boundary conditions, the
differential equation can be solved as was discussed in detail
recently {12}. With the corresponding solution the concentration
difference across the active layer is obtained and, therefore, the
osmotic pressure difference across the active layer. This can be
introduced into the corresponding equations. After some algebraic
rearrangement, the following final result is obtained:

$$q(1 + \sigma\ell_p\Pi''\delta/D_s) = \ell_p(\Delta P - \sigma\Delta\Pi) + \sigma\ell_p\Pi_o(\delta/D_s)\Phi_s \qquad (25)$$

where Π_0 is the osmotic pressure of a one molar solution of a non-
dissociating compound. For small salt fluxes the second term in
Equation (25) is of the order of $10^{-3}\sigma\ell_p$ and may, therefore, be
neglected. For simplicity only this case will be discussed here.

As can be seen then, one obtains a relationship between q and
ΔP and $\Delta\Pi$ which gives a linear dependence of q on ΔP but the slopes
of the corresponding straight lines depend upon the salt concen-
tration c_s'' (Π''). Therefore, the measured mechanical permeability
depends on the salt concentration c_s''. From Equation (25) the
following relation between measured permeability, $\overline{\ell}_p$, and intrinsic
permeability can be made:

$$\overline{\ell}_p = \ell_p/(1 + \sigma\ell_p\Pi''\delta/D_s) \qquad (26)$$

The corresponding volume flux is then given by:

$$q = \overline{\ell}_p(\Delta P - \sigma\Delta\Pi) \qquad (27)$$

Thereby, it is shown that one always measures the reflection coefficient of the active layer of the composite membrane as long as
one can assume that the reflection coefficient of the porous sublayer is zero or nearly zero. For asymmetric cellulose acetate

membranes as well as composite membranes this assumption is always in agreement with experimental findings.

The foregoing analysis can also be made if the volume flux, q, is reversed from phase (") to phase ('). In this case again a concentration profile within the porous sublayer is built up which leads to a larger salt concentration at the interface between the active layer and the porous sublayer and therefore, creating again an osmotic pressure difference across the active layer which opposes the original net driving force ΔP and/or $\Delta\Pi$ reducing the measured volume flux. The whole analysis shows that the osmotic pressure difference, developed across the active layer of the membrane, depends, in addition to, the volume flux essentially on the concentration c_s'' of that solution which is adjacent to the porous surface of the asymmetric membrane.

The theoretical results allow the following conclusions. As long as c_s' and c_s'' are equal or nearly equal the created osmotic pressure difference, $\Delta\Pi^m$, will be nearly independent on the orientation of the membrane. Thus, the same volume flux will be observed if the membrane is turned over. On the other hand, as soon as the difference in the two salt concentrations, c_s' and c_s'', starts to become large enough different values of the volume flux will be measured depending on the orientation of the membrane as was demonstrated by the corresponding reported measurements (Figure 12).

Therefore, one has to differentiate between three experimental situations. In the first case, no asymmetric behaviour of the membrane system will be observed. The system is considered to obey the restrictions of the linear range of the thermodynamics of irreversible processes. This will always be true for small concentration differences across the entire membrane and not too large volume fluxes. The asymmetric structure of the membrane is manifested by the strong concentration dependence of the corresponding transport coefficients. In the second case, one will also find linear relationships to hold between fluxes and forces but the measured fluxes and transport coefficients will depend upon the orientation of the membrane. This will mostly be true with larger concentration differences but small volume fluxes. One may call this regime of concentrations and fluxes the linear asymmetric regime. Finally, in the third case, no linear relationships between fluxes and driving forces will be observed in general and the system will exhibit highly asymmetric behaviour. This corresponds to the general case of large fluxes and concentration differences as is well known from hyperfiltration experiments with asymmetric membranes {20,21}.

There is one point which should be examined in somewhat more detail. With all permeability measurements reported in this paper the asymmetric membranes were supported by a porous metal plate.

Since concentration polarization in the porous support will never really be overcome by stirring, the question arises whether the observed effects may be partly or even entirely caused by the porous support instead of being due mainly to the porous sublayer of the asymmetric membrane. Thus, to make sure that the main reason for the concentration dependence of the fluxes as well as for the asymmetric behaviour is the porous sublayer, one should think about measuring without any support. The only quantity which can be measured without any support would be the membrane potential $\Delta\Psi$. Therefore, the membrane potential of an asymmetric cellulose acetate membrane annealed at 85°C was measured as a function of electrolyte concentration using NaCl, $CaCl_2$, and Na_2SO_4 solutions. Two different boundary conditions were used.

First, the active layer was adjacent to the solution with a constant salt concentration, which was 0.001 m NaCl. Second, the membrane was turned over and the porous surface was adjacent to the constant salt solution. The results obtained are shown graphically in Figure 14 {9}. As can be seen from this figure, a difference in membrane potential exists for an asymmetric membrane depending on the boundary conditions with regard to the salt solutions. If the membrane is turned over, so that the porous surface of the asymmetric membrane is contacting the solution of low salt concentration (0.001 m NaCl), the experimentally determined membrane potential deviates from the membrane potential determined with the active layer adjacent to the solution of low salt concentration. The difference is pronounced at high salt concentrations, c_s''. The fact that the deviation always starts at large salt concentration differences across the membrane indicates that a volume flow across the asymmetric membrane may cause the difference in membrane potentials at different boundary conditions.

The experimentally confirmed concentration gradient within the porous sublayer of an asymmetric membrane is connected with a diffusion potential. Thus, the membrane potential of an asymmetric membrane $\Delta\Psi_a$ is composed of the membrane potential of the active layer and the diffusion potential within the porous sublayer. If it is assumed that the membrane potential of the active layer is nearly the same in both cases, the following explanation can be given. In case 1 of Figure 14a, the active layer is adjacent to phase ('), which is the salt solution of constant concentration $c_s' = 0.001$ m NaCl. Here, the volume flow across the membrane produces a concentration gradient which leads to a smaller salt concentration c_{s1}^m at the interface between the active layer and the porous sublayer (Figure 15). The volume flux, q_1, in this case, is determined by the effective osmotic difference $\Delta\Pi_{eff}$ which is smaller than $\Delta\Pi = \Pi' - \Pi''$. The corresponding ratio c_s''/c_{s1}^m, which determines the magnitude of the diffusion potential within the porous sublayer, can be estimated using a simplified Nernst-Planck equation. As was reported recently {22} the result is as follows:

$$c_s''/c_{s_1}^m \;=\; \exp(q_1 \delta/D_s) \tag{28}$$

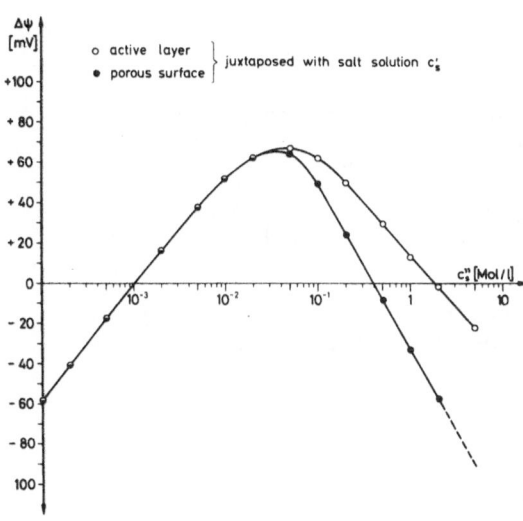

Figure 14a: Membrane potential, $\Delta\Psi$, as a function of NaCl concentration, c_s'', using an 85°C annealed asymmetric cellulose acetate membrane. The membrane potentials were measured by means of calomel electrodes. The data represented by the open circles were taken with the actice layer adjacent to the salt solution c_s' and the solid circles, with the porous surface adjacent to the salt solution c_s' (T = 298°K; c_s' = 0.001 m NaCl solution).

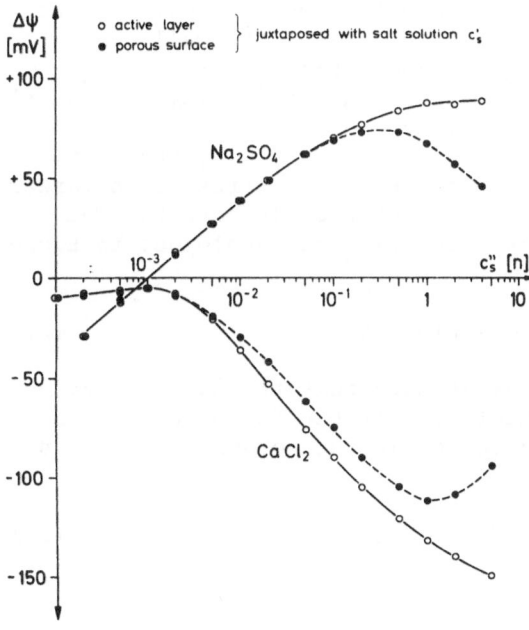

Figure 14b: Membrane potential as a function of $CaCl_2$ or Na_2SO_4 concentration, c_s'', respectively. With the sulphate solutions mercurous sulphate electrodes were used which possess a K_2SO_4 salt bridge. The data represented by the open circles were taken with the active layer adjacent to the salt solution c_s' and the solid circles (dashed lines), with the porous surface adjacent to the salt solution c_s' (T = 298°K; c_s' = 0.001 m electrolyte solution).

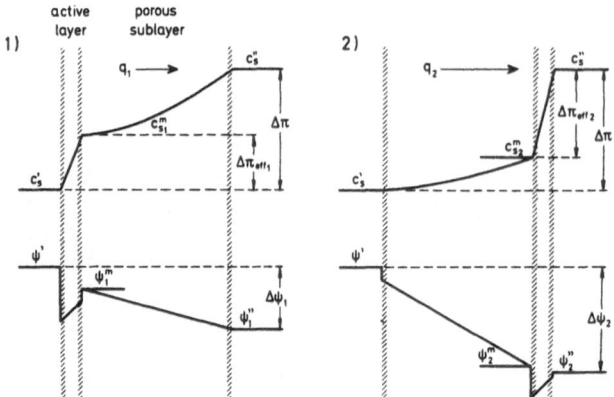

Figure 15: Diagrammatic presentation of the concentration and membrane potential profile for the two different cases of Figure 14.

In case 2 of Figure 14, where the membrane is turned over, the active layer is adjacent to phase ("), in which the salt concentration c_s'' is varied. There the volume flux q_2 also produces a concentration gradient within the porous sublayer. This leads also to a salt concentration $c_{s_2}^m$ at the interface porous sublayer/active layer which is larger than c_s'. Since c_s' itself is small compared to c_s'' at large values of c_s'', $c_{s_2}^m$ is also small compared to c_s''. Therefore, the osmotic difference across the active layer is nearly equal to the osmotic difference, $\Delta\Pi$, across the entire membrane. Because $\Delta\Pi > \Delta\Pi_{eff}$ of case 1, the volume flux $q_2 > q_1$ and thus, $c_{s_2}^m/c_s' > c_s''/c_{s_1}^m$. As a consequence of this, there results a larger diffusion potential within the porous sublayer in case 2. The ratio $c_{s_2}^m/c_s'$ is related to q_2 by a relationship analogous to Equation (28).

$$c_{s_2}^m/c_s' = \exp(q_2\delta/D_s) \tag{29}$$

In deriving this equation, it was assumed that the salt rejection, r, of the membrane is approximately 1. If the following common relationship for the diffusion potential in the porous sublayer is applied

$$\Delta\Psi_D = -(URT/F)\ln(c_s^a/c_s^b) \tag{30}$$

where R is the absolute gas constant; T is absolute temperature (^{O}K); F = Faraday number; $U = (D_{+} - D_{-})/(D_{+} + D_{-})$; D_{+}, D_{-} = diffusion coefficient of cation and anion, respectively, within the porous sublayer; c_{s}^{a}, c_{s}^{b} = concentrations at the corresponding boundaries; the two diffusion potentials within the porous sublayer can be estimated. Thus, the following relationships are obtained:

$$\Delta\Psi_{D_{1}} = -(URT/F)\ln(c_{s}''/c_{s_{1}}^{m}) \qquad (31)$$

$$\Delta\Psi_{D_{2}} = -(URT/F)\ln(c_{s_{2}}^{m}/c_{s}') \qquad (32)$$

With Equations (31) and (32) the following relationship is obtained for the difference of the diffusion potentials between case 2 and case 1 of Figure 14 which may also be taken as the difference in the entire membrane potential of the two cases if differences in the membrane potentials of the active layer are neglected:

$$\Delta(\Delta\Psi) \simeq \Delta\Psi_{D_{2}} - \Delta\Psi_{D_{1}} = -(URT/F)(q_{2} - q_{1})(\delta/D_{s}) \qquad (33)$$

Thus, the difference between the membrane potentials of an asymmetric membrane under the two different boundary conditions is due to the different magnitude of the corresponding volume fluxes. This asymmetric behaviour, with regard to the membrane potential, takes place far from the linear range. Thus it is a real non-linear effect, but the non-linear effects manifest the influence of the concentration profile within the porous sublayer on the transport behaviour of the asymmetric membrane much more strongly. Furthermore, the double layer model of the asymmetric cellulose acetate membrane is confirmed by these experimental findings.

The author is indebted to Professor R. Schlögl for his interest in this work. Furthermore, the author is very much obliged to Dr. H. B. Hopfenberg for the invitation to the ACS meeting in Los Angeles and also for reading the proofs. The work was supported by the "Bundesminister für Forschung und Technologie", Bonn, Germany.

REFERENCES

1. C. E. Reid and E. J. Breton, J. Appl. Polymer Sci. 1, 133 (1959).

2. S. Loeb and S. Sourirajan, Advan. Chem. Ser. 38, 117 (1962).

3. R. L. Riley, U. Merten, and J. O. Gardner, Desalination 1, 30 (1966).

4. K. S. Spiegler, J. electrochem. Soc. 100, 303 C (1953)

5. T. Foley, J. Klinowski, and P. Mears, Proc. R. Soc. London A 366, 327 (1974).

6. G. Schmid, Z. Elektrochem. 54, 424 (1950); ibid. 55, 229 (1951); ibid. 56, 181 (1952).

7. R. Schlögl, Stofftransport durch Membranen, Dr.-Dietrich-Steinkopff-Verlag, Darmstadt, 1964.

8. A. J. Staverman, Rec. Trav. chim. Pays-Bas 70, 344 (1951).

9. W. Pusch, Proc. Nato Advanced Study Inst. "Polyelectrolytes II", Forges-les-Eaux, 1973, E. Sélégny ed., Reidel Publishing Co., Dordrecht, Holland (in preparation).

10. W. Pusch, Chemie-Ingenieur-Technik 20, 1216 (1973).

11. W. Pusch, Proc. 4th Intl. Symp. on Fresh Water from the Sea, Vol. 4, p. 321/32, A. and E. Delyannis, published by the editors, Athens 1973.

12. W. Pusch, Proc. of a Symp. on "Structure of Water and Electrolyte Solutions", W. Luck ed., Verlag Chemie und Physik Verlag, Weinheim/Bergstr., Germany, 1974.

13. K. S. Spiegler and O. Kedem, Desalination 1, 377 (1966).

14. W. Pusch, Proc. 3rd Intl. Symp. on Fresh Water from the Sea, Vol. 2, 535/49, A. and E. Delyannis, editors, published by the editors, Athens, 1970.

15. K. S. Spiegler and Ch. P. Minning, "Streaming Potentials in Hyperfiltration of Saline Water", Ph.D. thesis of Ph. P. Minning at U.C.L.A., Berkeley, Sea Water Conversion Laboratory, 1973; Proc. Nato Advanced Study Inst. "Polyelectrolytes II", Forges-les-Eaux, 1973, E. Sélégny ed., Reidel Publishing Co., Dordrecht, Holland (in preparation).

16. W. Pusch and R. Riley, Desalination 14, 389 (1974).

17. G. Boari, C. Merli, G. Mossa, and R. Passino, Proc. 4th Int.
     Symp. on Fresh Water from the Sea, Vol. 4, p. 49/63, A. and E.
     Delyannis, editors, published by the editors, Athens, 1973.

18. J. Jagur-Grodzinski and O. Kedem, Desalination 1, 327 (1966).

19. U. Merten, Desalination by Reverse Osmosis, M.I.T. Press,
     Cambridge, Mass., 1966.

20. W. Banks and A. Sharples, The Mechanism of Desalination by
     Reverse Osmosis, and Its Relation to Membrane Structure, OSW
     Res. and Develop. Rep. No. 143, June 1965.

21. W. Pusch and R. Göpl, Desalination 8, 277 (1970).

22. W. Pusch, Reverse Osmosis Membrane Research, H. K. Lonsdale
     and H. K. Podall, editors, Plenum Publishing Corp., New York,
     1972.

II. Industrial Membrane and Barrier Film Applications

THE EFFECT OF TRANSPORT PHENOMENA ON THE SHELF LIFE OF A PLASTIC

CARBONATED BEVERAGE BOTTLE

G. A. Gordon and P. R. Hsia

Continental Can Company, Inc.

7622 S. Racine Avenue, Chicago, Illinois

A large new market, the beer and carbonated beverage bottle market, is about to open up for plastic materials. This paper will discuss the shelf life of a plastic bottle for carbonated beverages in terms of the various parameters, transport related and otherwise, which determine it, evaluating their relative importance in absolute terms and as a function of time.

The problem of the shelf life of a plastic bottle is one of critical importance for its success in this application. While different soft drink manufacturers state their shelf life criteria in different ways, one rule of thumb is that the package must not allow a pressure decrease of more than 15% in three months at room temperature. It is important to note that this criterion is stated in terms of total pressure loss, and not permeation loss alone. Permeation loss, the quantity of CO_2 that permeates through the bottle wall and is lost to the atmosphere, is only one factor in the total pressure loss and is small at short times, as will be shown. Other factors, more important at short times, are pressure loss caused by sorption of CO_2 into the bottle wall, pressure loss caused by volume expansion of the bottle due to creep of the bottle wall under internal pressure, and a pressure loss which we have attributed to some change in the packed beverage. This change might well be caused by consumption of head-space oxygen by the beverage, which would result in a total pressure reduction equivalent to the reduction in oxygen partial pressure. A side effect of oxygen consumption could be pH change in the beverage, which would lead to a change in CO_2 solubility, and therefore CO_2 pressure. Finally, leakage through the closure, losses related to the measurement procedure, and miscellaneous losses round out the material balance. With good technique, loss from these last three categories can be accounted for and/or minimized.

There are, therefore, six important factors which determine the total pressure loss from the pressurized plastic bottle as a function of time, though only the first five of these could occur in a commercial application:

1) permeation loss,
2) sorption into the bottle walls,
3) volume expansion due to creep
4) beverage change,
5) closure leakage, and
6) measurement related losses.

Factors 1) and 2) (permeation and sorption) can be evaluated analytically (1) by an iterative calculation based on the solution to Fick's Laws (2) if the diffusion coefficient and the sorption isotherm are known and if some simplifying assumptions are made regarding bottle geometry. Factor 3) (volume expansion) can also be evaluated analytically from first principles of rheology (1) if the proper data are available, but the properties involved are affected by thermal history and by the presence of permeant (3), and of moisture; therefore it can be difficult to obtain the appropriate data. Furthermore, the volume expansion of a pressurized bottle can be measured directly and the pressure loss due to such expansion can then be easily calculated. The latter approach was used in the calculations reported here. The magnitude of factor 4) (pressure loss due to beverage change) can be determined by the measurement of pressure changes as a function of time in glass control bottles, if CO_2 losses due to closure leakage (factor 5) and measurement losses (factor 6) are determined and subtracted from the total pressure change. Since the permeation loss is only an important factor at long times, and since at long times the amount of CO_2 sorbed in the bottle walls becomes constant, it should, in principle, be possible to determine the six different factors individually, solely on the basis of pressure and volume measurements. Unfortunately, however, uncertainty about such things as the effect of moisture on the transport and rheological properties of the resin makes this difficult to do with confidence. Nonetheless, the purpose of this paper is to describe the results of these calculations and to compare the pressure loss due to transport phenomena (permeation and sorption), as calculated from pressure and volume measurements,with that predicted by Fick's Laws.

EXPERIMENTAL

Steady state CO_2 permeability data were measured on a variable pressure type permeation cell like the one in Figure 1. The sample holder is stainless steel with a sintered bronze disc film support

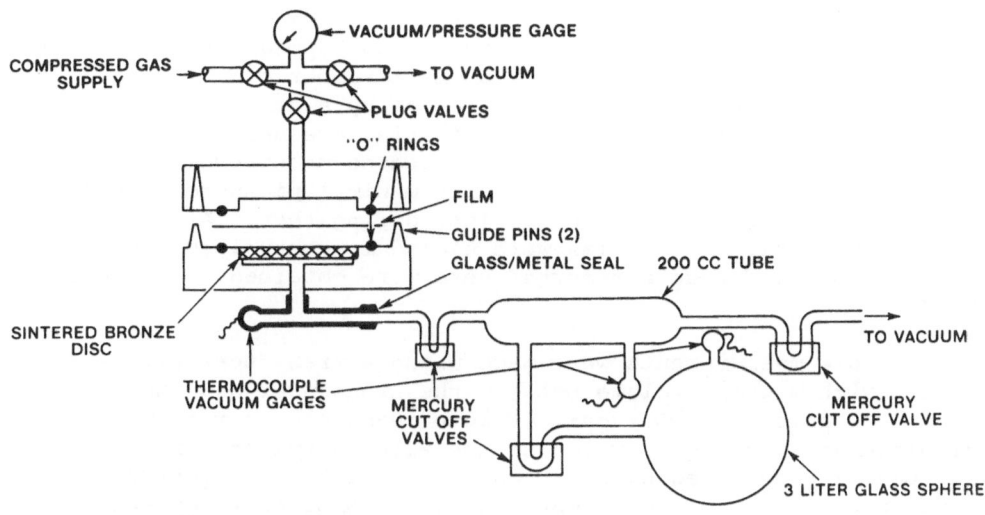

Figure 1.   Schematic drawing of a variable pressure type permeation
cell.

and "O" ring seals near the outer circumference of the film.   To
minimize leakage, all joints in the measuring section of the cell
are fused--either silver solder or glass--and mercury cut-off valves
are used to vary the receiving volume and to seal off the system from
the vacuum line to take a measurement.   Thermocouple gages calibrated
for the range of 1-20 millitorr (μ Hg) are used for pressure measurement.   Used with 5 mil thick compression molded films, this system
has a sensitivity on the order of 10^{-4} barrer (cc(STP) x cm x 10^{-10}/
cm^2 x sec x cm Hg), limited by leakage past the "O" ring seals.   With
some modification, this could probably be significantly improved.
However, this sensitivity has proved to be completely adequate for
even the demands of the high nitrile barrier films.   Using a 5 mil,
high nitrile barrier film, an individual measurement can routinely be
made in less than 1/2 hour, though a series of such measurements over
10-15 days is necessary to allow the system to come to steady state
before the true CO_2 permeability can be determined.

Two kinds of data taken on pressurized bottles are of interest
here:

1)   internal pressure measurements (for pressure loss data),
and

2)   total volume measurements (for volume expansion data).

The bottles were packed as follows: ten bottles for each variable

were filled with a carbonated beverage, then capped with a special
closure containing an Eaton valve, a small rubber plug designed to
allow repeated, pressure tight penetration by a small diameter probe.
Use of the Eaton valves allowed both initial pressurization of the
test bottles to a precise level, and repeated measurement of the
pressure level on the same bottles with minimal pressure loss (equivalent to about 0.1 psi per penetration) during test. After packing,
and between tests, the bottles were stored in controlled temperature-
humidity rooms. The data reported here were obtained at 73 deg. F.,
50% RH.

Pressure measurements were made using a transducer with a digital
readout, with minimal volume between the transducer and the sampling
probe. Precision was $\pm$0.1 psi. Volume measurements were made using
a displacement technique, weighing the water displaced from the spout
of a specially designed cylinder when the bottle was immersed. While
the reproducibility of this test was found to be operator sensitive,
a well-trained operator could expect to achieve reproducibility of
$\pm$0.2 ml, or about $\pm$0.07% for a 10 oz. bottle.

In choosing the appropriate solution of Fick's Laws for calculations, spherical geometry was assumed (even though a bottle more
closely resembles a cylinder than a sphere) because this greatly
simplified the calculations. The use of this simplifying assumption
introduces an error in the surface-to-volume ratio, which tends to
compensate for the error introduced by the non-uniformity of wall
thickness in a real bottle. The net effect of these compensating
errors, while changing some of the numerical results, should not
change the conclusions drawn from the analysis. The sorption isotherm used in the Fick's Law calculations was assumed to follow the
dual sorption model (4, 5, 6, 7), which postulates that the isotherm
can be resolved into two parts, one of which obeys Henry's Law while
the other resembles a Langmuir isotherm. The use of this model
instead of the simpler Henry's Law model allows for more accurate
calculation of the pertinent transport parameters.

CO_2 loss from the bottle closure was measured by isolation of
the neck/closure region of the bottle (by an "O" ring seal around the
neck) in a chamber equipped with valved ports and a septum for hypodermic needle entry. After packing with product and measuring the
initial pressure, a bottle was sealed into the chamber, the chamber
was flushed with helium, and gas samples were periodically drawn for
determination of CO_2 content on a gas chromatograph. The loss rate
is readily calculated from this information.

A small amount of CO_2, that contained in the pressure transducer
probe, was lost to the atmosphere after every pressure measurement.
Repeated pressure measurements on the same bottle over a short time
have shown that this loss is of the order of 0.1 psi per penetration

TABLE I

CO$_2$ Transport Properties of Barrier Resins and Other Materials at 25° to 30° C.

Material	Permeability, P, Barrers (a)	Diffusion Coefficient, D, cm^2/sec	Henry's Law Solubility, S, cc/cc atm	Langmuir Isotherm Coefficients	
				C$_h'$, cc/cc	b, atm^{-1}
Experimental Cycopac	0.02	2.0 x 10^{-10} (b)	0.49 (b)	6.0 (b)	0.55 (b)
Barex 210	0.008	5.2 x 10^{-11} (c)	2.43 @ 1 atm (c) 0.81 @ 5 atm (d)		
Experimental NR-16R	0.014	6.9 x 10^{-11} (c)	2.33 @ 1 atm (c)		
Lopac I	0.004	–	–		
Mylar	0.11	1.1 x 10^{-9} (e)	2.0 @ 1 atm (f) .38 (f)	5.3 (f)	0.44 (f)
SAN	3.2	7.5 x 10^{-9} (c)	2.2 @ 1 atm (c)		
HDPE (ρ = 0.96)	2.0	–	–		
HDPE (ρ = 0.95)	3.2	–	–		
LDPE (ρ = 0.92)	10.0	5.2 x 10^{-7}	–		

(a) 1 barrer = $1 \dfrac{\text{cc (STP) cm}}{\text{cm}^2 \text{ sec cm Hg}}$ x 10^{-10}

(b) Data courtesy of Borg-Warner Chemicals Div. of Borg-Warner Corp.
(c) Data courtesy of E. I. duPont de Nemours & Company
(d) Data courtesy of Mr. Jon Eilenberg, Rutgers University
(e) Data from Reference (8)
(f) Data from Reference (6)

of the Eaton valve. Since there may be as many as 15 to 20 penetra-
tions over a 4 month test period, the total loss in this time could
amount to 1.5 to 2.0 psi. Because the test schedule called for measurements after 1, 2, 3, 4, 8, 12, and 16 weeks, a graphical representation of these data appears as a straight line with a change in
slope at the 4 week point.

RESULTS AND DISCUSSION

Some typical data on the steady-state CO_2 transport properties
of high nitrile barrier resins and some other materials are shown in
Table I. It should be pointed out here that the data reported for
Cycopac were taken at 25 deg. C., and in such a way as to measure the
dual mode character of the sorption isotherm, while the other data in
Table I (except for the second set of Mylar data) reflect the assumption that the systems follow Henry's Law. This explains the comparatively low value for the Henry's Law solubility and might partially
explain the relatively high diffusion coefficient for Cycopac. The
discrepancy between the two results reported for Barex at different
pressures can also be understood in terms of a dual mode sorption
isotherm. Since, with the exceptions noted above, the Henry's Law
solubility coefficients of the barrier resins are of about the same
magnitude as for other polymers, the extremely low CO_2 permeabilities
must be attributed to the unusually low CO_2 diffusion coefficients in
these materials.

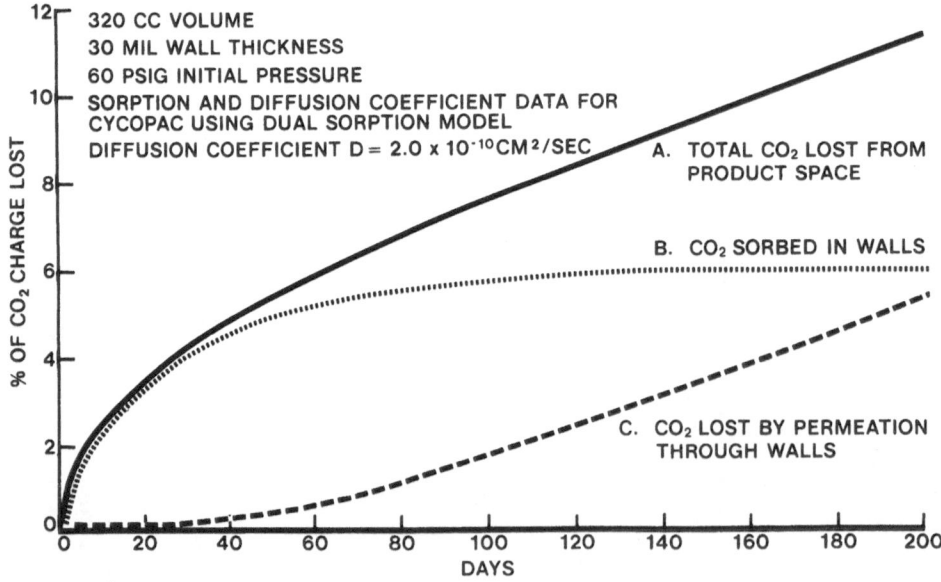

Figure 2. Calculated CO_2 losses from pressurized sphere-higher
           diffusivity resin.

Figure 2 compares the percentage of the original CO_2 charge which has left the product space (curve A) with the percentage of CO_2 which has been sorbed by the walls as a function of time (Curve B), as calculated from the Fick's Laws solutions. The difference between these curves (Curve C) is the percentage of CO_2 that has permeated out of the container; this curve shows that essentially all of the transport-associated pressure loss in the first several weeks after packing is due to sorption into the walls, while permeation loss does not become an important factor until sorption equilibrium has been attained, after several months. The importance of the magnitude of the diffusion coefficient is illustrated in Figure 3 where the same calculation has been made using a coefficient about one-third of that used in the calculations for Figure 2. Initial breakthrough of CO_2 lost by permeation through the walls (curve C), occurs only after a factor of three longer time (about 75 days compared with about 25 days in Figure 2).

Figure 4 shows typical volume increase data for bottles with and without application of a proprietary Continental Can Company process for dimensional stabilization, the CR (for creep roduction) process. While the major part of the volume increase takes place in the early stages, the bottle continues to expand throughout the test.

Figure 5 shows typical data for the experimentally determined total pressure loss as a function of time for untreated and CR treated bottles, and a similar curve for a set of glass control bottles tested at the same time.

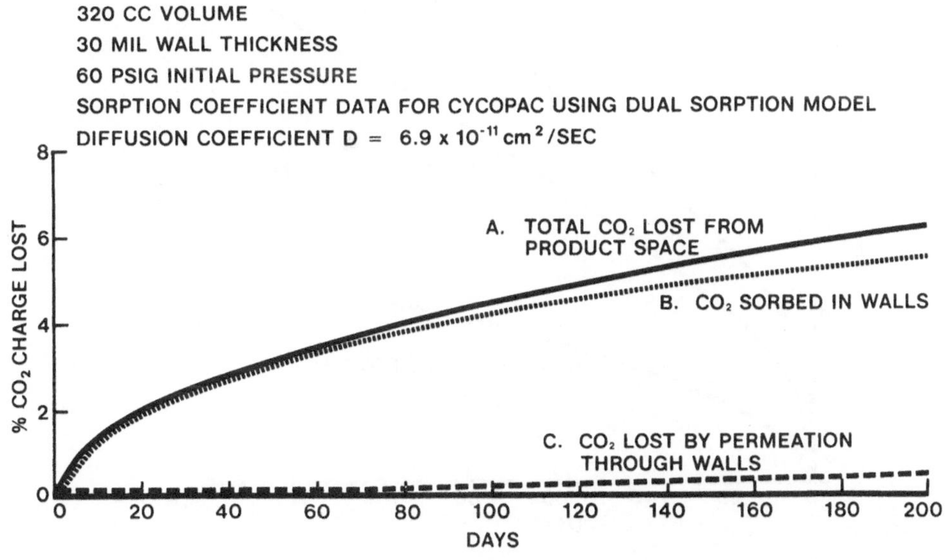

320 CC VOLUME
30 MIL WALL THICKNESS
60 PSIG INITIAL PRESSURE
SORPTION COEFFICIENT DATA FOR CYCOPAC USING DUAL SORPTION MODEL
DIFFUSION COEFFICIENT D = 6.9×10^{-11} cm^2/SEC

Figure 3. Calculated CO_2 losses from pressurized sphere - lower diffusivity resin.

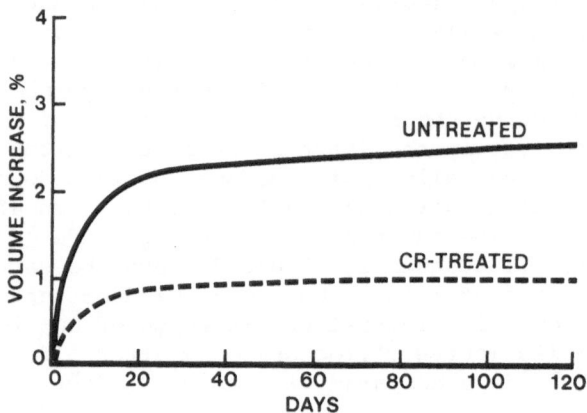

Figure 4. Typical volume increase of pressurized 10 oz. bottle.

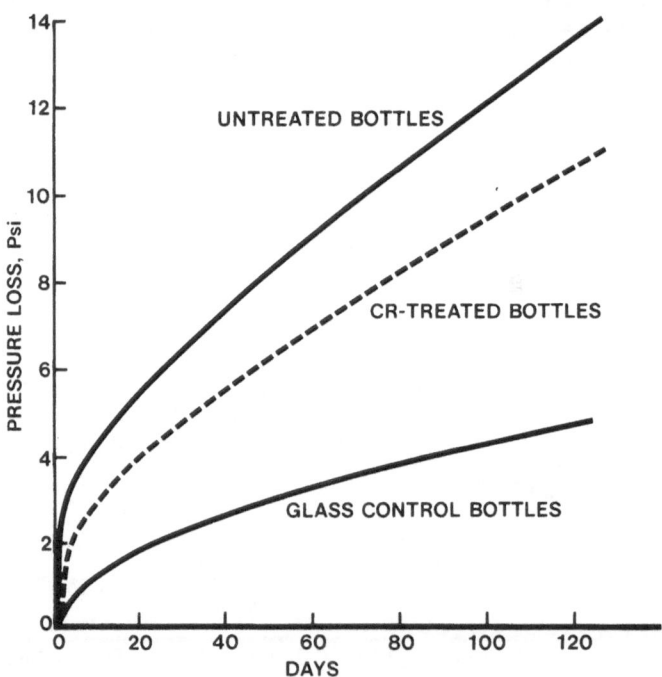

Figure 5. Typical total pressure loss from pressurized 10 oz. bottle.

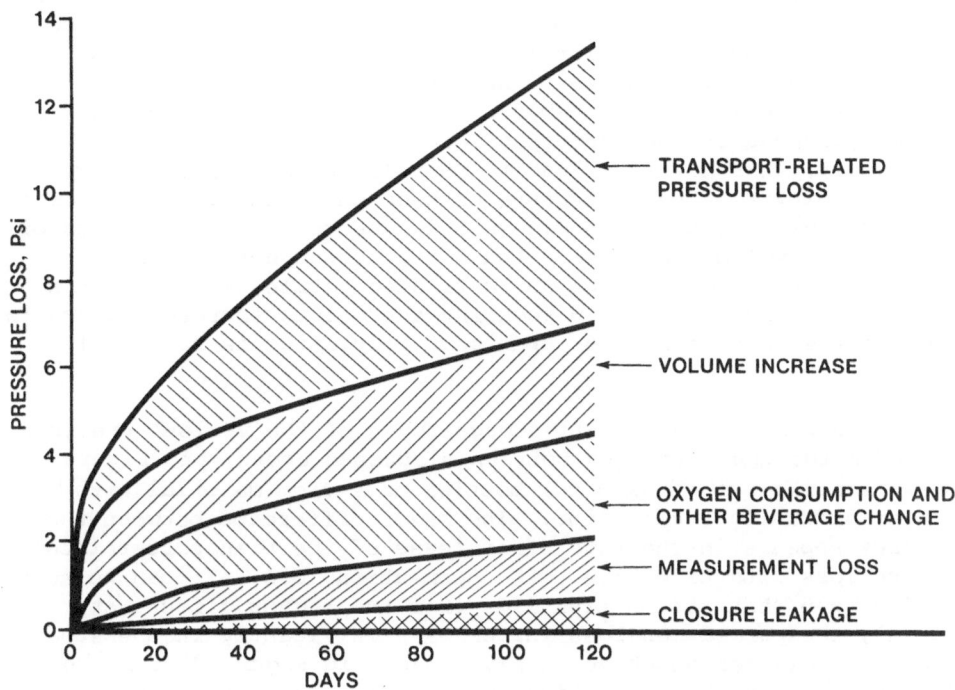

Figure 6.   Pressure loss modes for untreated bottles - higher
diffusivity resin.

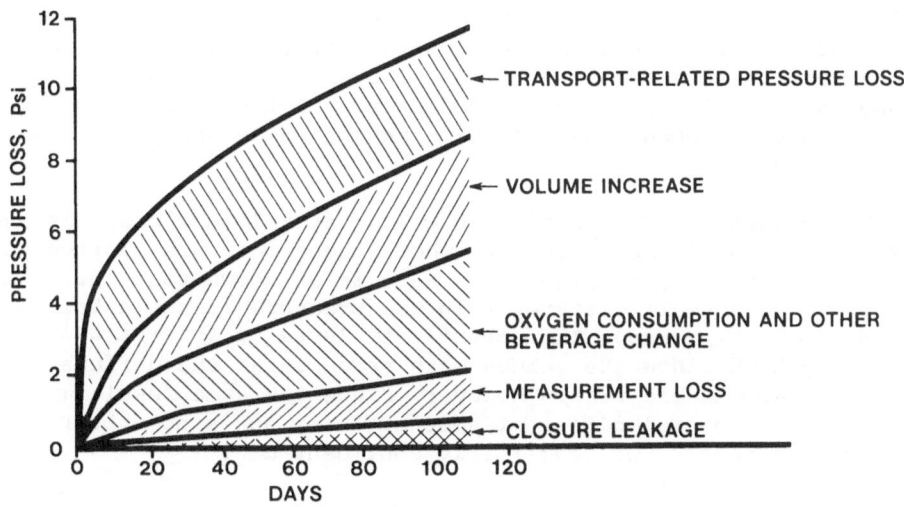

Figure 7.   Pressure loss modes for untreated bottles - lower
diffusivity resin.

The shape of the glass control curve and the magnitude of the difference between it and data (shown in Figs. 6, 7 & 8) for closure leakage and measurement loss both suggest that some additional mechanism must be found to completely explain the pressure loss in glass control bottles. At least part and perhaps all of the excess glass control pressure loss is probably due to consumption by the beverage of oxygen trapped in the head-space of the bottle when it was packed with product; this may also lead to pH related changes in the CO_2 pressure, however. The difference between the pressure losses from plastic and glass bottles in Figure 5 is due to combined effects of transport and volume increase. And most, though not all, of the difference between CR-treated and untreated bottles is due to the effect of the CR treatment in reducing the volume increase, since the effect of the CR treatment on transport properties is relatively small.

Figure 6 shows the relative magnitudes of the various pressure loss modes for the untreated bottles discussed above, and Figure 7 shows a similar set of data for bottles made from material with a 3x lower diffusion coefficient. The major difference between these two cases appears in the rate of increase of the transport-related pressure loss with time, i.e., the permeation loss. It can be seen from Figure 6 that for bottles made from the higher diffusivity material, the permeation loss increases significantly in the steady state period (after about 40 days), whereas in Figure 7, for the lower diffusivity material, permeation loss is negligible throughout the life of the test. (This is illustrated more clearly in Figures 9 and 10.)

Figure 8 shows the data of Figure 6 along with similar data for CR-treated bottles (indicated by cross-hatching), showing the effect of a reduction in volume increase on the various factors which make up the total pressure loss. It can be seen that the loss due to volume increase is markedly reduced by the CR treatment, and the loss due to transport phenomena (solution in the walls and permeation) is slightly reduced.

The effect of CR treatment on transport related losses is shown more clearly in Figures 9 and 10, where these data are plotted without the other components of Figure 8. Here it is shown that the CR treatment produces a small difference in the pressure losses attributable to transport phenomena. The uncertainty level of this calculation is relatively high since the values were produced as the difference of numbers of comparable size. However, since the same result was achieved with several different sets of data, and since theory would predict this result, we feel fairly confident of it.

The real purpose of these curves, however, is to show a comparison of the experimentally determined pressure loss due to transport

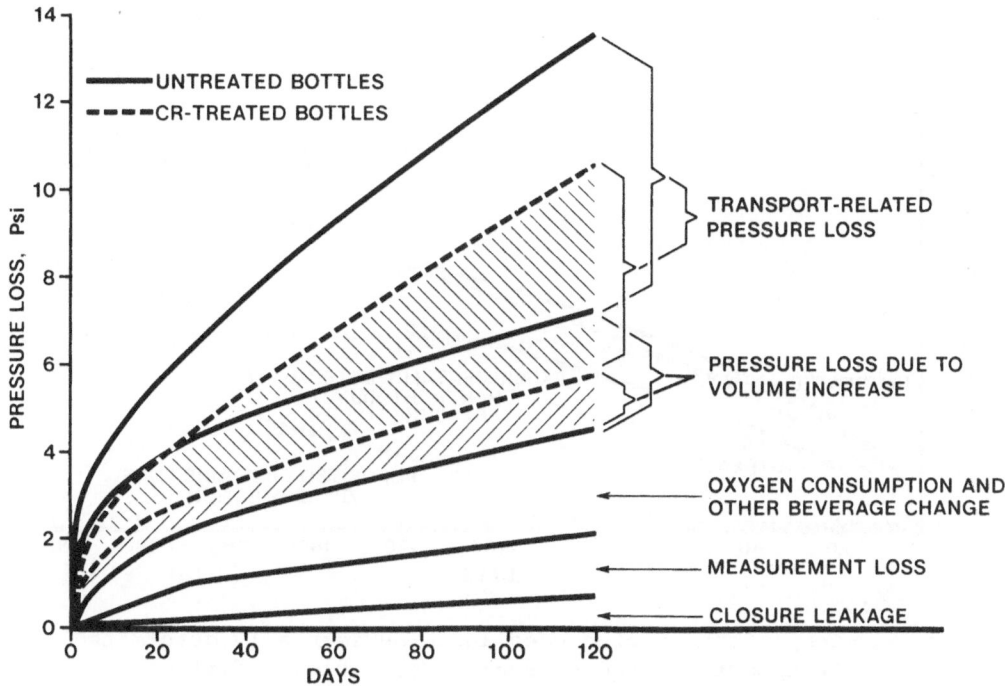

Figure 8. Pressure loss modes - comparison of CR treated and untreated bottles (data for CR treated bottles indicated
by cross-hatching).

phenomena, as evaluated in Figure 8, with the prediction from Fick's
Laws from Figures 2 and 3. Figure 9 shows this comparison for the
material of higher diffusivity. The comparable lines are those for
predicted total pressure loss and for untreated bottles (since the
transport parameters used in the computation were measured on untreated material). The agreement of these two lines is surprisingly
good, considering the simplified geometry assumed for the calculation
and the fact that the transport parameters were measured on dry
material, while the bottle tests, of course, were made with one surface in contact with an aqueous solution and the other at 50% RH.
In Figure 10, for the lower diffusivity resin, the agreement is not
quite so good, at least at short times, though the lines do approach
each other at longer times. The important factor, however, is not
the goodness of agreement of the lines as drawn, but the fact that
they are of the same general shape and order of magnitude, which
suggests that they do, in fact, represent alternative means of determining the same quantity.

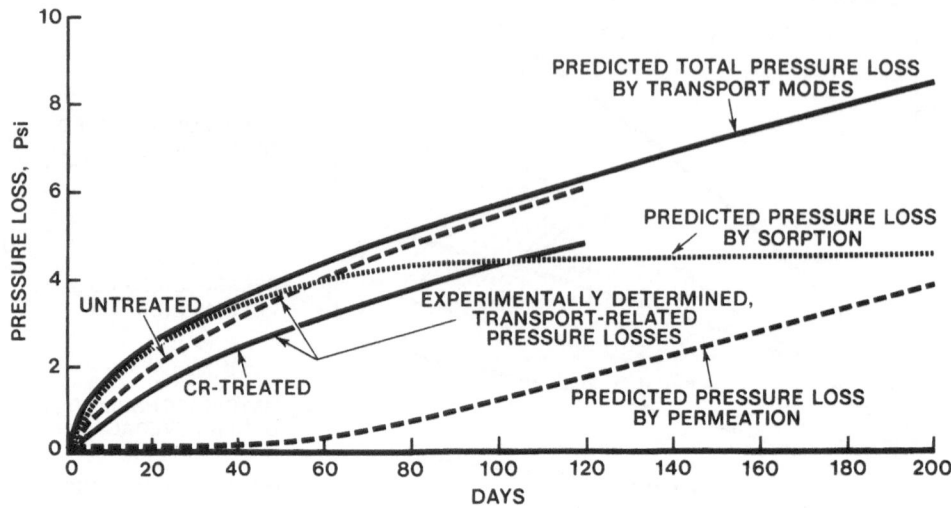

Figure 9.   Comparison of predicted and experimentally determined,
            transport-related pressure loss - higher diffusivity resin.

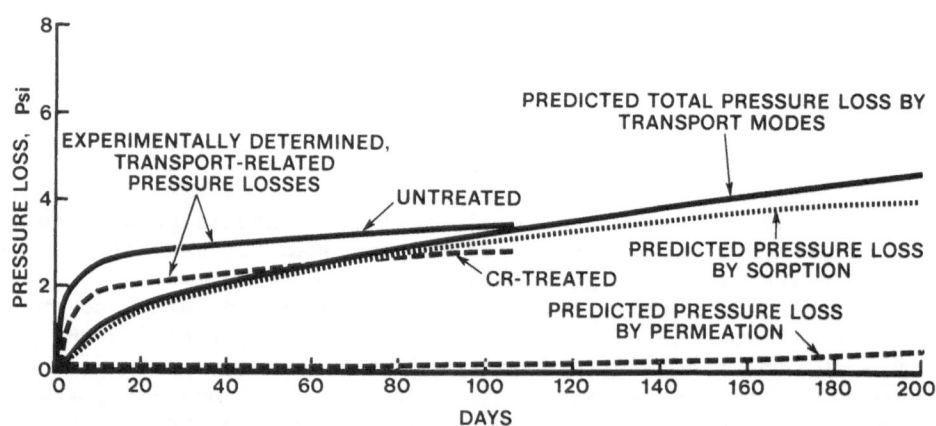

Figure 10. Comparison of predicted and experimentally determined,
            transport-related pressure loss - lower diffusivity resin.

CONCLUSIONS

1) Very long times, on the order of weeks or months, must pass before steady state conditions of mass transport are achieved through the wall of a carbonated beverage bottle made from acrylonitrile based barrier resins.

2) While solubility coefficients of CO_2 in these materials are of the order of magnitude of CO_2 solubility coefficients in other common polymeric materials, the CO_2 diffusion coefficients are orders of magnitude smaller than in other materials.

3) Both the unusually low steady state CO_2 permeability coefficients and the unusually long times required to reach steady state conditions are attributable to the low CO_2 diffusion coefficients of these materials. It is noteworthy that as a result of failure to recognize the long times necessary to reach steady state, erroneously low values of the CO_2 permeability for these and related materials have been published in both the patent and the scientific literature. (9, 10, 11)

4) The effect of diffusion coefficient on transport phenomena is reflected not only in theoretical calculations, but also in pressure loss measurements on bottles packed with a carbonated beverage.

5) The dual mode sorption isotherm model seems to describe the experimental facts more accurately than does the simple Henry's Law isotherm insofar as this information is used for Fick's Law calculations.

6) Fick's Law calculations can provide a useful means of predicting the pressure losses from a bottle due to transport phenomena, and of separating sorption losses from permeation losses.

7) These calculations indicate that in the early stages of the process the predominant mechanism for transport related pressure loss in the container is by sorption into the walls, and that only much later, when sorption equilibrium is approached, does permeation through the wall become an important factor in the pressure loss. And, finally,

8) Suitability of bottles made from acrylonitrile-based barrier resins for carbonated beverage application is determined not only by the transport properties of the resin, but also by its dimensional stability under long term stress and by such non-resin related factors as bottle design (to minimize the stresses which induce various kinds of bottle failure) and closure efficiency.

ACKNOWLEDGEMENTS

    Thanks are due to Mr. G. R. Johnson for preparation of the
computer program used for the Fick's Law caluclations, to Drs. K.
A. Boni and P. J. Fenelon for helpful discussions, and to the
Continental Can Company for permission to publish these findings.

LITERATURE CITED

1. Fenelon, P. J., paper presented at the Society of Plastic
    Engineers ANTEC in Montreal, May, 1973.

2. Crank, J., The Mathematics of Diffusion, Oxford University Press,
    London, 1956.

3. Machin, D. and Rogers, C. E., J. Polymer Sci: Polymer Phys.
    Ed., 11, 1535-1554 (1973).

4. Fenelon, P. J., personal communication.

5. Meares, P., Trans. Faraday Soc., 54, 40 (1958).

6. Michaels, A. S., Vieth, W. R. & Barrie, J. A., J. App. Phys.,
    34, 1 (1963).

7. Vieth, W. R. & Sladek, K. J., 39th National Colloid Symposium
    Preprints, (June, 1965), 484.

8. Michaels, A. S. Vieth, W. R. & Barrie, J. A., J. App. Phys.,
    34, 13 (1963).

9. Hughes, E. C., Idol, J. D. Jr., Duke, J. T., & Wick, L. M.,
    J. Applied Polymer Sci., 13, 2567-77 (1969).

10. Product brochure #v-10.006-2M - 2/69 for Barex 210, by the Vistron
    Corp., copyright 1969.

11. Isley, R. F., et al., U. S. Pat. #3,437,717 (April 8, 1969).

THE USE OF LOW PERMEATION THERMOPLASTICS IN FOOD AND BEVERAGE PACKAGING

Morris Salame

Monsanto Company

101 Granby Street, Bloomfield, Conn. 06002

Background

The use of synthetic polymers for food packaging is certainly not new. Many food and beverage items, over the years, have found their way into packages composed of polymeric materials, but no truly large scale inroads have been made by polymers into the majority of food packaging areas now using metal and glass. Perhaps the lack of barrier properties, or conversely the relatively high permeability of synthetic polymeric materials, as compared to glass and metal, is the most dominant reason for this lack of use. This study attempts to examine the applicability of various low permeation polymeric materials, taking only into consideration the permeability of the material, in the packaging of specific foods and beverages. It is not meant to be all encompassing nor absolute in its findings, since the ultimate acceptance of any polymer application for packaging foods will depend upon other parameters such as absorption and extraction in addition to permeation properties.

In the past, the phrase "barrier" polymer has been used to indicate a polymer of low permeability when associated with a particular use. In the case of food packaging, the phrase "barrier" polymer should only be used to indicate a polymer capable of <u>completely</u> protecting the food in all aspects of quality, purity and taste. Since this paper covers only permeability, the word "barrier" polymer will not (and should not) be used.

Scope of Study

The word "food" describes a diversity of materials, and a definition, for purposes of this study, should be made at the onset. In the area of food packaging studied here, only those foods requiring a relatively high degree of permeation protection are considered. In summary, these are those foods, beverages and other comestibles, normally stored at ambient temperature conditions, and normally expected to have a shelf life of at least a year - and in some cases more. Table I gives a list of the items studied along with a generalized degree of permeability protection needed by each (1,2). In addition, each one of these items carries with it a specific set of requirements as to method of preparation, method of packaging and method of post-treating of the package(3). This study does not attempt to grade the polymeric packaging material as to its acceptability on the food preparation and filling line. Most of the items listed in food groups 1, 2, 5 and 6 require high temperature autoclaving after filling, which many polymers are not able to survive.

After defining the area of foods of concern, the classes of polymer within which to work must also be described, since a study of all known materials would be beyond the scope of this report. Our study centered upon the use of low permeation thermpolastic polymers in the context of packaging the items listed in Table I. Combinations of materials, such as coated films, laminations, aluminum foil-polymer composites, etc., were not considered in this study, nor were materials that could not be formed directly into containers. We have, however, measured the permeation of these materials in the past(4), and films can be formed from some. The specific polymers covered in this report are given in Table II. All of these polymers possess a level of gas, vapor and moisture permeation below conventional polymeric materials(5). The object of the study was to produce a generalized matrix, taking permeability alone as a parameter, which paired each individual polymer with each food category.

Factors Affecting Packaging

Previous studies(6), have attempted to examine the specific properties of polymeric materials which have an affect upon food and beverage packaging. There are, in addition to permeation, other properties unique to polymers, which must be taken into account when studying the overall taste compatibility of a particular food/polymer pair. The various other interactions

TABLE I

Degree of Permeation Protection Required by Various Foods and Beverages (One year shelf life assumed at 25°C)

Food/Beverage	Estimated Max. Amt. of O_2 Gain Tolerable (ppm)	Other Gas Protection Needed	Estimated Maximum Degree of Water Gain or Loss Required	Requires High Oil Resistance	Requires Good Barrier to Volatile Organics
1. Canned Milk, Canned Meats, Canned Fish, Poultry	1-5	-	3% Loss	●	-
2. Baby Foods	1-5	-	3% Loss	●	●
3. Beer, Ale, Wine	1-5	< 20% CO_2 (or SO_2) Loss	3% Loss	-	●
4. Instant Coffee	1-5	-	2% Gain	●	●
5. Canned Vegetables, Soups, Spaghetti, Catsup, Sauces	1-5	-	3% Loss	-	-
6. Canned Fruits	5-15	-	3% Loss	-	●
7. Nuts, Snacks	5-15	-	5% Gain	●	-
8. Dried Foods	5-15	-	1% Gain	-	-
9. Fruit Juices, Drinks	10-40	-	3% Loss	-	●
10. Carbonated Soft Drinks	10-40	<20% CO_2 Loss	3% Loss	-	●
11. Oils, Shortenings	50-200	-	10% Gain	●	-
12. Salad Dressings	50-200	-	10% Gain	●	●
13. Jams, Jellies, Syrups, Pickles, Olives, Vinegars	50-200	-	3% Loss	-	●
14. Liquors	50-200		3% Loss	-	●
15. Condiments	50-200		1% Gain	-	●
16. Peanut Butter	50-200	-	10% Gain	●	-

TABLE II

Permeability Properties of Polymers Studied

Polymer	Gas Permeation Rates at 25°C, 50% RH ($\frac{cc\text{-mil}}{100\ in.^2\ day\text{-Atm.}}$)			Direct Liquid Permeation Rates at 25°C ($\frac{gm.\text{-mil}}{100\ in.^2\ day}$)			
	O_2	CO_2	SO_2	H_2O	C_2H_5OH	C_7H_{16}	$CH_3COOC_2H_5$
1. Poly(acrylonitrile co-styrene)*	1.10	3.50	0.15	1.45	<0.5	<0.5	<0.5
2. Poly(acrylonitrile co-acrylate) + Butadiene Graft**	1.10	3.75	0.20	2.15	<0.5 (+ haze)	<0.5	≤0.5
3. Poly(chlorotrifluoroethylene)***	3.01	12.2	0.50	<0.1	<0.5	≤0.5	<0.5
4. Poly(caprolactam)	3.05	6.00	1.00	4.70	∼5.0	<0.5	<0.5
5. Poly(ethylene terephthalate)	7.00	30.5	1.50	0.77	≤0.5	≤0.5	1.7
6. Poly(vinyl chloride)	8.02	20.5	1.15	0.60	<0.5	<0.5	∼5000 (+ attack)
7. Poly(acetal)	12.0	35.0	2.50	4.50	<0.5	≤0.5	12.0
8. Poly(methyl methacrylate) + Butadiene Graft****	40.0	130.	7.05	3.45	∼50. (+ attack)	<0.5	attack

*   Material from which Monsanto LOPAC® containers are made, AN range ≥60%.
**   Sohio Barex® Polymer.
***   Allied Halon® CTFE.
****   American Cyanamid XT® polymer.

are:

1. Chemical and Physical Interaction between food and
    polymer.
2. The ability of the polymer to interact with the
    food in such a manner as to "absorb", or "scalp"
    key flavor notes or other components of the food
    onto the polymer surface and into the matrix.
3. The level of migration of polymer or additive
    residuals within the polymeric matrix into the
    food or baverage.

This study will deal with the gas, vapor and moisture
barrier of the polymeric material as it affects the
various foods, and will not attempt to cover the other
parameters. For the purpose of this study we will
assume that there are no chemical or physiochemical
interactions between the food and polymer (other than
those related to permeability) which would adversely
affect the product. Secondly, we will assume that the
polymers are essentially pure and cannot cause taste
affects in the food by allowing off-taste residuals to
migrate into the food. Finally, we will assume that a
polymer of low permeability will not absorb the food
(or its components) to any great extent. Although we
also assumed that all the materials could survive the
post-packaging treatment required by the food, this is
the weakest assumption since many packages require post
filling autoclaving operations at temperatures over the
heat distortion temperatures of the thermoplastic.

Barrier Requirements of Specific Foods

The permeability protection requirement of a food
or beverage will, of course, depend upon not only the
susceptibility of the components in the food to change
but also to its expected or required shelf-life prior
to consumption. For all the items in Table I we have
assumed, for reasons of simplification, a one-year
shelf-life at ambient conditions (25°C, 50% RH).

There are various barrier needs that we can
quantify for each item. These are O_2, CO_2, SO_2, H_2O,
oil and volatile organic vapors. For this study, we
examined the specific need of each material as to its
probable maximum tolerance to changes which might be
caused by permeation. These changes are:

a) The gain of oxygen from ambient air.
b) The loss of carbon dioxide (carbonated beverages).
c) The loss of moisture (water based foods, beverages).
d) the gain of moisture from ambient air (dry foods and
    oil-based products).

e) The loss of ethanol (alcoholic baverages).
f) The resistance to oil migration (oil based products).
g) The loss of volatile organic vapors (food, beverage flavors).

Table I lists an estimation of the requirements needed by each food/beverage category in order to allow acceptable pallatability after the shelf-life storage. These estimates and levels are based upon each category's susceptibility to change, as well as to the composition of the food or beverage.
It should be mentioned that these levels are general, taking the foods as a class. There may be specific foods in each class which need more or less protection than indicated.

Permeability Properties of Polymers Studied
Table II lists the O_2, CO_2, SO_2, H_2O, ethanol, heptane (as indicative of oil protection) and ethyl acetate (as indicative of organic vapors) permeation of the eight polymers studied. Permeation determinations were made by conventional means as described previously(7), and represent true equilibrium values as far as can be measured. The gases were measured in contact with films at equilibrium with 50% RH moisture. The liquids were measured in direct contact with the film, where possible, by employing permeation cups with circulating 25°C air at 50% RH outside the cup or by using blow-molded containers. In some cases, where it was found that the liquid attacked the polymer (such as the PVC/ethyl acetate pair), diluted solutions of the organic in water or water/alcohol solutions were employed.
If we assume a given package of unit wall thickness and unit area, we can calculate the degree of barrier offered by each polymer for the various permeant on a comparative basis. Since all these materials could be formed into 10 ounce containers (area 40 in.^{2}) of 0.030" wall thickness, we used this model package as our unit. Based on the values in Table II, the degree of permeation for the unit package over the shelf life of one year, 25°C, 50% RH can be calculated. Numerical correlations based upon a previous study(6) were used. Using the proper correlations, the barrier properties of the eight packages are given in Table III.

Polymer/Food Pairs
By combining Table I (protection needed) with Tables II and III, a judgement can be made as to the

TABLE III

Permeation Effectiveness of Polymeric Materials in Unit Package Form

Package (See Table II)	ppm of O_2 Gain into Package	% CO_2 (or SO_2) Loss**	% Water*** Loss or Gain	Alcohol Barrier*	Oil Barrier*	Organic Vapor Barrier*
Polymer 1	5.4	5.4	2.2	A	A	A
Polymer 2	5.4	5.8	3.4	A	A	A
Polymer 3	15.	19.	<0.1	A	A	A
Polymer 4	15.	9.3	7.7	B	A	A
Polymer 5	34.	39.	1.2	A	A	(A)
Polymer 6	39.	26.	0.9	A	A	B
Polymer 7	59.	44.	7.3	A	A	B
Polymer 8	197.	~100.	5.6	B	A	B

*A = Acceptable.
 B = Unacceptable or borderline.
 ** = Includes gas lost by absorption into polymer.
*** = Assumes gradient of 100% moisture one side,
       50% moisture downstream.

Low Permeation Package/Food Packageability Judgement Matrix*

Food/Beverage (See Table I)	Package (See Table II)							
	P(AN/S)	Barex®	P(CTFE)	Nylon 6	PET	PVC	P(Acetal)	XT®
1	A	A	B	B	B	B	B	B
2	A	A	B	B	B	B	B	B
3	A	(A)	B	B	B	B	B	B
4	A	B	B	B	B	B	B	B
5	A	A	B	B	B	B	B	B
6	A	A	A	B	B	B	B	B
7	A	A	A	B	B	B	B	B
8	B	B	A	B	B	B	B	B
9	A	A	A	B	(A)	(A)	B	B
10	A	A	A	B	B	B	B	B
11	A	A	A	A	A	A	A	A
12	A	A	A	A	(A)	(A)	(A)	(A)
13	A	A	A	B	(A)	(A)	B	(A)
14	A	(A)	A	B	(A)	(A)	B	B
15	B	B	A	B	(A)	(A)	B	B
16	A	A	A	A	A	A	A	A

*A   = Full acceptance from a permeation aspect.
(A) = Qualified acceptance.
B   = Permeation too high for application.

Note: An "A" rating assumes acceptability merely from a
       permeation aspect. If organoleptic and other food
       quality tests were employed, many of these "A"
       ratings would be "B" or otherwise qualified for
       reasons stated in the test.

acceptance of each material in packaging the food or
beverage item, on a permeability basis. Table IV gives
the food/package matrix resulting from this treatment.
Note that there are 62 specific pairs out of the
possible 128 that look promising, from a permeability
vs. protection needed pairing. It is not surprising
that polymers 1 and 2 (the high acrylonitrile materials)
have so many acceptable pairings; polymer 1 was
designed for the specific purpose of packaging foods
and beverages(5).

Food/polymer pairs which were deemed acceptable
as well as some showing unacceptability from a permeability aspect were then subjected to actual filling,
storage and organoleptic tests where applicable(8).
Because of the physical property constraints of some
of the materials, certain pairs were not tested. In
general, the results agreed with the matrix except in
those cases where the assumption that polymer/food
interactions other than permeation (as previously
mentioned) did not apply. In those cases it was
difficult to judge the failure (when one occurred) as
to the specific cause.

A breakdown of all 66 rejected pairs indicates
that the majority were rejected due to the oxygen
barrier requirement, while only 18 were rejected due
to water permeation.

Conclusion
The feasibility of packaging the majority of foods
and beverages in polymeric packages, purely from a
permeation point of view, has been demonstrated.
Although low permeation is one of the chief parameters
leading to an acceptable package for foods and beverages,
other parameters such as absorption and extraction can
affect the overall quality of the food from a taste
aspect and should also be examined. The introduction
of low permeation polymeric containers, over the past
few years, possessing gas and vapor permeation rates
well below conventional polymeric packages, has opened
the door to the use of polymers as alternates to glass
or metal in the packaging of foods, beverages and other
products requiring a high order of protection.

References
1. N. W. Desrosier, "The Technology of Food Preser-
    vation", AVI Publishing, Westport, Conn., 1963.
2. J. Merory, "Food Flavorings", AVI Publishing Co.,
    Westport, Conn., 1960.
3. R. D. Buzzell and R. E. M. Nourse, "Product Inno-
    vation in Food Processing, 1954-1964", Harvard U.,
    Boston, 1967.
4. M. Salame, "Permeation and Absorption of Food
    Flavors in Plastics", Inst. of Food Technologists,
    Packaging Seminar, Dallas, March 1970.
5. M. Salame and E. J. Temple, "High Nitrile Copolymers
    for Packaging of Foods and Beverages", ACS Annual
    Meeting, Chicago, Aug. 1973.
6. M. Salame, "No Customer Complaints on Taste When
    You Pick The Right Plastic", Package Engineering,
    July 1972, pp. 61-66.
7. M. Salame, "Transport Properties of Nitrile
    Polymers", Journal of Polymer Science, Polymer
    Symposia 41, Wiley, 1973.
8. ASTM Manual in Sensory Testing Methods, STP 434.

EXPERIMENTAL SUPPORT OF "DUAL SORPTION" IN GLASSY POLYMERS

Paul J. Fenelon

Borg-Warner Chemicals

P. O. Box 68, Washington, West Virginia

INTRODUCTION

To explain equilibrium sorption isotherms for some gases in glassy polymers, Michaels, Vieth et.al. (1,2,3) postulated that sorption was occurring by two distinct modes. This postulate has since become known as the "Dual Sorption Model". Vieth et.al. (2,3) achieved a mathematical description of transient sorption behavior by further postulating that one of the sorption modes immobilizes the gas molecules while the other gives rise to the driving force for transport.

The above model leads to a modified form of Fick's second law and comprises a non-linear partial differential equation. Numeric solution of this equation by finite difference techniques provides a direct measure of absolute diffusion coefficients. A critical estimate of the validity of the "Dual Sorption Model" theory should therefore be provided by its capability of predicting steady state transport parameters from equilibrium and transient sorption behavior.

Experimental evidence in support of the above theory is presented for two polymeric materials, i.e., Mylar® Type A and an Experimental High Nitrile (HAN) polymer. The data obtained for Mylar® Type A is found to be in close agreement with published results. That obtained for the HAN polymer extends the applicability of the "Dual Sorption Model" to a new class of materials. These materials are presently of significant industrial importance in packaging applications. The practical significance of these results in terms of characterizing materials for specific industrial applications is also discussed.

BACKGROUND

The "Dual Sorption Model" applies to solutes that are sorbed according to two modes. Isotherms which follow this model consist of two distinct components. A linear component which represents ordinary dissolution and obeys Henry's Law and a non-linear component which represents "Hole Filling" and obeys Langmuir sorption behavior. The diffusion equation for the "Dual Sorption Model" is given by...

$$D \, \frac{\delta^2 C_D}{\delta x^2} = \frac{\delta C_D}{\delta t} \left[1 + \frac{C_H^1 \, b}{k_D \, [1 + \frac{b}{k_D} \times C_D]^2} \right] \qquad \text{Eq. 1}$$

where the terms are as defined in the appendix. For the special case of transport across a membrane, steady state is reached after an initial period of time. This time period will vary with the sorption and diffusion characteristics of the membrane. Once steady state is reached the concentration gradient of permeant within the membrane remains unchanged and the rate of appearance of the permeant on the low pressure side is constant. Under these conditions Equation 1 reduces to a simple differential equation where the flux rate is given by...

$$\overline{P} = D \cdot k_D \qquad\qquad\qquad \text{Eq. 2}$$

Equation 2 is valid only if the postulate that molecules which obey the nonlinear sorption component are immobilized. $\overline{P}$ is known as the permeability coefficient and is the flux rate at unit partial pressure differential and unit membrane thickness. Comparison of measured values of $\overline{P}$ against those calculated from Equation 2 should provide an estimate of the validity of the "Dual Sorption Model". Both diffusivities "D" and Henry's Law Constant "k_D" are independently obtained from pressure decay and equilibrium sorption data, respectively. Henry's Law constants are obtained by splitting measured sorption isotherms into two components. The procedure is outlined in reference 1. Diffusivities are obtained by scaling plots of ϕ versus $(\theta^1)^{1/2}$ to match those of ϕ versus $(\theta^1/D)^{1/2}$. The ϕ versus $(\theta^1)^{1/2}$ plots are obtained from numerical solution of Equation 1 whereas, ϕ versus $(\theta^1/D)^{1/2}$ plots are calculated from measured pressure decay data. The parameters ϕ and θ^1 are as defined in the appendix.

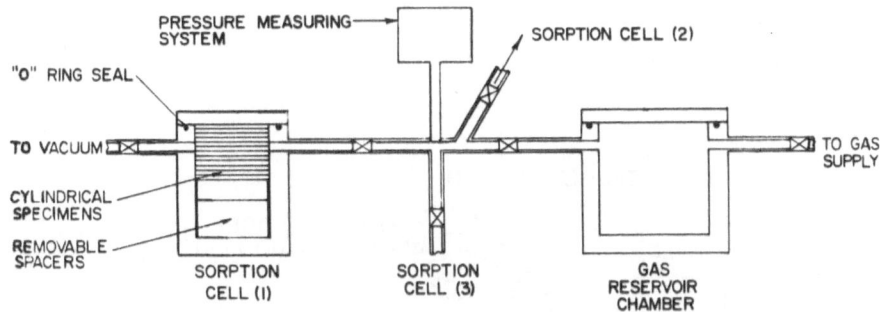

Figure 1.   Schematic of Sorption Apparatus.

EXPERIMENTAL DETAILS

Apparatus

The isotherm data were obtained by the static sorption method. A schematic of the apparatus is shown in Figure I. The basic elements of the apparatus are three sorption cells, a gas reservoir chamber, and a sensitive pressure measuring system. The pressure transducer with a 0-300 psia range (Model MPA-300) and the digital recording device (Model DM-100) were supplied by Tyco Instrument Division, Watertown, Mass. The gas reservoir chamber serves two functions: (1) it provides a source of thermally equilibrated gas at a pressure considerably closer to the working pressure than one would obtain from a gas cylinder, (2) it provides an accurate reference volume which allows repeated volume calibration of sorption cells and other line components. As shown, the pressure measuring system feeds all three sorption cells. Consequently, during a typical experimental run three isotherm data points and one pressure decay trace are generated. Prior to initiation of a test stacks of 1.5 inch diameter cylindrical specimens were loaded into the sorption cells and evacuated to a gas free state at 25°C. The steel spacers shown in Figure I were conveniently selected to give a suitable material volume to free space volume ratio.

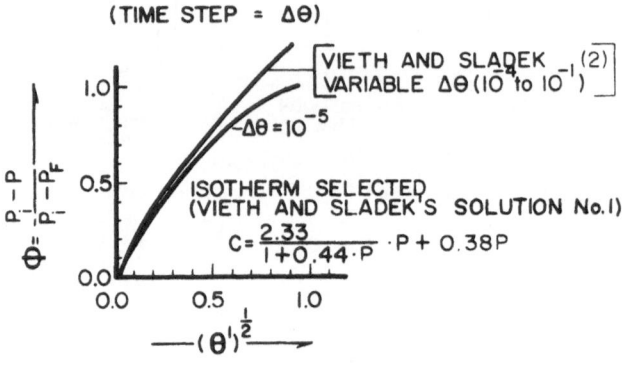

Figure 2.   Time Dependence of ϕ vs. $(\theta^1)^{1/2}$ Correlations.

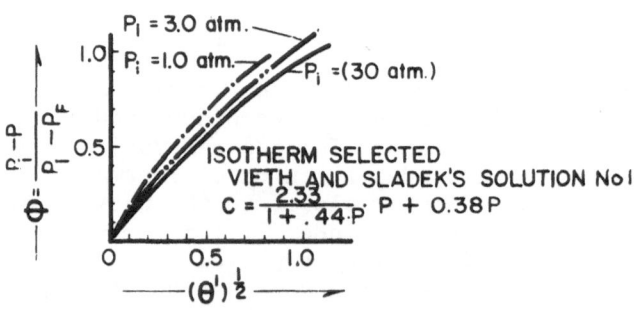

Figure 3.   Pressure Dependence of ϕ vs. $(\theta^1)^{1/2}$ Correlations.

Steady state CO_2 and O_2 permeation coefficients were measured on a modified Dow cell at 25°C in accordance with ASTM procedures. Pressure differentials which covered a range from 1 to 6 atmospheres were used. For comparison purposes, O_2 permeations were also measured on an Ox-Tran 100 at 25°C.

Materials

Sheets of Mylar® Type A, 4.7 mil thick, were used as received. Gravimetric measurements provided a density of 1.39 gram/cc. Thin strips 1.5 mil thick of the experimental HAN polymer were obtained by extrusion through a film die.

Data Analyses

The procedure used to analyze experimental data was essentially equivalent to that employed by Vieth and Sladek (2) with two notable exceptions. The size of the time step, $\Delta\theta$, was reduced to 10^{-5} and held constant throughout the iteration procedure. The influence of this on plots of ϕ vs. $(\theta^1)^{1/2}$ for Vieth's and Sladek's solution number 1 is shown in Figure 2. Secondly, and more important, experimentally determined ϕ versus $(\theta^1/D)^{1/2}$ relationships were not scaled against Vieth's and Sladek's proposed general correlation between ϕ and $(\theta^1)^{1/2}$. Rather, they were scaled against computed ϕ versus $(\theta^1)^{1/2}$ relationships where input parameters are set identical to those used in the experimental test run. As shown in Figure (3), the relationship between ϕ and $(\theta^1)^{1/2}$ is pressure dependent tending to asymptote at high pressures. Not unexpectantly, maximum dependence is observed at low pressures where "hole filling" predominates. The diffusion coefficient, D, is equal to the square of the reciprocal of the scaling factor and allowance must be made for all factors which influence the magnitude of this scaling factor. The pressure dependence of ϕ versus $(\theta^1)^{1/2}$ is such that diffusion coefficients calculated from Vieth's and Sladek's general correlation are considerably reduced. The author has found calculated diffusion coefficients to more than double when allowance is made for pressure dependence.

RESULTS

Equilibrium solubility data for Mylar® Type A are plotted together with similar data obtained by Michaels et.al. (1) in Figure 4. As shown, the isotherms generated in this study generally agree with those developed elsewhere. The agreement is closer for those materials with approximately equivalent densities. Equilibrium solubility data for Mylar® Type A and the experimental HAN polymer are plotted in Figure 5 along with

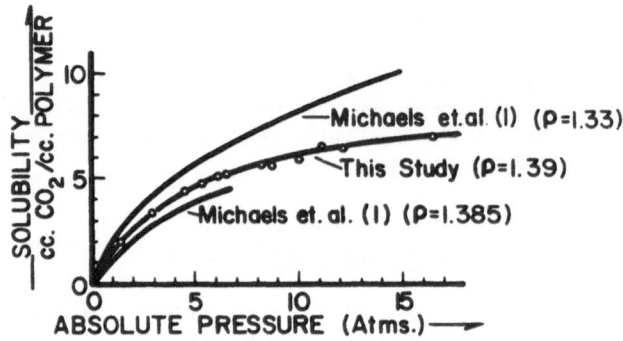

Figure 4. CO_2 Sorption in Mylar ® at 25°C.

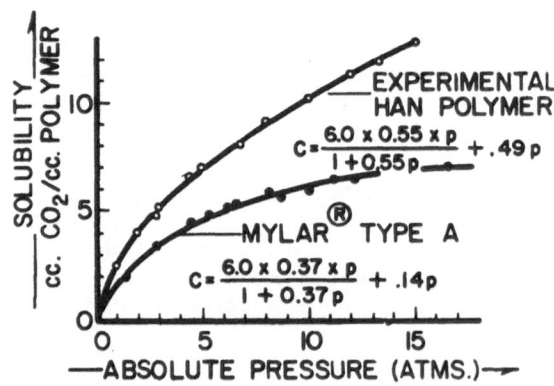

Figure 5. CO_2 Sorption Isotherms at 25°C.

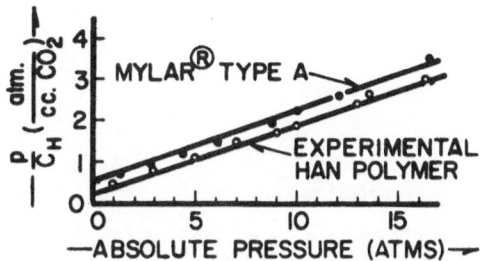

Figure 6.   Langmuir Plots for CO_2 Sorption by "Hole Filling"
            at 25°C.

isotherms which give the best fit to the experimental data.
These isotherms are:

    c = 3.3 p/(1 + 0.55 p) + 0.49 p, and

    c = 2.22 p/(1 + 0.37 p) + 0.14 p for

experimental HAN polymer and Mylar® Type A respectively. Langmuir
plots for CO_2 sorption by "hole filling" for these two materials
are shown in Figure 6. To calculate diffusion constants measured
values of p were converted into dimensionless pressure decay parameters, ϕ. These were plotted against...

$$\left(\frac{\theta^1}{D}\right)^{1/2} = \left[\frac{t}{L^2}\left[1 + \frac{C_H^1\, b}{k_D\,(1\,+\,bp)^2}\right]^{-1}\right]^{1/2}$$

using the appropriate values of C_H^1, b and k_D for the respective
materials. Scaling factors giving the best fit of these data to
calculated ϕ versus $(\theta^1)^{1/2}$ curves were used to estimate diffusion
constants. Calculated ϕ versus $(\theta^1)^{1/2}$ curves were obtained from
numerical solution of Equation 1. Numerical solution of Equation 1

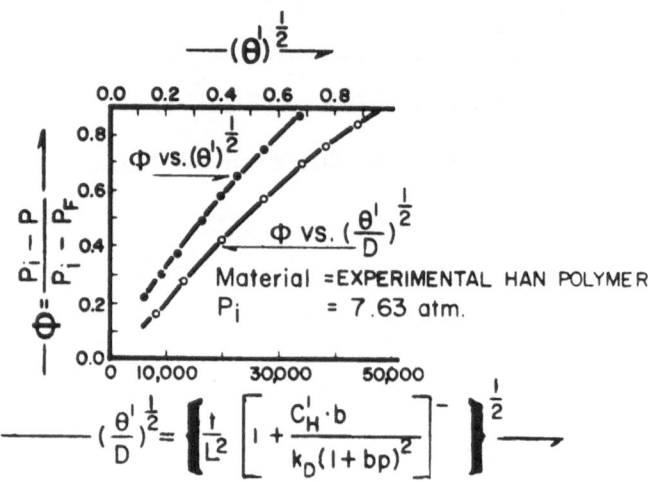

Figure 7.   Typical ϕ vs $(\theta^1)^{1/2}$ and ϕ vs $(\theta^1/D)^{1/2}$ Curves.

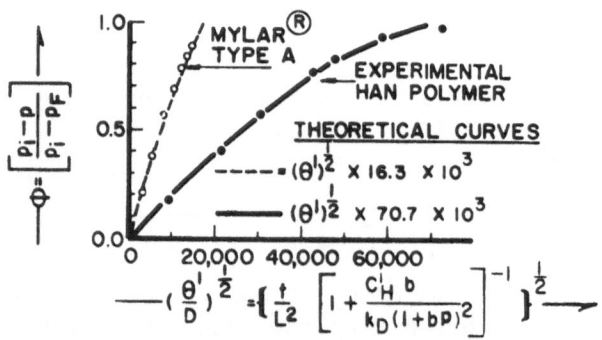

Figure 8.   Kinetics of CO_2 Sorption in Mylar® Type A and Experimental Han Polymer at 25°C.

TABLE I

Measured and Calculated Transport Parameters

for Mylar® Type A and Experimental HAN Polymer

Quantity	Method	Mylar® Type A	Experimental HAN Polymer
O_2 Permeation at 25°C (cc mil/24 hrs. atm. 100 sq. inches)	Dow Cell	4.4	1.6
	Oxtran 100	4.3	1.5
CO_2 Permeation at 25°C (cc mil/24 hrs. atm. 100 sq. inches)	Dow Cell (Upstream Press. Approx. 6 atms.)	12.7	2.6
	(Upstream Press. Approx. 3 atms.)	13.6	—
	(Upstream Press. Approx. 1.5 atms.)	11.3	—
	Calculated from $\overline{P} = k_D \cdot D$*	11.8	2.2
CO_2 Diffusivity at 25°C (cm^2/sec.)	"Dual Sorption" Analysis	$3.7_6 \times 10^{-9}$	2.0×10^{-10}
CO_2 Henry's Law Constant at 25°C (cc/cc atm.)	"Dual Sorption" Analysis	0.14	0.49

$$* \quad \frac{cc\ mil}{24\ hrs.\ atm.\ 100\ sq.\ ins.} = \frac{cm^2}{sec.} \times \frac{cc}{cc\ atm.} \times 2.19 \times 10^{10}$$

was carried out using the procedure outlined in the experimental section. Figure 7 shows typical calculated ϕ versus $(\theta^1/D)^{1/2}$ relationships for experimental HAN polymer. Figure 8 compares measured with scaled calculated plots for Mylar® Type A and HAN polymer, respectively. Comparison of calculated and measured plots yielded confidence limits on diffusion constants to be less than ± 15%. These limits apply over a wide range of tested experimental conditions. In the present study not less than three sorption pressure decay curves were analyzed for each material. Initial pressures fell within the range of 2 to 16 atmospheres

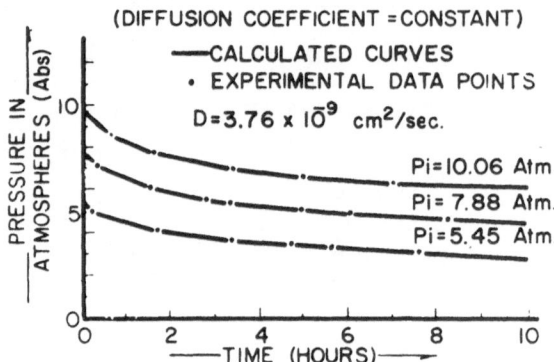

Figure 9.   Calculated vs. Measured Pressure Decay Curves for
Mylar® Type A at 25°C.

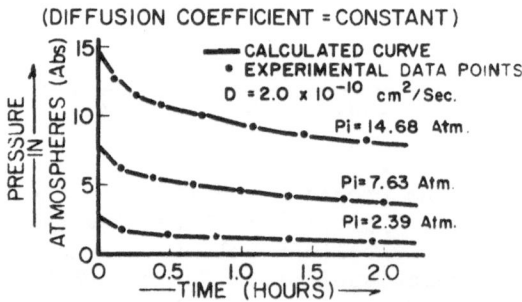

Figure 10.   Calculated vs. Measured Pressure Decay Curves for
Experimental Han Polymer at 25°C.

absolute. Calculated values of diffusivities were 3.76×10^{-9} cm^2/sec. and 2.0×10^{-10} cm^2/sec. for Mylar® Type A and experimental HAN polymer, respectively. Table 1 lists measures and calculated values of equilibrium CO_2 permeation coefficients. For comparison purposes, O_2 permeation coefficients are included.

DISCUSSION

The observed agreement between measured and calculated values of CO_2 permeation coefficients listed in Table 1 are excellent. They are within the experimental confidence limits of the various test procedures used. This evidence strongly supports "Dual Sorption Model" postulates. That is, the transport of gases through glassy polymers is controlled by the diffusion of penetrant molecules in amorphous regions with "Microvoids" acting as sinks to trap diffusing molecules and impede advancement of the diffusing front. The diffusion of molecules in amorphous regions appear to follow simple Fickian behavior, i.e., diffusivities are concentration independent and sorption isotherms in these regions are linear and obey Henry's Law. These latter assumptions are strongly supported by: (1) the concentration independence of measured equilibrium permeation coefficients for CO_2 in Mylar® Type A (see Table 1), and (2) the excellent agreement between measured and calculated pressure decay curves for Mylar® Type A and experimental HAN polymer as shown in Figures 9 and 10 respectively. It is worth noting that calculated curves are determined for a range of starting pressures with diffusion coefficients being held constant in each case. The affinity of the "microvoids" for CO_2 and for other "non-perfect" gases (3) appear to follow Langmuir type sorption behavior. This has now been demonstrated for CO_2 and CH_4 sorption in polystyrene (4,3), CO_2 sorption in a number of polyethylene terephthalates (1,5), and as shown here for CO_2 sorption in a high acrylonitrile polymer. The immobilization action of the nonlinear component molecules affect the kinetics of the transport phenomena only during the sorption process. For the case of transport across a membrane, this region is normally called the transient or non-steady state region.

It is of interest to compare CO_2 sorption isotherms for the experimental HAN polymer studied here and that established for polystyrene by Vieth et.al. (4). The closeness of the absolute magnitudes (see Figure II) of these isotherms is remarkable in light of the differences in chemical nature of the polymers involved. The observed closeness suggests to the author that sorption characteristics of amorphous polymers below their glass transition temperature are more related to the morphological nature rather than the chemical nature of the system. Employing the ideas of Meares (6), the glass state of an amorphous polymer as regards gas transport properties may be satisfactorily charac-

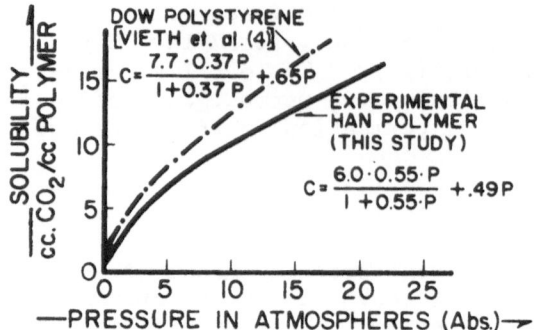

Figure 11. CO_2 Isotherms for Polystyrene and Experimental High Nitrile Polymer at 25°C.

terized through use of the "hole" theory of liquids. That is, as a polymer traverses the glassy state from the rubbery state segmental rotations are restricted so that pre-existing "holes" are immobilized and only interchain separation can adjust itself to minimize the free energy of the system. For polymers, the bulk volume occupied by "holes" should be equivalent at equivalent distances below their glass transition temperatures. The glass transition temperatures for polystyrene and the experimental HAN polymer discussed here are approximately equivalent.

Ratification and acceptance of the "Dual Sorption Model" theory is important for a number of reasons. Transport parameters such as diffusivities, Henry's Law constants, their respective activation energies and enthalpies are used as molecular probes to assist in molecular structure determinations. Historically these parameters are calculated by Time Lag Techniques (7). Analysis by Time Lag Techniques assume simple Fickian behavior. Therefore, actual diffusivities and solubilities may be respectively greater and smaller than those calculated from Time Lag Techniques if immobilizing sorption contributes significantly to the sorption process. Paul (8) has discussed this topic at length. Michaels, Vieth and Barrie (1) used this approach to explain why gas diffusion sonstants are abnormally low and activation energies are abnormally high in glassy polymers when compared to their values in typical rubbery polymers such as polyethylene. Evaluation of the various kinetic, equilibrium and their respective temperature dependent parameters which the "Dual Sorption Model" theory defines may provide some deeper insight into the molecular make-up of glassy polymers which cannot presently be characterized by other techniques. One area which may greatly benefit from a combination of this approach, fracture mechanic principles, and microscopic techniques is the phenomenon known as "environmental stress cracking" of polymers.

In a practical sense, a number of situations exist in which immobilization of the penetrant accompanies diffusion. They include binding of dyes to specific sites in fibers (9), clustering of water in some polymers (9), sorption of a penetrant onto a filler (10,11), and sorption of a penetrant onto a high energy site such as "microvoids" in glassy polymers (1,3,4,5). The realization that this phenomena occurs in so many cases in real life is important. Higuchi et.al. (11) made use of the dual sorption model concept to produce protective skin ointments with extremely long diffusion time lags. They showed that this procedure was preferred to developing ointments with reduced equilibrium permeation coefficients. Fenelon (12) showed by use of a hypothetical example that screening of thermoplastics as suitable candidates for the carbonated beverage container market must include evaluation of kinetic and equilibrium parameters which are defined by the "Dual Sorption Model". Since that time, his arguments have been supported by quoted pressure loss data for a specific container (13,14).

CONCLUSION

Evidence is presented which supports the postulates of the "Dual Sorption Model" theory. Confirmative evidence includes evaluating equilibrium permeation coefficients from transient and equilibrium sorption behavior for CO_2 transport in a Mylar® polyester film and an experimental high acrylonitrile polymer. The extension of this model to another class of glassy polymers strongly suggests its general applicability. The utility of using the concepts of this model to define molecular parameters and to characterize materials for use in commercial applications is also presented.

ACKNOWLEDGEMENT

The author gratefully acknowledges the permission of Borg-Warner Chemicals to publish this article.

REFERENCES

1. A. S. Michaels et.al., J. Appl. Phys., $\underline{34}$, 1 (1963).

2. W. R. Vieth and K. J. Sladek, J. Colloid Sci., $\underline{20}$,1014(1965)

3. W. R. Vieth et.al., Jour. Colloid and Interface Sci., $\underline{22}$, 454 (1966).

4. W. R. Vieth et.al., Jour. Colloid and Interface Sci., $\underline{22}$, 360 (1966).

5. W. R. Vieth et.al., J. Appl. Poly. Sci., Vol. 8, 2125 (1964).

6. F. Meares, Trans. Faraday Soc., $\underline{54}$, 40 (1958).

7. R. M. Barrer, "Diffusion in the Through Solids", London (1951).

8. D. R. Paul, J. Poly. Sci., A-2, $\underline{7}$, 1811 (1969).

9. J. D. Wellons and V. Stannett, J. Poly. Sci., A-1, $\underline{4}$, 593 (1966).

10. D. R. Paul and D. R. Kemp, Polymer Preprint, ACS, New York (1972).

11. N. I. Higuchi and T. Higuchi, J. Amer. Pharm. Assoc., Sci. Ed., $\underline{49}$, 598 (1960).

12. P. J. Fenelon, SPE ANTEC, Montreal (1973).

13. E. Merz, 75th Nat. Mtg. AIChE, Detroit, (1973).

14. G. Gordon, ACS Preprints, Los Angeles, April (1974).

® Registered trademark of E. I. DuPont deNemours & Co. (Inc.).

APPENDIX

D = diffusivity, cm^2/sec.

C_D = concentration of diffusing species, c.c. gas (STP)/c.c. solid.

t = time, sec.

x = distance, cm.

C_H^1 = hole saturation constant, c.c. gas (STP)/c.c. polymer.

b = hole affinity constant, atm^{-1}.

k_D = Henry's Law Constant, c.c. gas (STP)/c.c.-atm.

$\overline{P}$ = permeability coefficient c.c. gas (STP) mil/24 hrs. atm. 100 sq. ins.

L = half thickness of polymer sheet, cm.

p = pressure, atms.

ϕ = dimensionless pressure decay, $(p_i-p)/(p_i-p_f)$.

θ^1= correlating parameter.

TRANSPORT PROPERTIES OF ORIENTED POLY(VINYL CHLORIDE)

Thomas E. Brady, Saleh A. Jabarin, Gerald W. Miller

Owens-Illinois Technical Center, Toledo, Ohio 43666

INTRODUCTION

In commercial thermoplastic forming operations, poly(vinyl chloride), or PVC, is stretched to several times its original dimensions at fixed rates and temperatures. The stretching operation is usually followed by rapid quenching which freezes in the strain-orientation imparted to the polymer during forming. Since orientation is known to strongly influence physical properties, it was of interest to correlate several critical end-use properties with the range of processing conditions which PVC might experience during a typical forming operation. In addition to mechanical strength, gas and vapor transport are perhaps the most critical properties from the packager's point of view. This report describes the influence of 1) degree of biaxial orientation, 2) stretching rate, and 3) stretching temperature on the oxygen and water vapor transport parameters of non-impact modified PVC. The diffusion and solubility coefficients, as well as permeability, were determined for oxygen transport, while only equilibrium permeability was measured for water vapor. Structural changes were monitored by measuring both birefringence and density as a function of orientation. Since processing temperature was found to have a pronounced effect on properties, separate annealing studies were carried out, where variations in the transport properties were correlated with structural changes as measured by both x-ray diffraction and differential scanning calorimetry, DSC.

Other investigators have examined the effect of orientation on permeability,[1-8] but have not attempted to correlate a range of processing (stretching) conditions with permeability, P, diffusivity, D, and solubility, S. The permeabilities of both crystalline and non-crystalline polymers have been shown to decrease with increasing orientation. Among those which exhibit this behavior are polyethylene,[1-3] polypropylene,[1,2] polyvinyl chloride,[1] polycarbonate,[4] and polyethylene terephthalate.[1,2,8] Although the permeability decrease for polyethylene was shown to depend upon the method of orientation,[1] it has not been firmly established whether or not diffusivity always decreases proportionately with the degree of orientation.[1] There is some evidence, however, that the mode of orientation (uniaxial versus biaxial) affects the relationships between transport parameters and orientation,[1] and at temperatures near Tg, the permeability of polyethylene terephthalate actually increases at low elongation levels.[8] A comprehensive study of the effects of processing variables, including both temperature and stretch rate, was reported for cellulose acetate reverse osmosis membranes,[6,7] but the emphasis in that study was on percent salt rejection and flux.

EXPERIMENTAL

Preparation of Oriented Samples

American Hoechst Company PVC calendered sheet ($\bar{M}_w$ = 104,000) was cut into four-inch square samples (15 mils thick) and stretched biaxially at temperatures ranging from 82°C to 130°C and at velocities ranging from 0.02 in/sec to 2.00 in/sec. The stretching was done on a Long Extensional Tester(R) (LET). This device is a commercial film stretching machine which stretches films both uniaxially and biaxially and in either a sequential (transverse direction first, then machine direction) or a simultaneous mode. Both equal and nonequal degrees of biaxial stretching were studied, although the maximum strain attained for PVC was <300%, even at optimum stretching conditions.

With the wide range of processing conditions to be evaluated, a point-to-point determination of all the property-processing relationships was not possible. Therefore, an orthogonal type statistical design was used to select the minimum number of experimental trials which would allow at least a quadratic fit of the property-processing relationships. The data consisted of twenty-seven designed experimental trials incorporating various strains, temperatures, and stretch rates. These data were supplemented by constant strain and constant temperature runs.

Preparation of Annealed Samples

Hoechst PVC sheet, 5.6 mil thick, was biaxially restrained and placed in an oven. Temperature of the oven was varied from 80°C to 155°C, but the insertion of the large metal restraining frame caused the temperature to drop by as much as 15°C. Consequently, times as long as one-half hour were required to again achieve the desired annealing temperature. After this temperature was reached, the sample was kept in the oven for five minutes, then removed and cooled in air. These samples were designed to simulate unstretched samples with the same heat history as those samples stretched in the LET. Those samples used for the x-ray diffraction and DSC structural studies were annealed without restraint and the temperatures and times for those samples were carefully controlled.

Oxygen Transport Measurement

The permeability and diffusion coefficients were determined at 23°C and 100% RH using the Modern Controls Corporation OXTRAN 100 coulometric oxygen permeability tester. Permeability was calculated from the equilibrium oxygen transmission rate, and diffusivity was calculated by applying the method of Pasternak, Schimscheimer and Heller[9] to the nonequilibrium portion of the data. Solubility was calculated from the relationship[10] P = DxS. A reproducibility study showed approximate errors of ±4% for permeability, ±10% for diffusivity, and ±13% for solubility, but the constant strain and constant temperature data showed variations well within those limits, and the empirical model was able to account for 80% of the data variation.

Water Vapor Transport Measurement

Equilibrium water vapor permeabilities were determined at 100°F (38.5°C) and 90% RH using the Modern Controls Corporation IRD-2C infrared diffusometer. Reproducibility was determined to be ±5% of the measured value.

Density Measurement

Densities were measured at 30°C in a gradient column. Commercially available glass beads were suspended in a density gradient of aqueous calcium nitrate solution. Accuracy was within ±0.0002 gm/cc.

Birefringence

Following the method of Okajima and Koizumi[11] as described by Schael,[12] birefringence was calculated from the refractive indices measured in all three principal directions using an Abbe-3L refractometer equipped with a polarizing eyepiece. Cargille Laboratories fluid (n_D = 1.580) was used as the immersion liquid.

Wide Angle X-Ray Diffraction

A North American Phillips X-Ray Diffractometer was used to obtain the wide angle x-ray spectra. The diffractometer was magnetically interfaced with a pulsed stepping motor so that step scanned intensity data could be collected automatically. The data were collected from 6° to 57° (2θ) in increments of 0.2°, and in 5° intervals up to 80°. CuK$_\alpha$ radiation and a nickel β filter were used in combination with Sollen receiving slits.

Thermal Measurements

The thermal properties were obtained using a Perkin-Elmer DSC-2 instrument. Sample weights ranged between 3.5 and 4.5 mg. In all DSC measurements, a heating rate of 20°/min was used.

RESULTS

The "Planar Strain" Parameter

Property data for biaxially oriented systems are usually represented by plotting the property value on the abcissa versus the machine direction strain on the ordinate, thereby producing a family of curves, with each representing a different transverse direction strain.[13] Two disadvantages to this method of two-dimensional data representation are: 1) that each data point represents strain in only one direction; and 2) that uniaxial, uniaxial restrained, and pure biaxial deformations may occur by fundamentally different mechanisms, resulting in quite different property value correlations.

We used a parameter designated "Planar Strain" as a way of representing two-dimensional strain with only one number. Planar strain, ε_p, is the areal strain A_f/A_o where A_o is the original area and A_f is the final area of the sheet after deformation. ε_p can then be expressed as either the product of (MD strain) x (TD strain) or as the ratio of initial to final thicknesses, assuming that the density (volume) change during deformation is negligible. Since density changes of the order of 0.5% are usual, this is a good approximation. For example, a 1 inch x 1 inch square sample stretched biaxially 100% in both directions is characterized by a planar strain of four since 2 x 2 = λ_{MD} x λ_{TD} = 1/(1/4) = t_o/t_f = 4, where the area of the sheet is increased four times and the thickness is reduced by a factor of four. This parameter adequately represents biaxial as well as uniaxial film orientation without the difficulties mentioned above. Permeability, which is measured perpendicular to the plane of orientation, and density, which is a bulk property, both correlate well with planar strain since machine direction (MD) strain and transverse direction (TD) strain are no longer separate strain parameters.

Empirical Models

The technique used to model the data was a conventional stepwise multiple linear regression analysis in which the most significant parameters were selected to describe the transport data. Between 70% and 80% of the data variation were accounted for by these models, and while this is not really good enough for detailed prediction, the models do show how each processing variable affects the property response. The models for oxygen permeability, diffusivity, and solubility are given below:

$$P = -5.09 + 0.052(T) + (1.04/\epsilon_p) + 0.002(T) \log (R + 0.001) + 23631.4(T^2) - 18.06 \log (R + 0.001)/(T).$$

where

P = oxygen permeability, cc·cm/cm^2·cm(Hg) x 10^{12}
ϵ_p = planar strain, in^2/in^2
R = stretch rate, sec^{-1}

The standard deviation for this fit is $\sigma = 0.19$ and the most significant term is $1/\epsilon_p$, while the effects of both T and R are minor. Log (R) best describes the rate dependence, but rate is not defined at $\epsilon_p = 1$ since R = 0. The (R + 0.001) term provides for a continuous relationship at $\epsilon_p = 1$.

Because the behavior at $\epsilon_p = 1$ (no stretch) was completely temperature dependent, it was not possible to fit diffusivity and solubility with a single equation over the entire strain range. The models given below define the behavior of D and S for the oriented samples, i.e., $\epsilon_p > 1$, while the data for the unoriented samples as a function of annealing temperature are plotted directly in Figures 10-12.

$$D = 2.06 + 1.24/(\epsilon_p) + 0.00012(T^2)/(\epsilon_p{}^2)$$
$$S = -1.37 + 0.0165(T) - 0.43/(\epsilon_p) + 9129.2/(T^2)$$

where

D = oxygen diffusivity, cm^2/sec x 10^9
S = oxygen solubility, cc/cm^3·cm(Hg) x 10^3

Standard deviations are $\sigma = 0.30$ for D and $\sigma = 0.13$ for S, and in both cases $1/\epsilon_p$ is dominant. The effect of stretch rate, which was small for permeability, was indeterminate for both D and S.

For water vapor transmission, the permeability model is:

$$P = 9.633 + 61827/(T^2) - 1137/T + 0.000716(T) \; (\ell nR)$$

where

P = water vapor permeability, $gm \cdot mil/in^2 \cdot yr$

$\sigma = 0.17$

Birefringence

Birefringence was used as a quantitative measure of molecular orientation in the stretched PVC samples.

With respect to the principal directions, z, y, and x in the film sample (indicated in Figure 1), the birefringence may be defined by:[14]

$$\Delta_z = n_y - n_x \tag{1}$$

$$\Delta_y = n_x - n_z \tag{2}$$

$$\Delta_x = n_z - n_y \tag{3}$$

where Δ is the birefringence and n is the refractive index.

For biaxial orientation, two birefringence values in two planes are required to specify orientation while in the case of uniaxial orientation, only one birefringence is needed, since

$$n_y = n_x$$

Figure 2 is a plot of birefringence versus planar strain for the three orthogonal planes of samples stretched at 90°C. The solid lines represent birefringence for samples stretched in an equal biaxial mode, while the broken lines represent birefringence for nonequal biaxial samples. Figure 2 demonstrates the utility of the planar strain correlation, where anisotropic properties can be correlated with isotropic properties by referencing the anisotropic values to a common transverse strain rather than to zero strain as is done for equal-biaxial results. Common transverse strains of x1.5 and x2.0 are represented in Figure 2, but birefringence values for other common transverse strains fall on similar curves.

In the case of the equal biaxial stretching, nearly zero birefringence, Δ_x, demonstrates a truly equal biaxial process. The

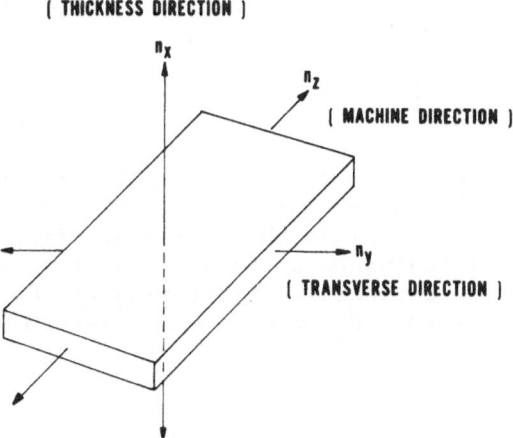

Figure 1. Relationship between principal refractive indices and film direction.

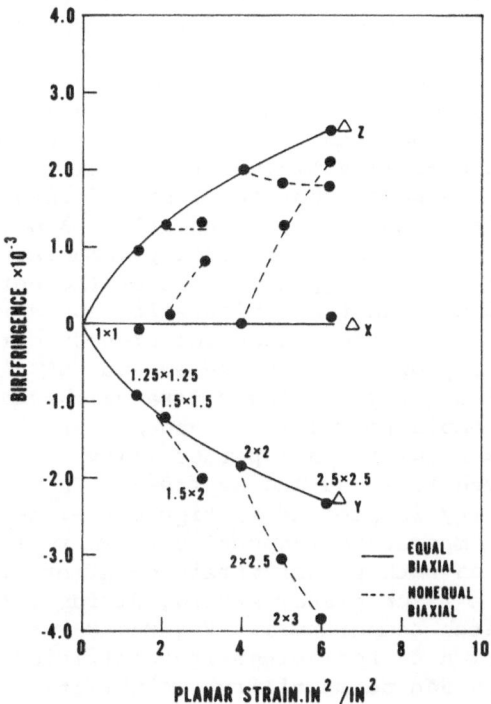

Figure 2. Birefringence versus planar strain for samples stretched at 90°C.

birefringence in the MD and TD stretching directions changes continuously with increasing planar strain.

For nonequal biaxial stretching the absolute value of birefringence in the transverse direction (Δ_z) decreases compared to the equal biaxial case, while the primary stretch direction birefringence (Δ_y) and the thickness direction birefringence (Δ_x) both increase in absolute value as indicated in Figure 2. Such a plot makes it easy to assess rapidly the different effects of equal and nonequal biaxial stretching on molecular orientation. Although similar plots were determined for stretching at 100°C and 110°C, no significant temperature dependence was observed in this temperature range.

Oxygen Transport

Figures 3-5 illustrate the effects of orientation on permeability as a function of planar strain, ε_p, stretch temperature, T, and stretch rate, R, respectively. Permeability, P, decreases rapidly for planar strains <2. At higher strains, permeability continues to decrease more gradually, leveling off to between 75% and 80% of the value for unoriented PVC. Permeability also decreases with increasing stretch temperature (for a fixed planar strain), reaching a minimum at T = 96°C when stretch rate R = 1.0 sec^{-1}. When R = 0.05 sec^{-1}, this minimum is shifted to T = 105°C. Apparently below 90°C, permeability is independent of stretch rate. The shift of the permeability minimum to higher temperatures at lower stretch rates suggests that the measured changes in permeability are not strictly caused by orientation, since strain relaxation should increase with both increasing temperature and with decreasing strain rate. We hypothesize that the permeability decrease is the combined result of orientation and structural changes which occur as a result of the annealing process during stretching. This hypothesis is supported by our x-ray and DSC results (to be presented), as well as by preliminary data which suggest that allowing the stretched sample to relax (or anneal) at the stretch temperature results in a lower final permeability than does rapid quenching even though rapid quenching freezes in more strain orientation. Diffusivity is plotted in Figures 6-7 as a function of planar strain and temperature respectively, where it is shown that diffusivity decreases with planar strain even more rapidly than does permeability. For a fixed planar strain, diffusivity is almost independent of stretch temperature. Figures 8-9 demonstrate the effects of orientation on the solubility coefficient, S. In contrast to diffusivity and permeability, solubility increases rapidly with increasing stretch up to $\varepsilon_p \approx 2$, leveling off at higher planar strain values. Figure 9 shows that even though solubility increases with stretching, there is a minimum at a stretch temperature of

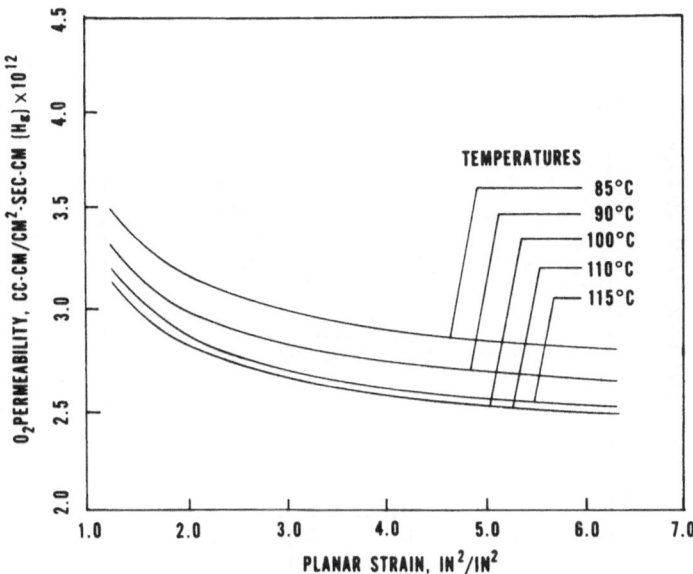

Figure 3.   Oxygen permeability versus planar strain for samples stretched at a strain rate of 0.05 sec^{-1}.

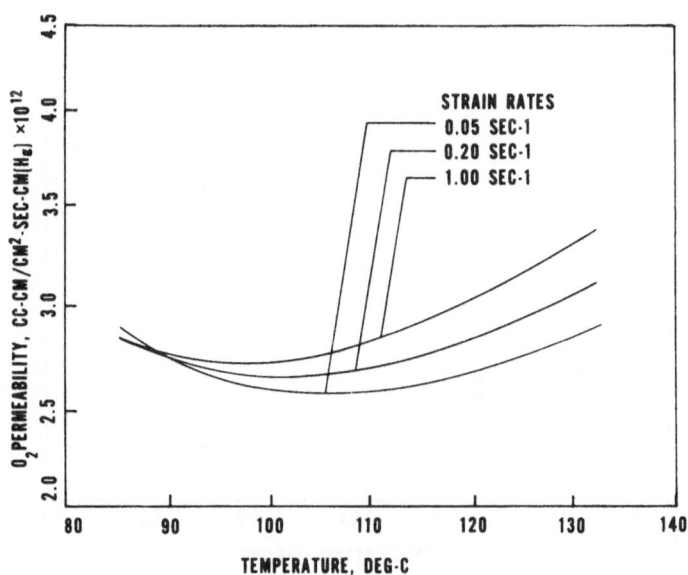

Figure 4.   Oxygen permeability versus stretch temperature for a fixed planar strain of 4 in^2/in^2.

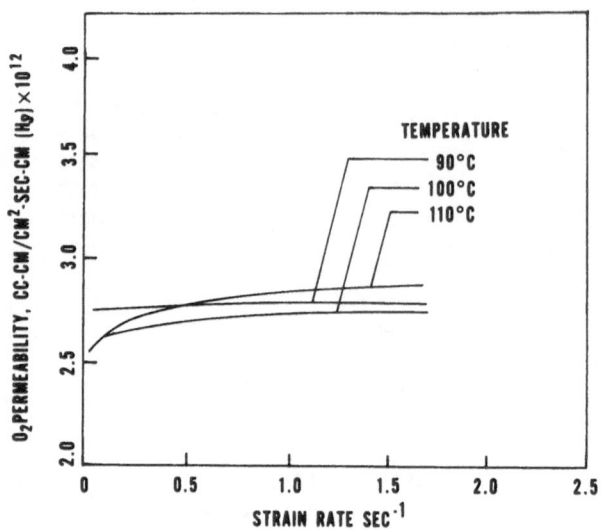

Figure 5.   Oxygen Permeability versus stretching strain rate for a fixed planar strain of 4 in²/in².

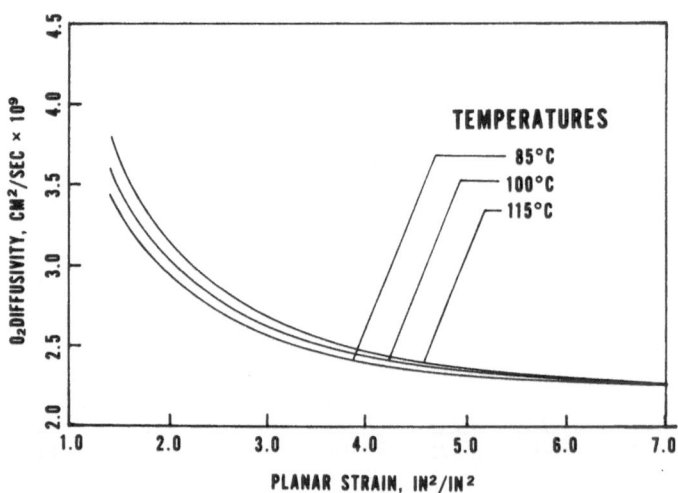

Figure 6.   Oxygen diffusivity versus planar strain.

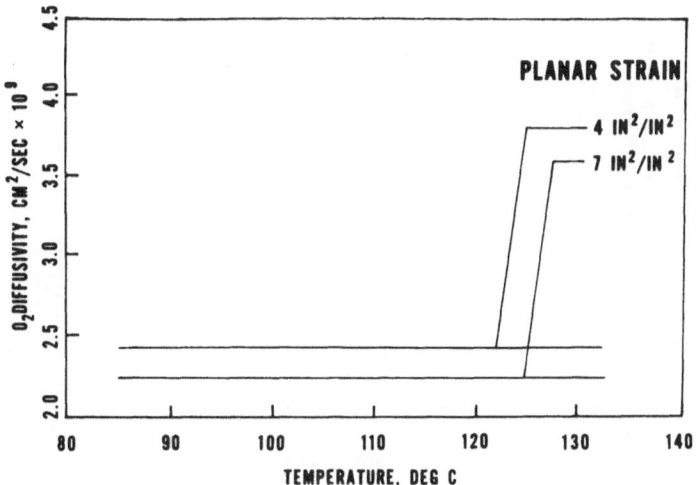

Figure 7. Oxygen diffusivity versus stretch temperature.

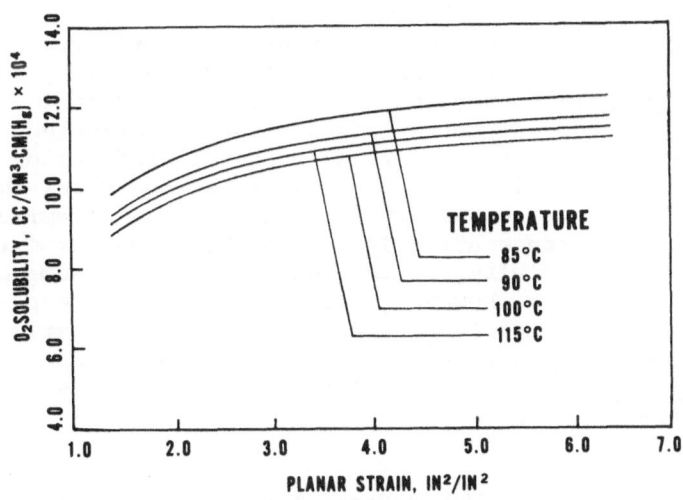

Figure 8. Oxygen solubility versus planar strain.

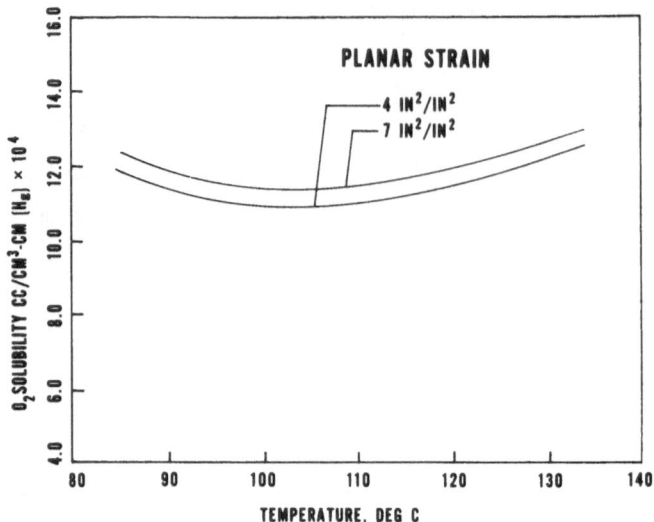

Figure 9. Oxygen solubility versus stretch temperature.

∿100°C for any fixed planar strain. Since no minimum was observed
for the diffusion coefficient, these results suggest that for oriented samples the permeability minimum at 100°C is controlled by a
similar minimum in solubility.

     The effects of annealing (in contrast to orientation) are
shown separately in Figures 10-12 where permeability, diffusivity,
and solubility are plotted as a function of annealing temperature.
Permeability again goes through a minimum somewhere in the 110°C
range, but in contrast to the results for the oriented samples,
both diffusivity and solubility also exhibit inflection points.
Diffusivity is a minimum at 100°C, which accounts for the permeability minimum, and solubility appears to reach a maximum at 110°C,
even though the data scatter is much higher than for either permeability or diffusivity, making this observation somewhat less
reliable. Certainly, however, solubility does not experience a
minimum with annealing temperature as it does with stretching temperature. These results suggest that while both annealing and
stretching cause permeability to decrease, they do so by different mechanisms.

Water Vapor Transport

     At 100°C and 90% RH, the decrease in water vapor permeability
(WVP) with planar strain is very rapid, even at strains as low as
1.2. The value for the unoriented sheet is indicated in Figure 13

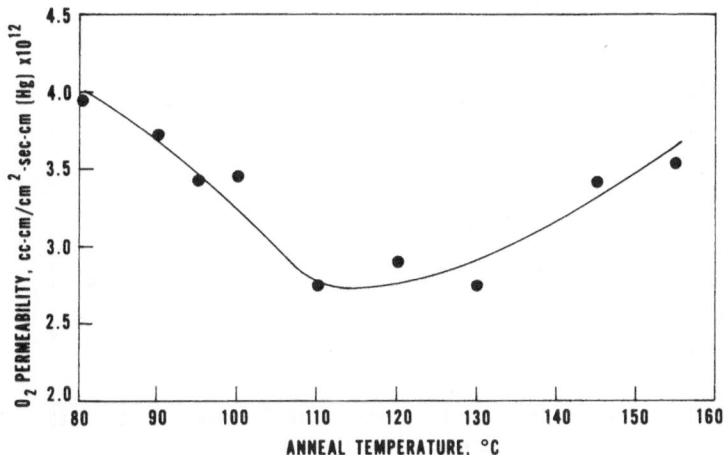

Figure 10.   Oxygen permeability versus annealing temperature for unoriented PVC.

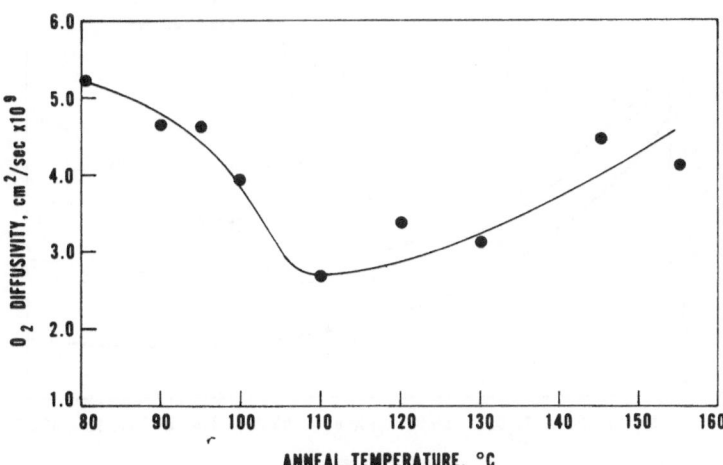

Figure 11.   Oxygen diffusivity versus annealing temperature for unoriented PVC.

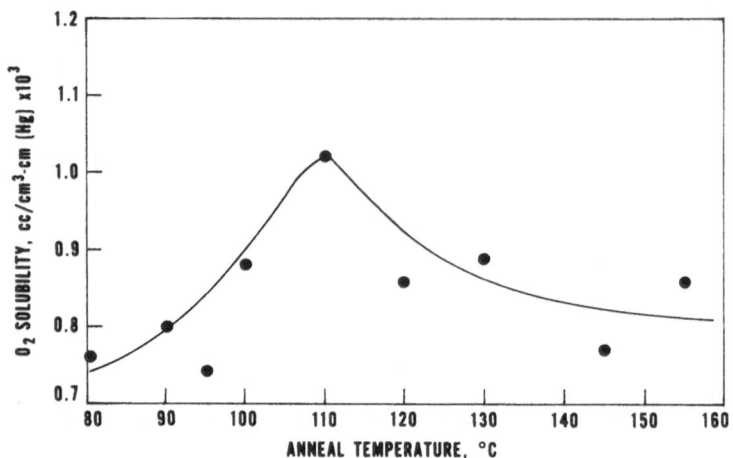

Figure 12.   Oxygen solubility versus annealing temperature for
unoriented PVC.

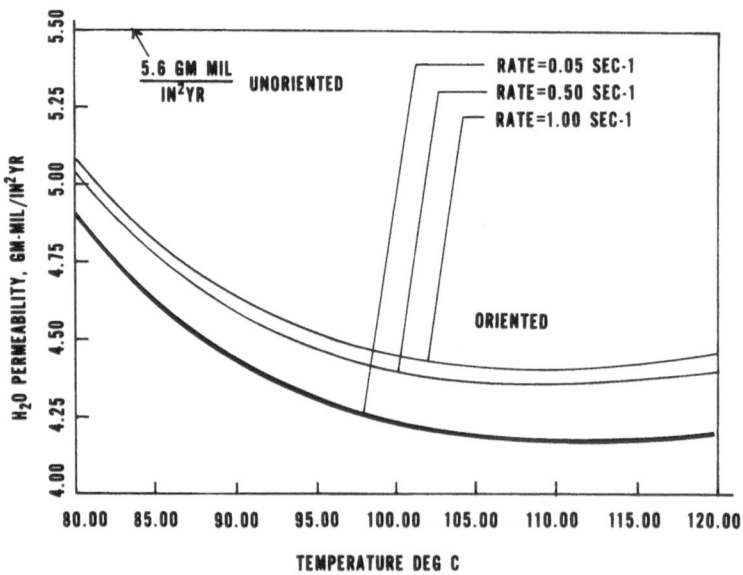

Figure 13.   Water vapor permeability versus stretch temperature.

while the plotted curves represent the temperature and rate dependence of WVP for the oriented sheet. As was found for O_2 transport, at high planar strains, WVP decreases to about 75%-80% of the value for the unoriented sheet.

Density

The density versus planar strain plot in Figure 14 demonstrates that density increases 0.2%-0.4% with orientation and that it correlates well with planar strain. For a fixed planar strain, equal and nonequal biaxial stretching result in sheets of slightly different densities. All of these densities can be correlated with degree of orientation in a single two-dimensional plot if the nonequal biaxial samples are referenced not directly to the unoriented sample but to each other through a common transverse strain. As indicated in Figure 14, a common curve exists for all samples of equivalent transverse strains. Since the density increases in nearly the same way that permeability decreases, there is a definite cross-correlation between the two.

Differential Scanning Calorimetry

Annealing an unoriented sample above the glass transition temperature, Tg, produced an endothermic peak in the DSC curve. The position as well as the area under the endotherm were dependent upon the annealing temperature, suggesting an increase in molecular "order" with annealing. As a measure of the degree of order, the endothermic peak area (enthalpy) was used since all the measurements were made under similar conditions. Figure 15 gives ΔH in cal/mole versus annealing temperature where it is apparent that a maximum occurs at 100°C, generally correlating with the maxima and minima observed in the transport parameters upon annealing in the 100°-110°C range. Our results are consistent with a similar study by Illers[15] who also reported a ΔH increase with annealing of the same order of magnitude.

The effect of annealing time on the degree of molecular order as measured by DSC is shown in Figure 16 where ΔH is plotted as a function of annealing time at 100°C. ΔH increases rapidly at short times, supporting the hypothesis that annealing times of less than ten minutes can cause significant changes in molecular order.

Wide Angle X-Ray Diffraction

The wide angle x-ray scattering data (WAXS) reported here were corrected for background, absorption polarization and Compton factors according to the procedures outlined by Lippert.[16]

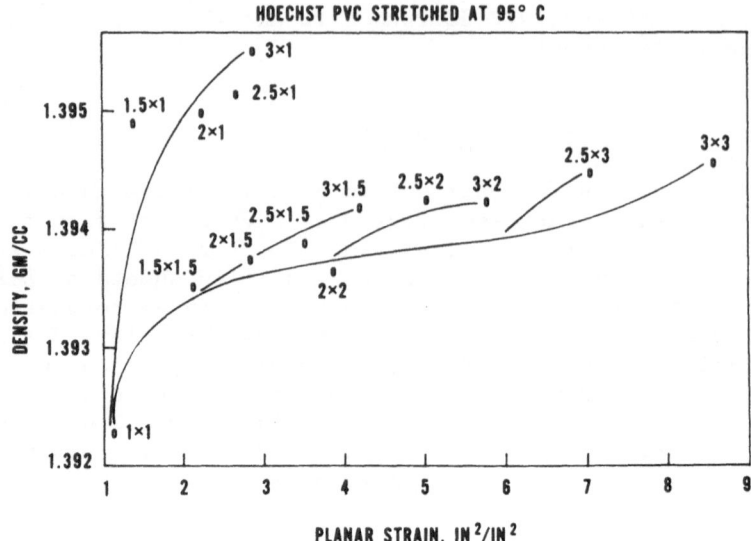

Figure 14. Density versus planar strain for samples stretched at 95°C.

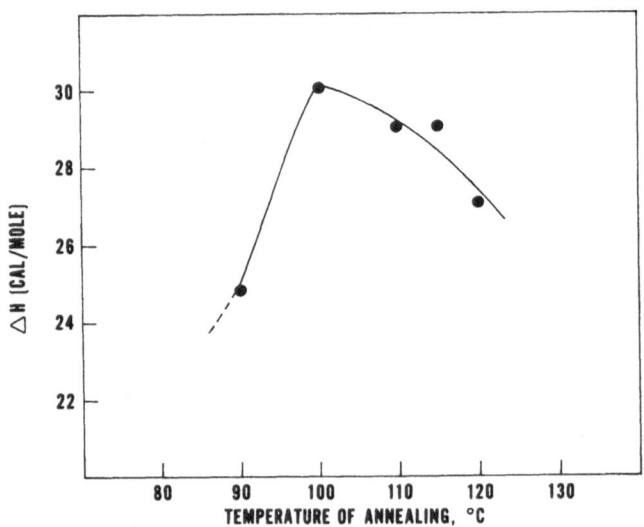

Figure 15. Heat of fusion versus annealing temperature.

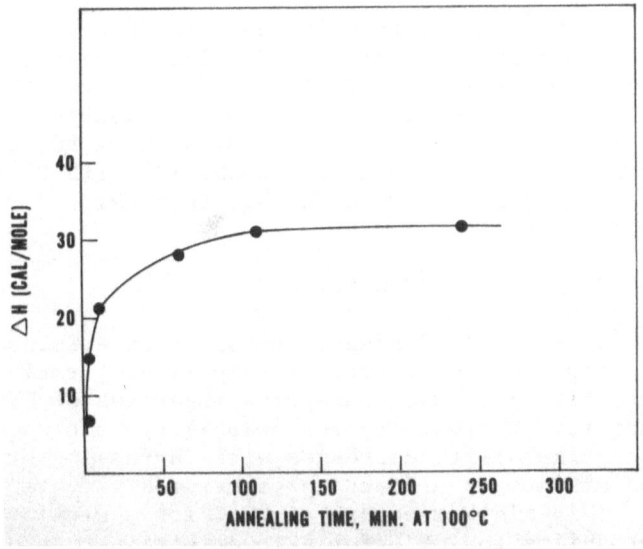

Figure 16.   Heat of fusion versus annealing time at 100°C.

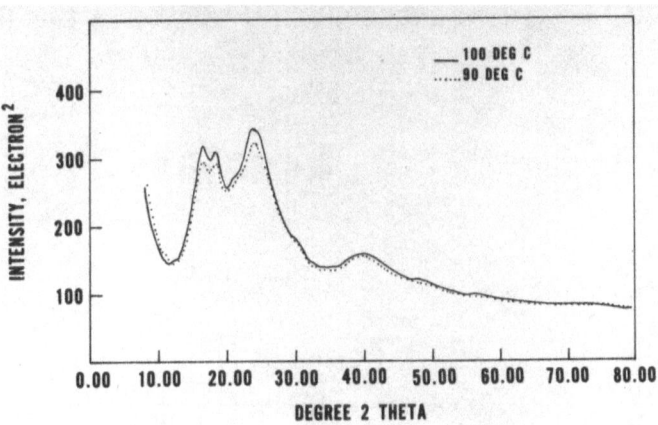

Figure 17.   Influence of annealing on the wide angle x-ray scattering of PVC.

Changes in the x-ray diffraction patterns for samples annealed at 90° and 100°C are shown in Figure 17. There are three major diffraction peaks occurring at 17.0°, 18.8° and 24.4° (2θ). The intensities of these three peaks depend upon annealing temperature. Figure 18 gives the peak height at 2θ = 24.4° as a function of annealing temperature, where a maximum in the 100°–115°C range is evident. This is consistent with the DSC results.

CONCLUSIONS

Using a statistically designed number of experimental trials, we empirically modeled the effects of temperature, rate and degree of biaxial orientation on the transport properties of PVC including oxygen permeability, diffusivity and solubility, and water vapor permeability. Permeability decreases with increasing orientation, but exhibits a minimum as a function of stretch temperature and stretch rate. This minimum occurs at 90°C for a strain rate of 1.0 sec^{-1}, but shifts to 105°C for a lower strain rate of 0.05 sec^{-1}. Since the direction of this shift is opposite to what one would predict from orientation relaxation effects, we suggest that microstructural changes caused by annealing play a strong role in determining the final permeability level.

Although diffusivity decreases with increased orientation, little change in diffusivity is observed with stretch temperature.

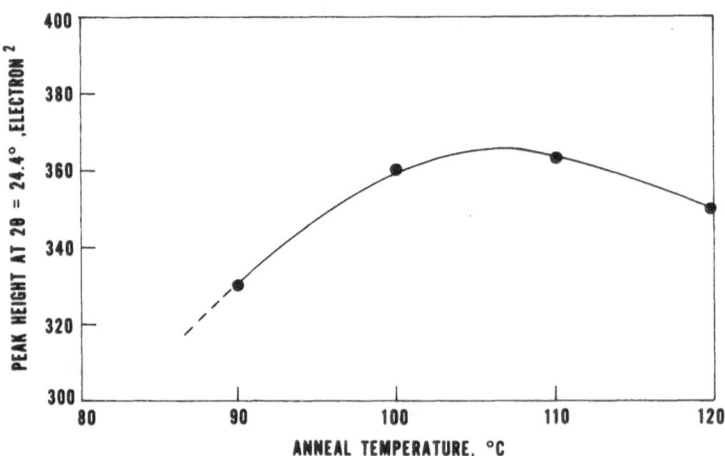

Figure 18. Peak height at 2θ = 24.4° versus annealing temperature.

Solubility, on the other hand, increases with increasing orientation but exhibits a minimum (for a fixed degree of orientation) at stretch temperatures of ∿100°C, correlating with the observed minimum in permeability in that temperature range.

Our hypothesis that annealing during the stretching operation can significantly affect the transport behavior is consistent with the results of the annealing studies, where both x-ray diffraction and thermal techniques were used to measure an increase in molecular "order" with annealing temperature and time. Both techniques confirmed a maximum change in the 100°-110°C range consistent with the results of the transport studies where permeability, diffusivity and solubility all exhibited inflection points in that temperature range.

Density and birefringence both increase with increasing orientation, suggesting that for a fixed stretch temperature and stretch rate, there is a direct correlation between these parameters and permeability.

The concept of planar, or areal, strain was introduced, providing an easy technique for correlating both equal and nonequal biaxial stretching in the same two-dimensional graph since both MD and TD strains are included in a single planar strain value.

ACKNOWLEDGMENTS

The experimental aid of F. R. Marsh for the permeability measurements, N. J. Curto for preparation of stretched films, Dr. E. L. Lippert for the x-ray measurements, and Dr. G. N. Lauer for the statistical data analysis are gratefully acknowledged. The authors also thank Owens-Illinois, Inc., for permission to publish this work.

REFERENCES

1. Y. Ito, Kobunshi Kagaku, 18, 6 (1961).

2. S. W. Lasoski and W. H. Cobbs, J. Polymer Sci., 36, 21 (1959).

3. A. S. Michaels and R. W. Hausselein, J. Polymer Sci., Pt-C, 10, 6 (1965).

4. Y. Ito, Kobunshi Kagaku, 19, 412 (1962).

5. W. W. Brand, J. Polymer Sci., 41, 415 (1959).

6. E. A. Meinecke and D. V. Mehta, ACS Polymer Preprints, 12 (2), 104 (Sept. 1971).

7. Ibid., 12 (2), 111 (Sept. 1971).

8. W. R. Vieth, E. S. Matulevicius and S. R. Mitchell, Kolloid-Z, 220 (1), 49 (1967).

9. R. A. Pasternak, J. F. Schimscheimer and J. Heller, J. Polymer Sci., Pt-A2, 8, 467 (1972).

10. J. Crank and G. S. Park, ed., Diffusion in Polymers, Academic Press (New York, 1968).

11. S. Okajima and Y. Koizumi, Kogyo Kagaku Zusshi, 42, 810 (1939).

12. G. W. Schael, J. Appl. Polymer Sci., 8, 2717 (1964).

13. K. Matsumoto, S. Utsuromiya and R. I. Mamura, Sen-I-Gakkaishi, 26, 303 (1970).

14. R. S. Stein, Newer Methods of Polymer Characterization, Chapter 4, B. Ke, ed., Wiley Interscience (New York, 1964).

15. K. H. Illers, Makromol. Chem., 127, 1 (1969).

16. E. L. Lippert, Jr., Paper presented at the American Physical Society Meeting, San Diego, California, March, 1972.

BARRIER PROPERTIES OF EARTH LINING FOR POLLUTION CONTROL

Kenneth J. Brzozowski and Charles A. Kumins

Tremco Incorporated,

10701 Shaker Blvd., Cleveland, Ohio   44104

Introduction

Environmental considerations prompted by governmental regulations require industries and new housing developments to treat waste liquid effluents to either de-toxify or sequester harmful matter prior to release into the surroundings. Treatment generally takes place while the liquids are in storage. Because of the large volume of liquids involved, tanks and other types of preformed containers are extremely expensive. A less costly solution to this problem has been taken by some sanitary engineers who have simply bulldozed large holes in the ground and then lined the walls and bottoms with what was expected to be impervious coatings or films. This approach, however, leads to further considerations because of the organic matter in the ground. If the H_2S and methane constantly being generated cannot be dissipated through the lining, the resulting pressure causes the films or coatings to balloon and eventually burst.

This paper describes the transmission characteristics of various water proofing compounds towards hydrogen sulfide, water vapor, and hydrocarbons. Their solubilities and diffusion coefficients are given as well as their concentration dependence when applicable. The compositions studied are a tar based material modified with an epoxidized urethane and crosslinked with a polyamide amine, a moisture-cured urethane modified tar, and a pigmented, urethane based water proofing cured by moisture.

Materials and Procedure

All films were cast and allowed to cure under ambient conditions. Prior to testing, samples were placed in a vacuum

oven at 75°C. for 72 hours, then mounted in a nichrome wire
sample holder, and finally set into the test apparatus. The
apparatus employed for absorption and desorption studies
consisted of a Cahn Electrobalance, Model RG suitably adapted
to allow evacuation by means of a Welch Duo-Seal vacuum pump.
The entire system was constructed from glass, and pressure
measurement was done with an open-end mercury manometer. After
the sample was positioned in the testing apparatus, the entire
system was pumped overnight or until the sample weight had
stabilized. Test gases were introduced by means of hand-made
glass bulbs. The water used in the experiments was degassed
by several freeze-pump-thaw cycles. The Cahn balance was
connected to an Omni Scribe, Series 5000 recorder from Houston
Instruments. This allowed for continuous monitoring of sample
weight gain and loss. All studies were performed at 77°F. Film
thicknesses were from 0.40 to 0.70 mm.

Results

The commonly used forms of Fick's first and second laws
of diffusion are[1].

$$P = -D \frac{dc}{dx} \qquad (1)$$

$$\frac{dc}{dt} = D \frac{d^2c}{dx^2} \qquad (2)$$

where P is the rate of transfer per unit area of section, D is
the diffusion coefficient, c is the concentration of the
diffusant, x is the space co-ordinate measured normal to the
section, and t is the time. The boundary conditions for equation
2 are $c=c_1$ at x=o and x=L (L=film thickness) for all t; c=o at
any point within the film at t=o.
Using equations 1 and 2, Barrer[2] has developed an equation
for the quantity of material transferred (Q)

$$Q = L \left[\frac{c_1 + c_2}{2} + c_0 \right] \left[1 - \frac{8}{\pi^2} \sum_{m=o}^{m=\infty} \frac{1}{2m+1} \exp\left[\frac{-D (2m+1)^2}{L^2} \pi^2 t \right] \right] \qquad (3)$$

where
   c_1 = Vapor concentration in film at x=0
   c_2 = Vapor concentration in film at x=L
   c_0 = Vapor concentration in film at t=0
   m = 0, 1, 2, 3......
In our experiments $c_1 = c_2$ (since the gas dissolves in the
film from both sides) and $c_0 = 0$. Further, when equilibrium is
reached, the quantity of material absorbed (Qe) is equal to Lc_1.
Assuming these conditions and ignoring all terms but the first

(m=0) in equation 3 (possible because the series converges rapidly), Long has simplified this equation to

$$\frac{d \log (Qe - Q)}{dt} = \frac{-\pi^2 D}{2.3L^2} \qquad (4)$$

where Qe equals the equilibrium amount of vapor sorbed and Q equals the amount of vapor sorbed at any time, t. Equation 4 was derived for the situation where D is constant.

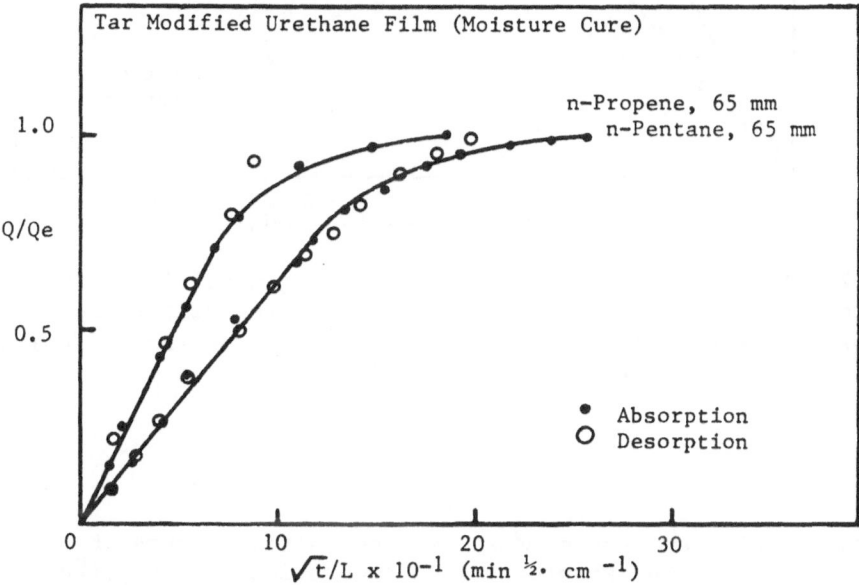

Fig. 1.- Typical curves showing fractional amounts of n-propane and n-pentane absorbed or desorbed, plotted against square root of time over film thickness. Initial straight lines indicate Fickian diffusion. Parallel lines for absorption and desorption indicate a concentration independent diffusion coefficient.

Long [3] has also examined the case in which D is concentration dependent. Based on the Boltzmann solution [4] of Fick's law, Long gives the following equation for the initial stages of Fickian diffusion:

$$\frac{Q}{Qe} = K_s \frac{\sqrt{t}}{L} \qquad (5)$$

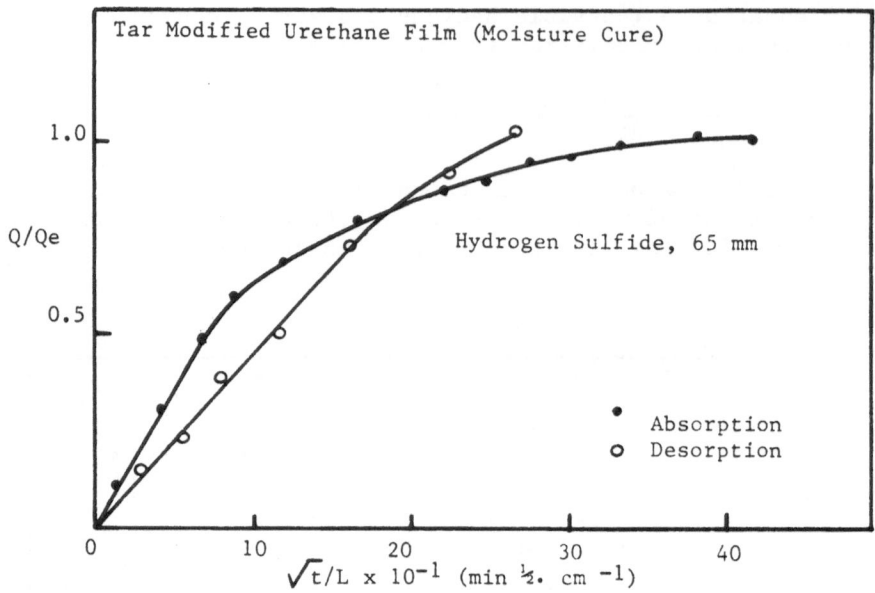

Fig. 2. –Absorption and desorption of hydrogen sulfide. Initial straight lines indicate Fickian diffusion. The differences between the absorption and desorption curves mean the diffusion coefficient is concentration dependent.

If Q/Qe is plotted versus $\sqrt{t}/L$ and the initial slopes, K_s and K_d, obtained for the absorption and desorption process, the integral diffusion coefficient, $\bar{D}$, may be calculated from the following equation:

$$\bar{D}\ (c) = \frac{1}{c_t} \int_{c_0}^{c_f} D\ (c)\ dc = \frac{\pi}{32} \left[K_s^2 + K_d^2 \right] \qquad (6)$$

In equation 6, c_t = concentration of vapor within the film at time t, c_0 = concentration at t=0, and c_f is the equilibrium or final concentration.

When a plot of Q/Qe vs. $\sqrt{t}/L$ does not give an initial straight line, this is termed anomalous or non-Fickian diffusion.

It has been shown by Crank[5] that solution of Fick's differential equations with the proper boundary conditions gives:

$$P = DS \qquad (7)$$

where P and S are the permeability and solubility coefficients.

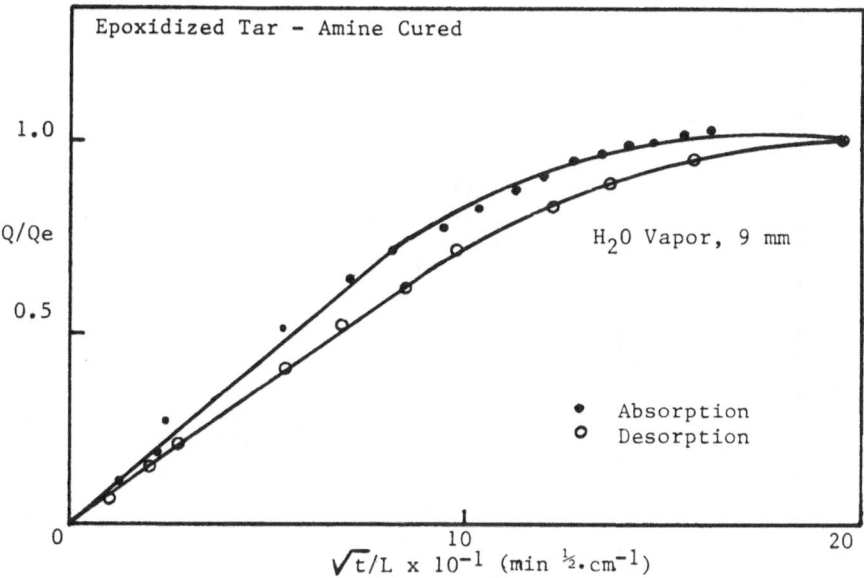

Fig. 3. —Absorption and desorption curves showing Fickian diffusion and concentration dependence of the diffusion coefficient for water vapor.

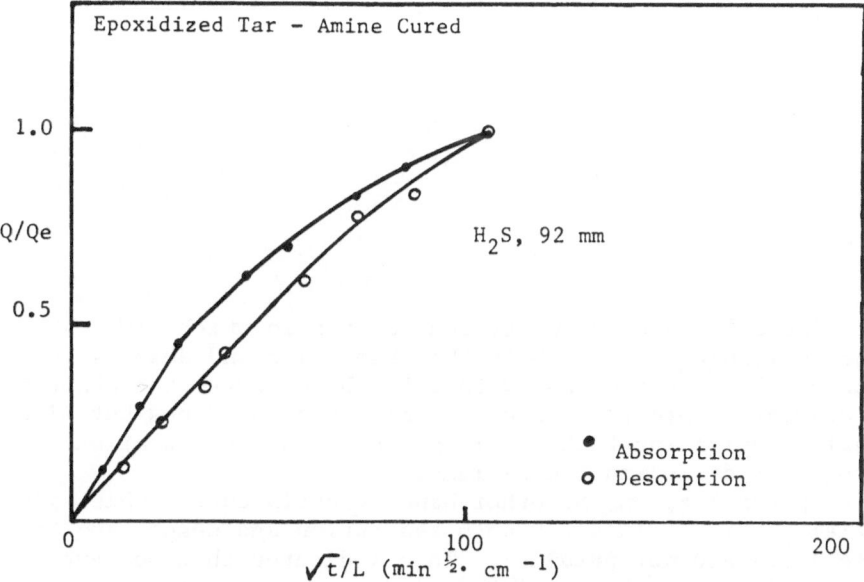

Fig. 4. —Absorption and desorption curves showing Fickian diffusion and concentration dependence of the diffusion coefficient for H_2S.

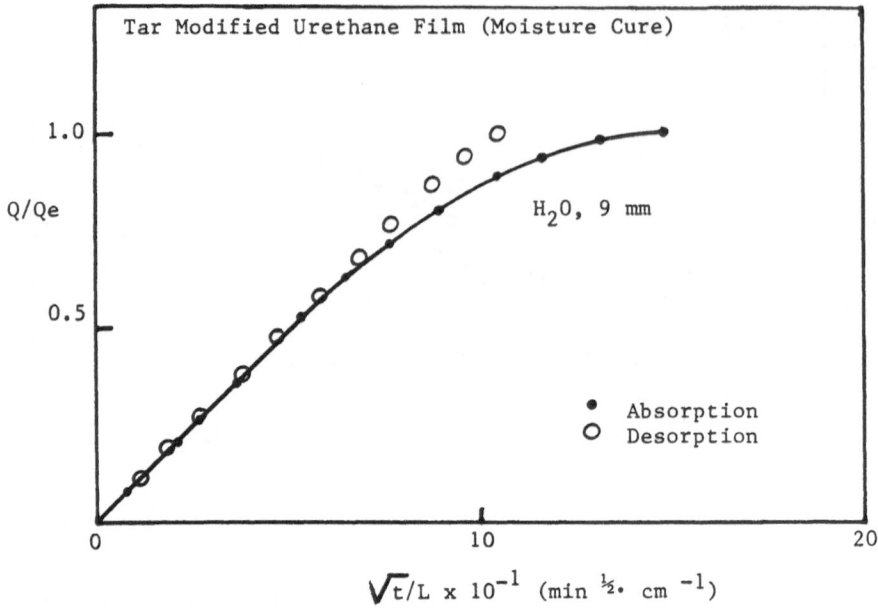

Fig. 5. –Absorption and desorption of water vapor.   Slight
concentration dependence is observed.

Figure 1 contains two typical curves in which Q/Qe is
plotted against $\sqrt{t}$/L for both the absorption and desorption
process. It should be noted that for both curves the absorption
and desorption processes are not only straight lines but also
parallel, indicating Fickian diffusion and a concentration
independent diffusion coefficient.

Figures 2–4, on the other hand, contain curves which give
initial straight lines for both absorption and desorption but
these lines are not parallel. This indicates that we have
Fickian diffusion, but that the diffusion coefficient is
concentration dependent.

Table I lists the values of the solubility coefficients,
diffusion coefficients, and permeabilities obtained.

Table I

Film	Gas	Press. (Torr.)	Diffusion Coeff. $(cm^2/sec \times 10^7)$	Permeability $\dfrac{cm^3 \cdot cm}{cm^2 \cdot sec \cdot cm\ Hg} \times 10^{10}$
Moisture Cure Tar Urethane	H_2O	9	2.6	5.3
Moisture Cure Tar Urethane	H_2S	65	0.85	0.85
Moisture Cure Tar Urethane	CH_4	745	No Absorption	No Absorption
Moisture Cure Tar Urethane	$n-C_3H_8$	65	3.0	2.3
Moisture Cure Tar Urethane	$n-C_5H_{12}$	65	1.1	7.0
Epoxidized Tar	H_2O	9	2.2	8.5
Epoxidized Tar	H_2S	92	5.3	10.7
Epoxidized Tar	CH_4	745	No Absorption	No Absorption
Pigmented Urethane	H_2O	9	5.5	11.0
Pigmented Urethane	H_2S	92	18.0	18.0
Pigmented Urethane	CH_4	745	No Absorption	No Absorption

Discussion

An examination of Table I indicates that the permeability
values obtained in this study are very low. They may, for example,
be compared to the permeability of poly (vinylidene chloride)
to H_2O (0.5 x 10^{-10} $cm^3 \cdot cm/cm^2 sec \cdot cm$ Hg)[6] and H_2S (0.03 x 10^{-10}
$cm^3 \cdot cm/cm^2 \cdot sec.$ cm Hg)[7] or the permeability of polypropylene to
H_2O (51 x 10^{-10} $cm^3 \cdot cm/cm^2 \cdot sec.$ cm Hg)[8]. It is apparent, from
this comparison, that the materials listed in Table I would be
excellent for water proofing applications -especially when one
considers that they are typically applied in a thickness of 60
mils in actual usage. It is also apparent that the other
permeability values listed in the Table are low. This fact would
have to be taken into account if any of the films were used as
pond liners. Specifically, a suitable venting system would have
to be incorporated to prevent gas build up between the film and
the earth surrounding the sewage pond. This precaution would
eliminate bursting of the liner and pollution of the area
surrounding the pond.

Further examination of Table I shows that the permeabilities
of the films generally increase as we go from the moisture cure
tar urethane to the epoxidized tar to the pigmented urethane.
This phenomenon occurs for at least two reasons, crosslink density
and pigment flocculation. It is likely that in the moisture
cured tar urethane and the epoxidized tar, some reactions take
place between the nitrogen and sulfur moieties of the tar and the
film-forming polymer. This, of course, would result in a tighter
film and reduce both the solubility factor and the diffusion
coefficient. With the pigmented urethane, on the other hand,
the high filler content would lead to some flocculation in the
final film, and this would produce paths for the diffusing
substance to travel through.

With the exception of Figure 1, all the absorption-desorption
curves presented in this paper show some degree of concentration
dependence. In the case of the tar modified urethane film, it
can be seen that initial desorption of both the H_2O and H_2S
indicates that there is some interaction with the polymer or
tar portions of the film. As stated earlier, this is not
unusual considering the polar nature of both these film components.
Desorption from the expodized tar, however, takes place more
rapidly than absorption. This could be due to the fact that
the amine curative is added as a separate component and reacts
or interacts with the polar moieties of the tar and prevents
absorption onto these sites.

The lack of absorption of methane by any of the films even
at pressures of approximately one atmosphere is thought to be
due to the fact that we are substantially below the critical
pressure of methane (46.5 atmospheres at - 82.3°C.)

References

1. J. Crank and G. S. Park, "Diffusion in Polymers," Academic Press, 1968, p. 1-2.

2. R. M. Barrer, "Diffusion In and Through Solids," Cambridge Unit Press, 1941, p. 17, equations 60-63.

3. F. A. Long and L. J. Thompson, J. Polymer Sci., 15, 413 (1955).

4. J. Crank, "Mathematics of Diffusion," Oxford Univ. Press, London, Secs. 4.32, 11.66, 1956.

5. Ibid., chapter 4.

6. A. W. Meyers, J. A. Meyer, C. E. Rogers, V. Stannett, and M. Szwarc, TAPPI, 44, 58 (1961).

7. W. Heilman, V. Tammela, J. A. Meyer, V. Stannett, and M. Szwarc, Ind. Eng. Chem., 48, 821 (1956).

8. A. W. Meyers, J. A. Meyer, C. E. Rogers, V. Stannett, and M. Szwarc, TAPPI, 44, 58 (1961).

Synopsis

A study was conducted to relate the potential performance of sewage-pond lining materials to their permeability characteristics. Permeabilities and diffusion coefficients were measured for several film/diffusing-substance combinations. Measurements were made on a Cahn Electrobalance (sensitivity 10^{-8} grams) using a weight gain/weight loss technique. The permeability and diffusion coefficient values obtained were quite low in almost all cases. This indicated that the films studied were excellent water barriers, but that some type of venting system would be required to prevent bursting of the liner because of gas build up between it and the earth. Some explanations are given for the low permeability values measured and for the concentration dependent diffusion coefficients.

SPONTANEOUS POLYMERIZATION OF 4-VINYL PYRIDINE

FOR HOLLOW FIBER ION EXCHANGERS[*]

A. Rembaum, S.P.S. Yen and C. Robillard

Jet Propulsion Lab., Calif. Institute of Technology

Pasadena, California 91103

Introduction

It was previously shown[1] that 4-vinyl pyridine (4-VP), polymerized spontaneously in presence of alkyl halides. The growing active species in this polymerization system exhibited high specificity. It was reacting only with monomeric vinyl pyridinium salts and not with other vinyl monomers including 4-VP. Therefore, in presence of excess of 4-VP, the polymerization was arrested after the alkyl halide was used up in the quaternization reaction. It was also observed that free radical inhibitors had no effect on the polymerization rate.

In view of these facts, Kabanov[1] postulated a "zwitterion" mechanism shown in Fig. 1. The failure of addition of I to II (Fig. 1) was explained by the relatively higher charge density and therefore more negative electrical character of the double bond in 4-VP than in the quaternary salt.

However, it was later found that the quaternization of 4-VP with a weak nucleophile (e.g., p-toluene sulfonate) or with a dilute acid, a different polymerization mechanism is operative leading to a different structure of the polymer.[2]

---

[*]This paper represents one phase of research carried out at the Jet Propulsion Laboratory, California Institute of Technology, under Contract No. NAS7-100, sponsored by the National Aeronautics and Space Administration.

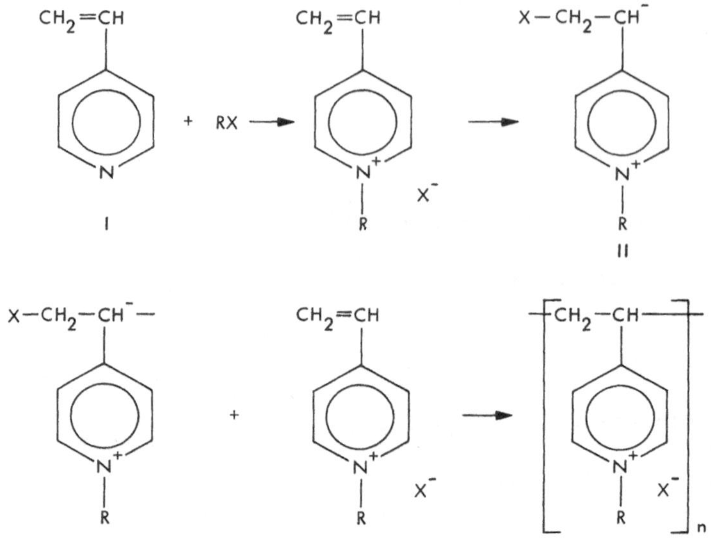

Figure 1.   Mechanism of Spontaneous Polymerization
of 4-Vinyl Pyridine

The alternative mechanism which does involve the addition of a neutral molecule of 4-VP to the quaternized salt is shown in Fig. 2. This mechanism postulates a hydrogen transfer to the negatively charged carbon atom in III and formation of ionene polymers (IV) where the positively charged nitrogen is located in the backbone of the chain.   Thus, the understanding of 4-VP polymerization can be summarized by means of Fig. 3.

In our investigations, 4-VP was reacted under a variety of conditions with α,ω dihalides yielding highly crosslinked polymers. Most of our findings can be accounted for on the basis of Kabanov's mechanism.

The use of dihalides involves a second quaternization reaction of the same molecule and brings up the problem of the effect of the first positive charge on the second quaternization reaction.

It was previously shown[4] that a series of dicationic crosslinking agents (Fig. 4) could be prepared and isolated. All the dicatonic crosslinking agents yielded crosslinked resins on addition of redox initiators.

Because of the spontaneous polymerization of quaternized 4-VP the isolation of a dicationic crosslinking agent was not successful with any α,ω dihalides investigated including dibromethane. The latter, when reacted with two molecules of dimethylamino ethylmethacrylate

(DEMA), yielded pure V indicating that the second quaternization was prevented by the proximity of the positive charge.

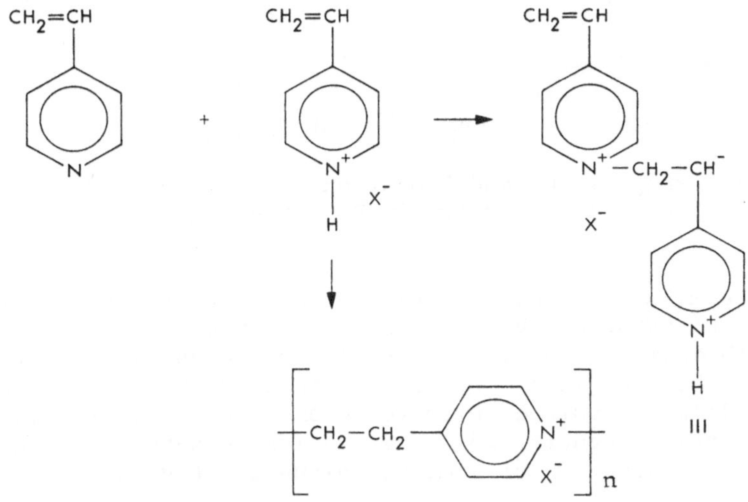

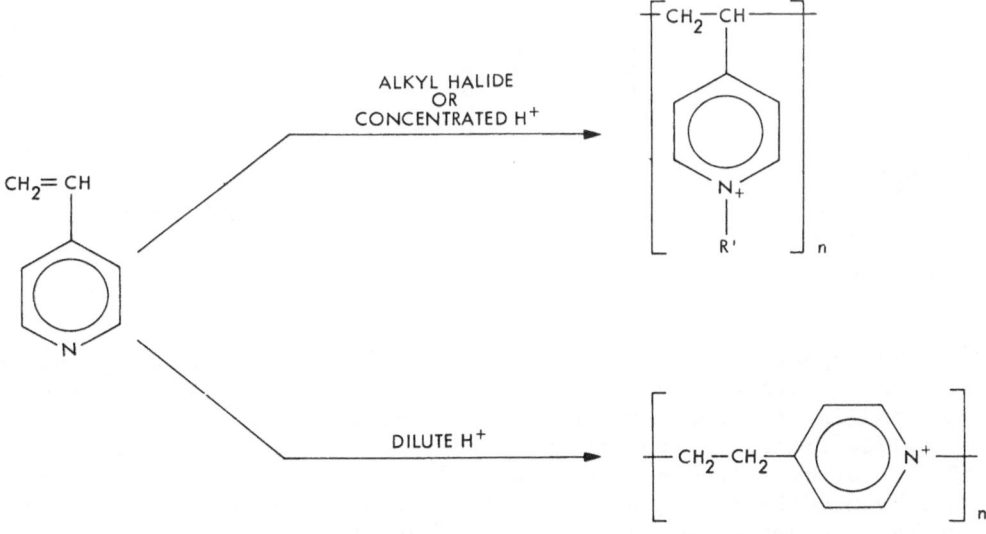

Figure 2.   Mechanism of Polymerization of 4-Vinyl Pyridine
with Weak Nucleophiles or Dilute Acids

Figure 3.   Postulated Structures of Quaternized Poly(4-Vinyl
Pyridine)   R' = alkyl group or H

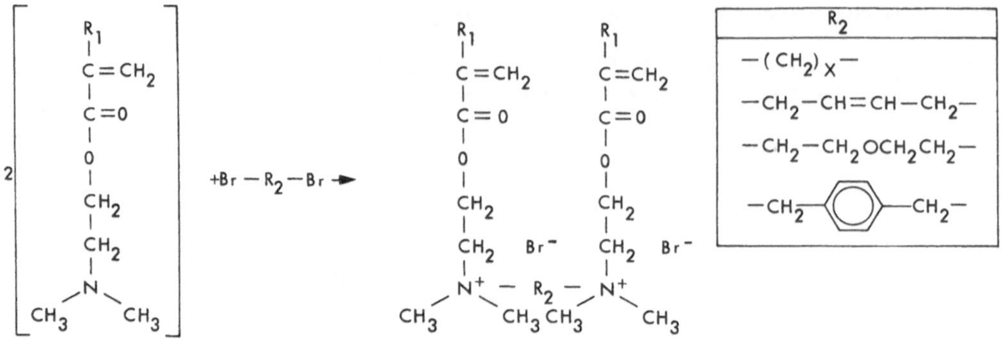

Figure 4.   Preparation and Isolation of Dicationic Crosslinking
Agents   X = 2 to 16   R$_1$ = H or CH$_3$

For similar reasons, we were unsuccessful to synthesize ionene polymers[5] of structure VI, and the reaction of tetramethylamino methane with dibromethane yielded small molecules instead of VI. Furthermore, the crosslinked resins obtained by the reaction of 4-VP with dihalides contained unreacted bromine end groups. In order to determine the reason for this observation we have investigated the rate of quaternization of a model compound (Structure VII) with pyridine.

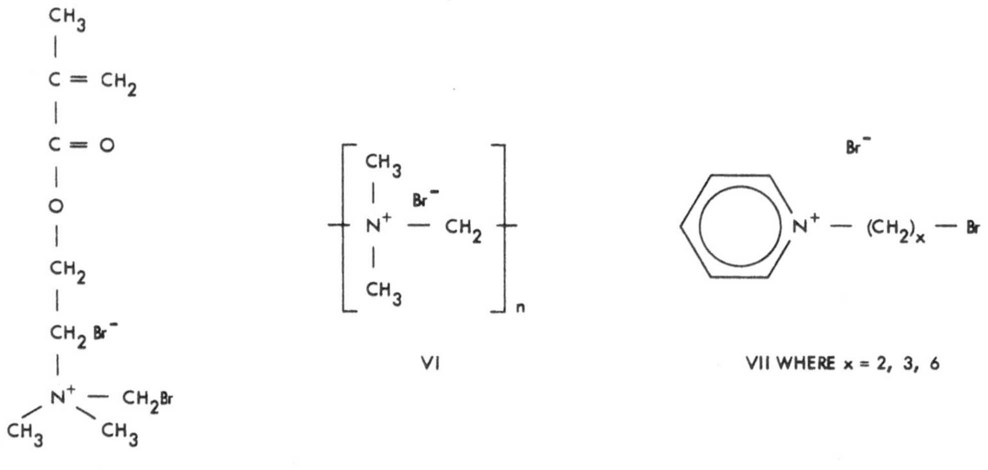

VI

VII WHERE x = 2, 3, 6

V

The spontaneous formation of high ion exchange capacity resins at room temperature in presence of a variety of materials on the surface and in situ lead to the synthesis of these resins inside the walls of hollow fibers and to the examination of the transport properties of the latter. It was found that polyacrylonitrile hollow fibers can be conveniently impregnated with 4-VP crosslinked resins and used to remove dichromate from aqueous solutions as well as from industrial effluents.

Experimental

a. <u>Synthesis</u>. Freshly distilled 4-VP was mixed with dihalides
(Molar ratio; 2:1) in bulk or in solution. The polymerization pro-
ceeded to virtual completion in bulk or solvents (benzene, methanol,
dimethylformamide (DMF) or mixtures of DMF with methanol) and in presence of e.g., silica gel, carbon black, sand, porous materials, e.g.,
paper cloth, polyacrylonitrle fibers, etc. The rate of polymerization was significantly enhanced in absence of air or by Coγ irradiation. The mixtures of 4-VP and dihalides were left at room temperature for periods of 5 days during which time the color changed from
colorless to pink or red. The resin was isolated by addition of acetone and washing with acetone. After drying it was obtained in the
form of a light yellow powder in yields of 70 to 100% of the theoretical amount.

b. <u>Exchange Capacity</u>. Approximately 1 g. of resin vacuum dried
at 100°C overnight and sieved to give a mesh size of 250-500, was
placed in a burette. 3N NaOH (100 ml) was then added to the column
and eluted with distilled water. The elutant was neutralized to pH 6
with N/10 HNO_3 and diluted to 250 ml.

A 30 ml aliquot was analyzed for Br-by the Mohr method. Exchange
capacity = meq/g of dry resin.

c. <u>Relative Swelling</u>. Samples of dry resins (dried at 100°C
were placed in containers at 100% humidity. The increase of weight
was measured after 120 hours.

d. <u>Preparation of Model Compounds and Rate Studies</u>. Freshly
distilled pyridine was mixed with α,ω dibromides (excess) and reacted
at room temperature (7 days). Compounds VIII, IX and X were obtained
in quantitative yields based on the amount of pyridine, and their
structure was ascertained by NMR analysis.

VIII   IX   X

The rate of quaternization of VIII,IX and X was studied in excess of
pyridine in DMF Methanol (1:1 by volume) at 25° and 50°C by following
the increase of ionic bromine concentration with time.

e. <u>Impregnation of Hollow Fibers</u>. Polyacrylonitrile hollow fibers fabricated at Gulf South Research Institute were used. Their
hydraulic permeability was 9 x 10^{-5} cm/sec atm., the wall thickness
50 μ, the inside diameter 200 microns and the wall micropore diameter

about 100 $\overset{\circ}{A}$. Hollow fibers (150) assembled in bundles with a total surface area of 140 cm^2 were washed first with water, and then with methanol and dried by passing nitrogen gas through them for one hour. They were immersed in a mixture of 4-VP and α,ω-dihaloalkane (2:1 Molar). The reaction was permitted to proceed for 10 days in case of dibromo ethane and 2 days in the case of dibromohexane.

f. <u>Ion-Exchange</u>. Removal of dichromate from aqueous solution is achieved by using the Donnan pumping principle, e.g., dichromate ions are pumped against their concentration gradient through the ion exchange hollow fibers by a second ion (of the same charge sign) present at a much higher concentration on the other side of the hollow fiber wall. In this manner polluting dianions can be concentrated in the pumping ion solution and the cleaned up water can be reused or discharged.

Ion-exchange experiments were run in the crossflow mode using 1 N NaCl solution to clean up wastewater which was initially either 10^{-3} or 10^{-4} M $Cr_2O_7^=$ solution. The fibers were bathed in the stirred wastewater while the sodium chloride solution was circulated continuously through the fibers. For wastewater containing 10^{-4} M $Cr_2O_7^=$, the initial dichromate content of the pumping ion solution (NaCl) was varied from 0 to 10^{-2} M $Cr_2O_7^=$.

Results

a. <u>Yields, Exchange Capacities and Swelling</u>. The yields (Table I) of resins varied with the method and temperature of preparation. The exchange capacities were found to be below the theoretical values. However, in most cases the exchange capacities were comparable to those of commercial strong base ion exchange resins determined by the already outlined procedure.

Table I

Yields, Exchange Capacities and Swelling Properties of
4-VP Resins (Room Temperature)

Dihalide	Polymerization Method	Yield W. %	Exchange Capacity meq/g	Relative Swelling %
1,2-Dibromoethane	Solution*/Bulk	70/51	4.9/ -	- / -
1,3-Dibromopropane	Solution/Bulk	97/70	- /4.52	- /67.1
1,4-Dibromobutane	Solution/Bulk	94/80	- /4.59	- /58.0
1,6-Dibromohexane	Solution/Bulk	100/79	- /3.11	- / -
1,8-Dibromooctane	Solution/Bulk	100/90	- /2.37	- /47.8

*DMF, Methanol (1:1 by volume)

b. <u>Halogen Analysis</u>. The total halogen content of 4-VP resins
was found to be greater than the amount of ionic halogen (Table II)
indicating presence of unreacted bromine end groups. After further
reaction with trimethylamine the ionic bromine content was increased
(Table II).

Table II

Theoretical and Observed Ionic Bromide Content of 4-VP Resins

α,ω dihalide	% Br⁻ Theoretical	% Br⁻ Found	% Br⁻ After Further Reaction
1,2-Dibromoethane	40.2	35.3	-
1,3-Dibromopropane	38.44	31.1	36.5
1,4-Dibromobutane	36.85	28.4	-
1,8-Dibromooctane	31.58	19.4	26.5

c. <u>Rate of Quaternization of Model Compounds</u>. Examination of
rates of quaternization of VIII, IX and X using a large excess of
pyridine yielded straight lines when $\log \frac{a}{a-x}$ was plotted versus time
(a and x = initial bromide concentration and concentration at time t,
respectively). The first order rate constants are recorded in Figs.
5 and 6 for reactions carried out at 25° and 50°C, respectively.

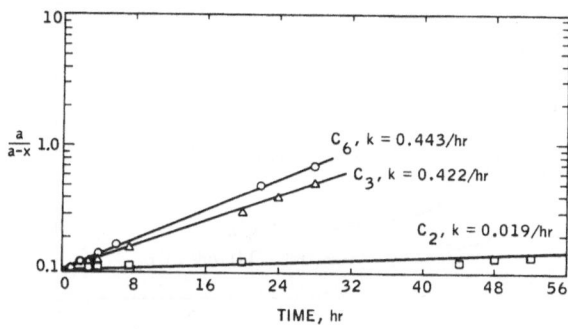

Figure 5. Rate of Reaction of 1-Pyridinium-2-bromoethane (C_2), 1-
Pyridinium-3-bromopropane (C_3), and 1-Pyridinium-6-bromo-
hexane (C_6) with Pyridine at 25°C.

d. <u>Removal of Dichromate by Means of Acrylonitrile Hollow Fibers
Impregnated with 4-VP Resins</u>. The experimental results of the ion-
exchange experiments are shown in Fig. 7. The laboratory experiments
were run in batches, but any plant design would provide for continuous
countercurrent flow of both wastewater and pumping-ion solution.

A. REMBAUM, S. P. S. YEN, AND C. ROBILLARD

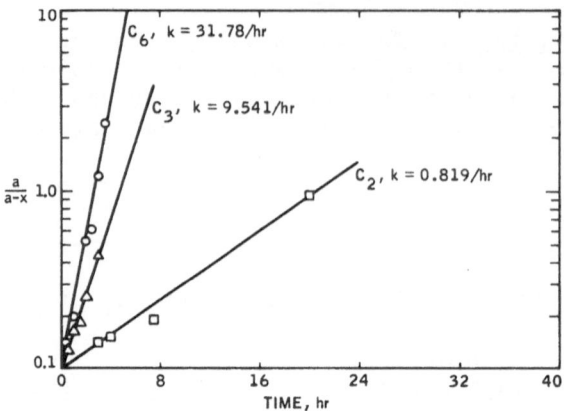

Figure 6. Rate of Reaction of C_2, C_3 and C_6 with Pyridine at 50°C.

Figure 7 suggests that wastewater dichromate concentration can be re-
duced from 10^{-4} to $2\text{-}5 \times 10^{-6}$ $MCr_2O_7^=$ (10 ppm to 0.2-0.5 ppm of chro-
mium) using countercurrent flow to concentrate dichromate up to 1000
ppm in the pumping solution.

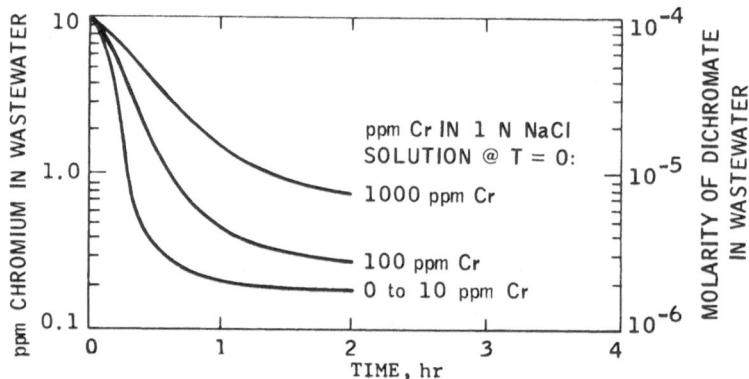

Figure 7. Dichromate Reduction as a Function of Time and of Dichro-
          mate Content of Pumping Ion Solution (765 ml of Wastewater;
          1100 ml of Pumping Ion (NaCl) Solution; Cl⁻ Leak Rate =
          0.0006 Moles per Hour per Fiber Bundle).

DISCUSSION AND CONCLUSIONS

The mechanism of the spontaneous polymerization of 4-VP in presence of α,ω-dihoalkanes may be represented by the scheme shown in
Fig. 8.

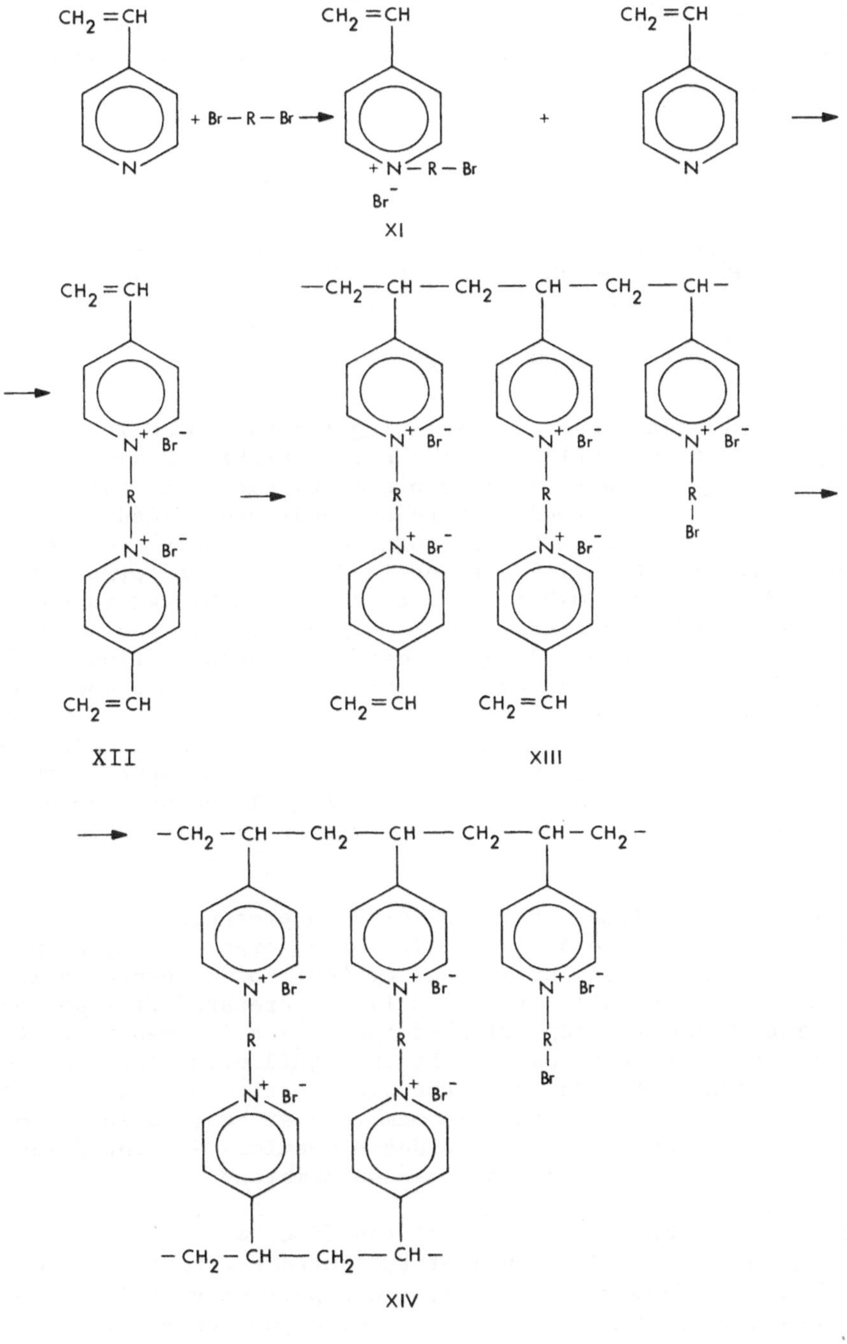

Figure 8. Tentative polymerization mechanism of 4-VP in presence of α,ω-dihaloalkanes.

Structures similar to XI have been isolated and characterized (VIII, IX, & X). Structure XII could not be isolated with 4-VP
because of the fast crosslinking reaction leading to networks represented by XIV.

XV

However, we have succeeded in isolating a substituted intermediate
XV by the use of 2-methyl-5-vinyl pyridine (MVP). The structure of
this bis-quaternary salt (XV) was proven by elemental analysis and
NMR. The isolated intermediate was unstable and formed an insoluble
resin on standing at room temperature, most probably because of the
polymerization reaction illustrated in Fig. 8. The formation of
crosslinked resin from MVP at a fast rate was achieved in solution
by the use of redox initiators. The higher stability of MVP than 4-VP
was attributed by Kabanov,[1] to the decreased polarization of the double bond in MVP. For the same reason 2-vinyl pyridine when reacted
with dihalides was found to be even less reactive than either 4-VP or
MVP and yielded only low molecular weight soluble materials under the
already described experimental conditons. It thus appears that the
polymerization mechanism illustrated in Fig. 1 can be adapted to the
4-VP, dihalide system and the mechanism shown in Fig. 2 does not apply
here.

The ease of polymerization of 4-VP in presence of dihalide unin-
hibited by the presence of impurities or a large variety of polymeric
materials led to the investigations of 4-VP ion exchange resins formed
in the pores of the walls of hollow fibers prepared from polyacrylonitrile. The transport properties of the fibers impregnated with 4-VP
ion exchange resins are governed by the equilibrium expressions derived by Donnan[6] which describe the distribution of ions on either
side of a perfectly semipermeable membrane at equilibrium. The attainment of this equilibrium provides a chemical driving force for
the movement of counter ions across the membrane.

To examine the permeability of the fiber wall with respect to
anions, $Cr_2O_7^=$ was used because of the current interest in removing
divalent anions from water wastes. In a perfectly semipermeable
system the number of milliequivalents of $Cr_2O_7^=$ permeating the wall
would be exactly equal to the number of milliequivalents of Cl^- moving in the opposite direction. In the actual experimental set-up,
this is not the case. The number of transferred dichromate ions is

significantly less than the number of chloride ions. However, the results obtained so far indicate that ion exchange hollow fibers may be
used successfully for the removal of unwanted dianions from wastewater
when the experimental condition of Donnan equilibrium are applied.
The potential applications of this system and the recent report by
Dressner[7] pointing out the requirements for economical Donnan soffen-
ing indicate a need for further investigations of the transport
properties.

On the basis of the results obtained, so far, the following conclusions can be reached.

1. The reaction of 4-VP with α,ω dihalides yields highly crosslinked
   anion exchange resins, the structure of which can be systematically varied by using a variety of dihalides.

2. The first quaternization reaction activates the double bond of
   4-VP leading to an addition polymerization of 4-VP monomers containing quaternary nitrogens. This conclusion is confirmed by
   the fact that the crosslinked resin is free of unquaternized pyridine rings.

3. The rate of formation of resin is determined by the reaction of
   the second bromine end group of the dibromide after the first end
   group has reacted. This second reaction increases as the distance
   between positive charges increases, i.e., as the number of CH_2
   groups in the dibromide increases. This is probably due to the
   fact that a close approach of positive charges increases the cou-
   lombic repulsion and the free energy of formation of diquaternary
   salts.

4. The principle of formation of hollow fiber ion exchangers by spontaneous polymerization of 4-VP in the walls of the fibers was demonstrated.

5. Hollow fiber ion exchangers can be used in the continuous process
   utilizing the Donnan pumping principle to remove unwanted dianions
   from an aqueous solution.

Additional results on the mass transport properties of 4-VP ion
exchange hollow fibers will be reported separately.[8]

References

1. V. A. Kabanov, K. V. Aliev, O. V. Vargina, T. I. Patrikeeva and
   V. A. Kargin, J. Polym. Sci., C, 16, 1079 (1967).

2. J. C. Salamone, B. Snider and S. L. Fitch, J. Polymer Sci., A-1,
   9, 1943 (1971); I. Mielke and H. Ringsdorf, Makrom. Chem., 153,
   307 (1972).

3. J. C. Salamone, E. J. Ellis, C. R. Wilson and D. F. Bardolivallan, Macromolecules, $\underline{6}$, 475 (1973).

4. A. Rembaum, S. Singer and H. Keyzer, J. Polym. Sci., B, $\underline{7}$, 395 (1969).

5. A. Rembaum and H. Noguchi, Macromolecules, $\underline{5}$, 261 (1972).

6. F. G. Donnan Chem. Rev. $\underline{1}$, (1924).

7. L. Dressner I & EC Process, Research and Develop. $\underline{12}$, 148 (1973).

8. A. Rembaum, S. P. S. Yen, E. Klein and J. Smith, "Polyelectrolytes and their Applications," D. Reidel Publ. Co., Dordrecht Holland eds: A. Rembaum and E. Selegny (in press).

PARTICLE COLLECTION FROM AQUEOUS SUSPENSIONS

BY PERMEABLE HOLLOW FIBERS

Daniel P. Y. Chang* and Sheldon K. Friedlander

W. M. Keck Laboratories, California Institute of Technology

Pasadena, California 91109

INTRODUCTION

Permeable hollow fibers can be utilized as a filter medium to increase the collection efficiency of conventional depth filters. The permeability of the wall permits modification of the chemistry and hydrodynamics of the fiber-suspension interface, thereby increasing the transport and attachment of particles to the filter. Modification of the interface is accomplished by induction of a minute flow of liquid into the fiber interior and/or by the diffusion of a destabilizing chemical from the fiber interior. The flow of liquid into the fiber surface reduces the viscous drag on the particle as it nears the surface of the filter thereby increasing the rate of transport. The destabilizing chemical reduces repulsion (electric double layer) between particle and fiber surfaces, increasing the fraction of particle-fiber contacts resulting in attachment.

The novel use of hollow fibers described above differs from ultrafiltration since only a small fraction of the suspending liquid is processed through the fiber. Primary transport of particles to the filtering surfaces results from the flow around the fibers. The experimental investigation of single hollow fibers operated in the above described mode is the subject of this research.

----

*To whom all correspondence should be addressed. Current mailing address:

        Department of Civil Engineering
        University of California, Davis
        Davis, California 95616

THEORY

It is convenient to view the initial stage of filtration as a process involving two distinct steps: a transport step in which particles are carried to the vicinity of a collector surface, and an attachment step in which particles "collide" and attach themselves to the collector. The interpretation of the effects of hollow fibers on particle collection are complicated by the different transport mechanisms which dominate over various ranges of experimental parameters (particle size and density, fiber size, velocity, etc.). For this reason particle diameter and density and fluid velocity were selected so as to result in interception being the dominant transport mechanism.

In filtration terminology interception is said to occur when a particle, free to follow an undisturbed fluid streamline, is carried within one particle radius of the collecting surface. A dimensionless transport efficiency, η_{GIM} , can then be defined as the ratio of the rate of collision of particles with the fiber to the rate at which particles would be carried through the projected area of the fiber if it were not present. For a given flow field η_{GIM} can be computed. The experimental conditions approximated that of uniform flow past a circular cylinder at low Reynolds number $(N_{Re} < 1)$. In that case the transport efficiency, η_{GIM} , for spherical particles is given by:

$$\eta_{GIM} = 2A_f(\frac{d_p}{d_f})^2 , \tag{1}$$

where d_p and d_f are the particle and fiber diameters and A_f is a parameter characteristic of the flow,

$$A_f = \frac{1}{2 \, [2.00 - \ln (N_{Re})]} , \tag{2}$$

(Friedlander, 1967). A corresponding experimental collection efficiency, η_{EXP} , can be calculated from the fluid velocity and the rate of deposition of particles on the fiber. If particle transport occurs by interception alone, the relationship between η_{EXP} and η_{GIM} is given by:

$$\eta_{EXP} = \alpha\eta_{GIM} , \tag{3}$$

where α , the collision efficiency, is the fraction of collisions resulting in particle capture.

The large viscosity and short "mean free path" of liquids compared to gases, leads to an appreciable viscous resistance to motion as the gap between the particle and collector surface narrows. This causes the particles to deviate from the fluid streamlines and results in a lower transport efficiency than that given by η_{GIM} . In fact Spielman and Goren (1970) have shown that the rate of collisions would be zero in the absence of an external force which can accelerate the particles through the viscous layer! The external force which is believed to be responsible for the majority of particle contacts is the van der Waals attraction. Spielman and Fitzpatrick (1973) have determined the rate of collisions of spheres with a cylindrical collector by solving the "creeping flow" equations of fluid motion and numerically integrating the particle trajectories subject to the van der Waals attraction. The collision efficiency so computed is termed the hydrodynamic retardation model efficiency, η_{HRM} , in this paper, and its ratio to the geometric interception efficiency, η_{GIM} , is shown in Figure 1 as a function of the adhesion number, N_{Adf} , a dimensionless group characterizing the van der Waals attraction. The portion of the curve falling below unity represents the effect of the viscous drag on the collision rate. The permeability of the hollow fibers allows the "viscous layer" to be "sucked" away and brings the particle into the effective range of the van der Waals attraction. Thus, the ratio η_{HRM}/η_{GIM} should be raised to a value of about unity and an increase in the rate of collisions should occur for the same adhesion number. It should be noted that primary transport of particles to the surface is still the result of the flow around the fiber, not the result of flow into the fiber as in ultrafiltration.

Forces other than viscous drag and van der Waals attraction may act upon the particle near the fiber surface. According to the Derjaguin - Landau - Verwey - Overbeek (DLVO) theory of the diffuse electrical double layer, the total repulsive energy of interaction between similarly charged double layers can be reduced by the addition of electrolytes into the solution (Verwey and Overbeek, 1948). Most naturally occurring surfaces are negatively charged near neutral pH. One common practice to flocculate colloidal suspensions is to add electrolyte (alum or ferric salts) to the suspension, or more recently to use polymers and poly-electrolytes to bring about destabilization.

Diffusion of destabilizing chemicals from the interior of hollow fibers can produce the same results more efficiently. Under the proper operating conditions the necessary electrolyte concentration can be maintained in a narrow concentration boundary layer surrounding the fiber where it is most effective. This is shown schematically in Figure 2. It should then be possible to reduce the total amount of destabilizing chemical added to the system.

The parameter which compares the relative effectiveness of addition of destabilizing agent through the fiber to addition

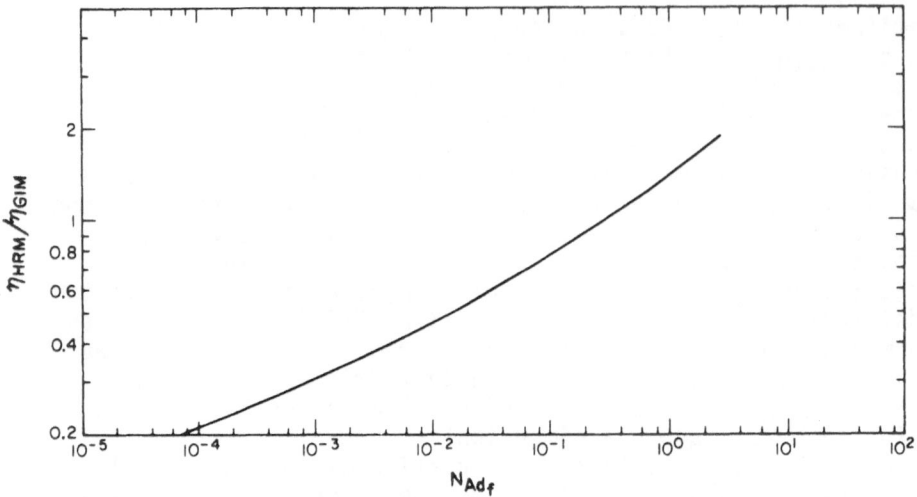

Figure 1.   Ratio of HRM Efficiency to GIM Efficiency versus
            Adhesion Number

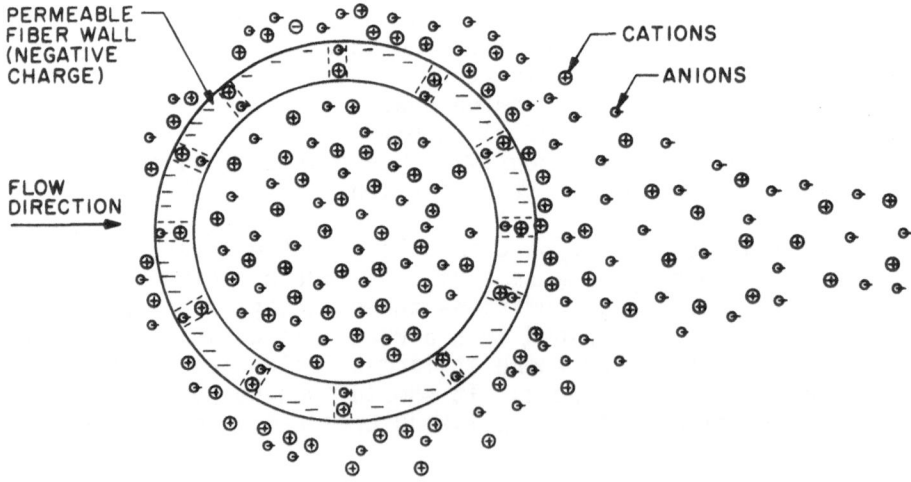

Figure 2.   Schematic of Concentration Boundary Layer Growth

to the bulk solution is the ratio of the average mass transfer coefficient of particles from the suspension to the surface, k_p , to the average mass transfer coefficient of destabilizing agent from the surface to the suspension, k_s . The mass transfer coefficient for particles is closely related to the transport efficiency, η_{GIM} . The mass transfer coefficient of diffusing species can be estimated from concentration boundary layer theory (Friedlander, 1957). The ratio has been estimated by Chang (1973) for the case where spherical particles are transported by the interception mechanism and is given by:

$$\frac{k_p}{k_s} \; \alpha \; \frac{d_p^2 A_f^{2/3} U^{2/3}}{d_f^{4/3} D_S^{2/3}} \quad , \tag{4}$$

where d_p and d_f are the particle and fiber diameters, A_f is defined by Equation (3), U is the velocity of the fluid far from the fiber surface and D_S is the diffusion coefficient of the destabilizing chemical. If the ratio k_p/k_s is large, then addition of electrolyte through the fiber should be more efficient than addition to the bulk solution. It can be seen that smaller diffusion coefficients and fiber diameters lead to larger ratios. Therefore larger molecules, e.g. low molecular weight polyelectrolytes rather than simple ions, and smaller fibers are predicted to be desirable filter characteristics.

EXPERIMENTAL PROCEDURE

Latex spheres of narrow size distribution, ranging in size from 2.02 to 25.7 μm geometric mean diameter, were suspended in 0.45 μm millipore - filtered distilled de-ionized water, filtered electrolyte solutions, or filtered tap water. The dilute suspensions of particles flowed past a single hollow fiber whose axis was mounted transverse to the direction of flow in a transparent plexiglass chamber (Figure 3).* The flow in the chamber approximated that of uniform flow past a cylinder of infinite length so that the geometric interception efficiency could readily be calculated from the measured flowrates. The flowrate was maintained constant by connecting a constant-head tank and throttling valve in series with a rotameter. A peristaltic pump was used to re-circulate the suspension to the constant head tank from a slowly stirred reservoir.

---

*Details of the experimental design and apparatus are described elsewhere by Chang (1973).

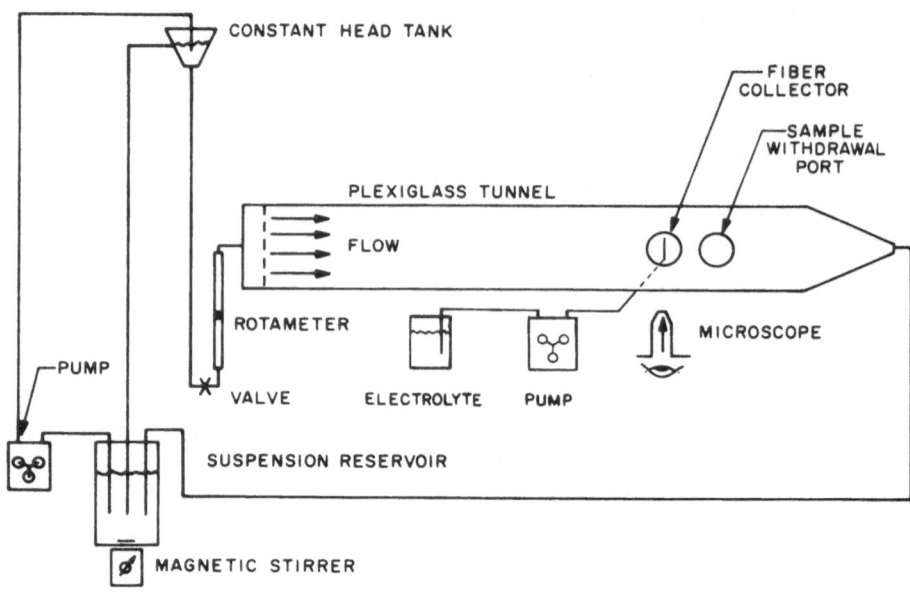

Figure 3.   Schematic of Experimental Apparatus

The number of particles deposited on the fiber was recorded as a function of time. An optical microscope whose axis was transverse to both the fiber axis and flow direction was used to count the particles. Samples of the suspension were withdrawn from the center of the flow chamber through a hypodermic syringe and the particle concentration determined with a Coulter Counter.

The hollow fibers were sealed between stainless steel tubes and mounted in the test section with their axes transverse to the flow. A peristaltic pump connected to one stainless steel tube support was used to circulate various solutions through or to apply mechanical suction to the fiber interior. Chemical destabilizing agents used included: simple electrolytes - NaCl , $CaCl_2$, an hydrolyzable metal ion - $Al(NO_3)_3$; and low-molecular weight polyelectrolytes - 1, 2 - dimethyl -5- vinyl pyridinium bromide $\sim 1.2 \times 10^4$ MW (DMVPB), Cat-Floc ~25,000 MW and Ionene $\sim$ 4000-14,000 MW*. The molecular weight cutoffs of the commercial

---

*Cat-Floc supplied courtesy of Dr. Frank Mangravite, Calgon, Corp. Ionene supplied courtesy of Dr. Alan Rembaum, Jet Propulsion Laboratory. DMVPB supplied courtesy of Dr. Dennis Kasper, University of Arizona.

cellulose acetate fibers used in these experiments were 200 MW and 30,000 MW[+]. These fibers had nominal outside diameters of 230μm with a wall thickness of 25μm. When immersed in water they became transparent, making it possible to count the particle deposit over the entire surface of the fiber.

RESULTS AND DISCUSSION

Initial experiments were directed at demonstrating the feasibility of electrolyte addition through the fiber wall to increase particle adhesion. Figure 4 illustrates the differences observed when suction is applied to the fiber interior, with and without electrolyte addition (CA-C fiber).

The deposition rate increased over the sides and rear stagnation region of the fiber when suction was applied. However, most of the particles caught on the rear stagnation region were released when the applied suction was removed. Adding an electrolyte, $CaCl_2$ in this example, before releasing suction, resulted in much higher retention of particles, approaching 100%. (The initial decrease in the number of particles retained is an artifact induced by an air bubble preceding the $CaCl_2$ solution).

These data indicated that a repulsive energy barrier existed at the fiber surface which prevented particles from attaching to the wall even though the applied suction reduced hydrodynamic drag. Introduction of simple electrolytes (NaCl , $CaCl_2$) reduced the repulsion and permitted irreversible attachment of particles to the surface. The behavior of the barrier qualitatively agreed with that predicted by the diffuse electrical double layer theory of Verwey-Overbeek-Derjaguin-Landau for similarly charged particles of like sign in aqueous suspensions, except that some particles did attach themselves to the hollow fiber before chemical addition. The anomolous deposition has been observed on many types of surfaces, solid and porous, no matter how carefully the surface has been prepared and is probably the result of sub-microscopic heterogenities of the surface (Marshall and Kitchener, 1966; Hull and Kitchener, 1969; Fitzpatrick and Spielman, 1973; Chang, 1973). Significantly these data demonstrated that addition of chemical destabilizing agents through the fiber wall is feasible and can result in increased collection efficiency.

[+]Hollow-fibers CA-A (200 MW cutoff) and CA-C (30,000 MW cutoff) manufactured by Dow Chemical Company.

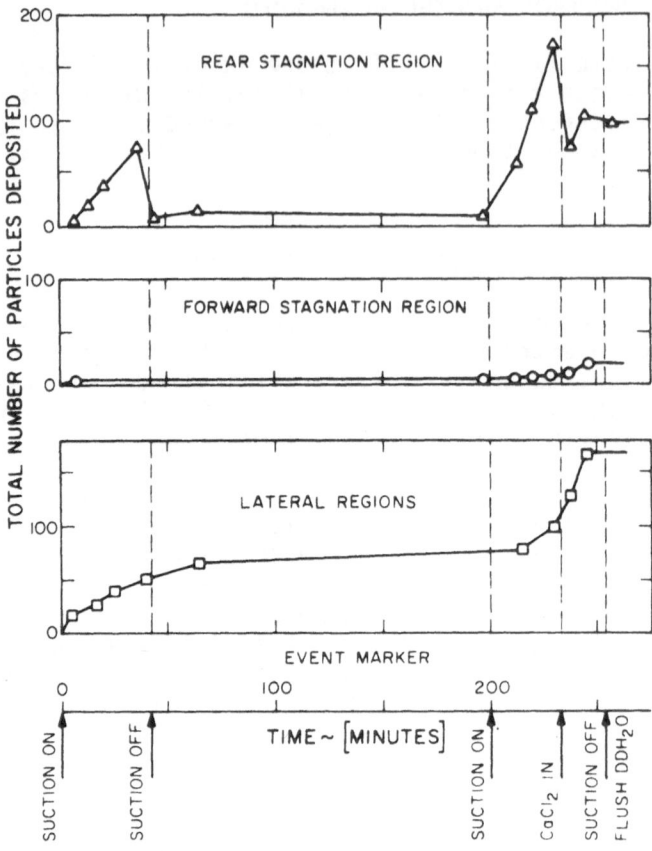

Figure 4.   Effect of Slight Suction and Chemical Addition on
Particle Capture by 30,000 MW Hollow Fiber

        Cellulose acetate (CA-A) fibers with a nominal molecular
weight cut off of 200 were used in a series of experiments to
study the effect of varying internal ion concentration on
collection efficiency. Unbuffered sodium chloride solutions
flowed through the fiber interior while latex particles in de-
ionized distilled water were circulated past the fiber exterior.
When calcium chloride solutions were used as the electrolyte, the
flow of solvent into the fiber rapidly diminished signaling block-
age of the membrane pores. This "blinding" was irreversible and
did not occur with the sodium chloride solutions.

        Figure 5 shows the ratio of the measured single fiber
collector efficiency (η_{EXP}) to that predicted by filtration
theory for removal by interception (η_{GIM}) versus the electrolyte
concentration (NaCl) . Not all of the captured particles adhered

to the fiber surface. Some were held solely by the osmotic flow
into the fiber. When the driving force was removed, that is, when
the osmotic flow was stopped by flushing the interior of the fiber
with distilled water, particles were released by the fiber.
Figure 6 shows the fraction of particles retained on the fiber
after flushing versus the original electrolyte concentration.

The high ratio, n_{EXP}/n_{GIM} , observed in Figure 5 suggested
that the flow induced by the osmotic pressure gradient across the
fiber wall transported particles to the surface, and dominated
transport by the undisturbed stream. However, particles could not
attach themselves until the surface ion concentration became
sufficiently high to compress the double layer as seen by the
sharp increase of particles retained as electrolyte concentration
increased (Figure 6). It is worthwhile noting that the suction of
fluid into the fiber interior may be the result of either a
mechanically or chemically induced pressure gradient and that this
must be taken into consideration in the proper design of a hollow
fiber filter. Otherwise the fibers would be operated less efficiently in an ultrafiltration mode.

The effect of varying the external ion concentration while
keeping the internal ion concentration fixed (~ 1.0 M) is shown
in Figure 7. The ratio n_{EXP}/n_{GIM} varied as expected for a
decreasing osmotic pressure gradient with increasing external ion
concentration. The experimental collection efficiency approached
the theoretical transport efficiency as the osmotic pressure
gradient decreased indicating that almost all particles striking
the fiber were captured, i.e., $\alpha \sim$ 100%.

Qualitative experiments involving low molecular weight
cationic polyelectrolytes and an hydrolyzable metal ion Al^{+3}
diffusing from the CA-C fiber indicate that collision efficiencies
close to unity can be achieved by allowing dilute solutions of
these substances [order of 5×10^{-6} M for DMVPB and 6×10^{-5} M
$Al(NO_3)_3$] to diffuse through the fiber wall. The larger molecules
diffuse from the surface region less rapidly, are effective at
lower concentration and in some cases appear to adsorb to the
fiber surface as evidenced by continued particle collection long
after the polyelectrolyte had been flushed from the fiber.

A separate experiment to demonstrate that suction on the
fiber can reduce the viscous drag on the particle was performed
(Figure 8). The permeability of the CA-C fibers was measured and
a known pressure drop applied to the fiber. The fluid velocity into
the fiber could then be calculated and the particle deposition rate
measured. The dashed line represents the rate of deposition with
no suction applied. The dot-dashed line represents the anticipated
deposition rate accounting for the additional particles carried

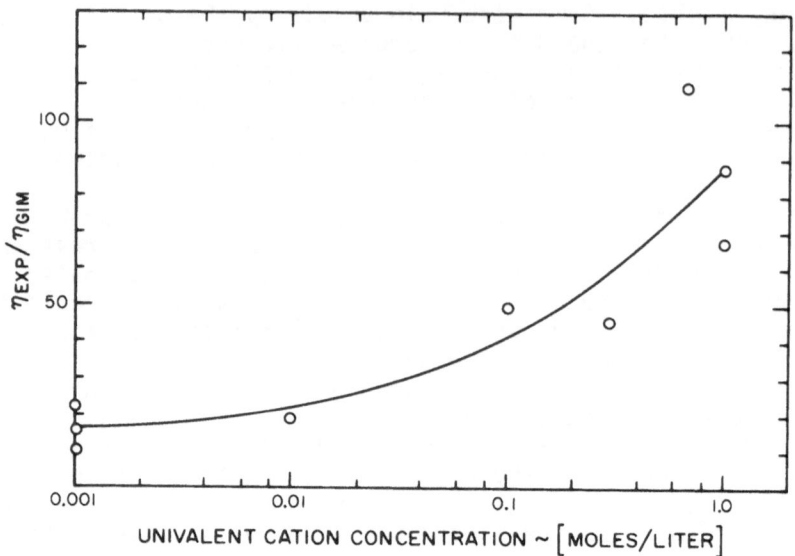

Figure 5. Effect of Internal Cation Concentration in 200 MW Fiber
on Initial Capture of 5.7 μm Latex Suspended in De-ionized
Distilled Water

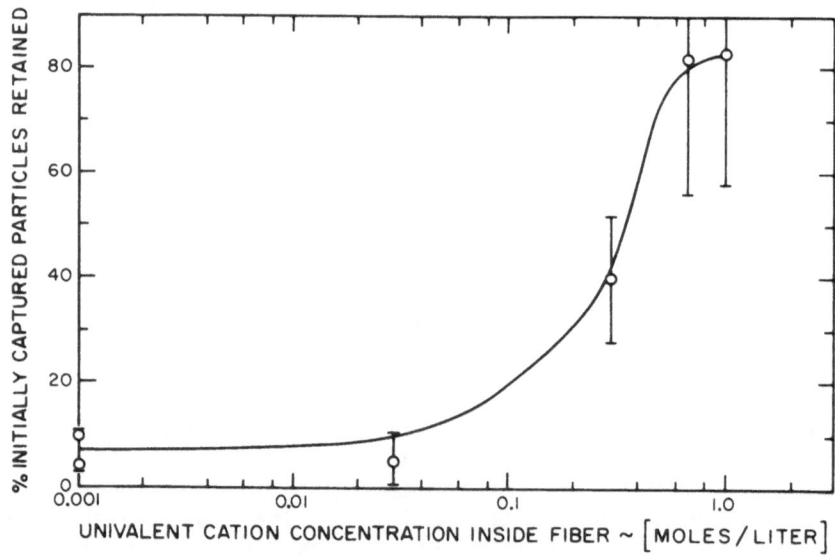

Figure 6. Effect of Internal Cation Concentration in 200 MW Cutoff
Fiber on Particle Deposit Retained After Flushing Interior
with De-ionized Distilled Water

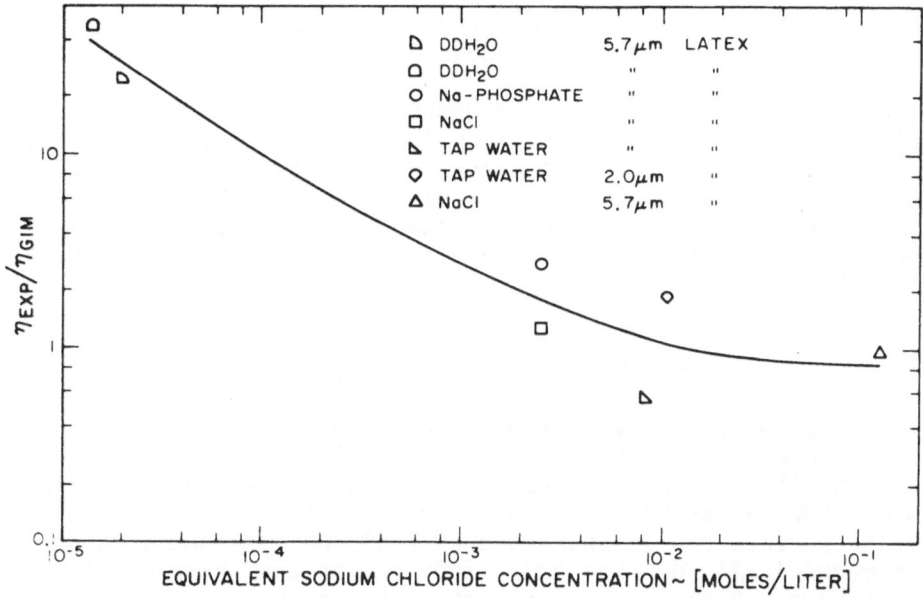

Figure 7. Effect of External Ion Concentration on Initial Capture Efficiency of Latices with Fixed Internal Ion Concentration 1M NaCl

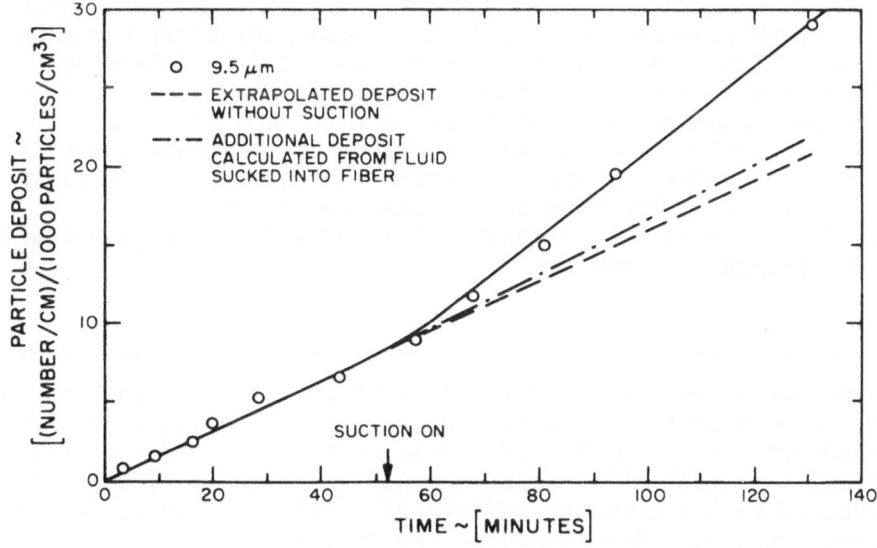

Figure 8. Effect of Slight Suction Under Constant Osmotic Pressure on Capture of 9.5 μm Latex Particles

toward the fiber by the flow into the fiber. The solid line
represents the experimentally observed deposition rate.

The increase in the observed deposition rate could not be
explained completely by the additional convection toward the fiber.
The difference between the two curves represents the effect of
reducing viscous drag. An analysis of the experimental error
prevents one from positively concluding that the reduction in
viscous drag is real, but the data strongly suggests that this is
the case.*

CONCLUSIONS AND RECOMMENDATIONS

Addition of chemical destabilizing agents through hollow
fibers has been demonstrated to be a feasible and effective means
of reducing the repulsive forces between particles and the filter
surface. Furthermore the rate of transport of particles to the
filtering surface can be increased by applying a slight suction,
either mechanically on chemically induced, to the interior of the
hollow fibers. A dimensionless parameter, k_s/k_p , can be used
to estimate the conditions under which hollow fibers can be
effectively used as a filter medium to minimize the amount of
chemical added.

Further experiments to determine compatible polyelectrolyte -
hollow fiber combinations are needed. Experiments with packed
beds of hollow fibers should also be conducted to determine
whether single fiber predictions are applicable to real filters
and to see if additional beneficial effects are produced. The
ability to modify the flow and chemistry of the fiber-suspension
interface provides a degree of control which can, in principle,
be used either to increase or decrease the deposition of
particles. This may have application in blood filtration and in
the coalesence of emulsions.

----

*The number of particle collisions with the fiber is governed by a
Poisson distribution, the size of the error varying with the
square root of the number of particles captured. Unfortunately
the accuracy with which deposited particles could be counted
decreased with increasing numbers of particles. There was an
optimum deposit which minimized the experimental error.

ACKNOWLEDGEMENTS

Messrs. Dwight Landis, Robert Jung and Chi-Ngong Pow are thanked for their part in assisting with the collection of data. Messrs. Elton Daly and Robert Greenway constructed the plexiglass flow chamber. Part of the research was performed while one of the authors (D. Chang) was a Public Health Service Trainee. The generous support of the Ametek Corporation is also acknowledged.

REFERENCES

1. Chang, D. P. Y., (1973), "Particle Collection from Aqueous Suspensions by Solid and Hollow Single Fibers." Unpublished Doctoral Dissertation, California Institute of Technology.

2. Fitzpatrick, J. A. and L. A. Spielman (1973), J. Coll. Interface Sci., 43, 350.

3. Friedlander, S. K., (1957), A.I.Ch.E. Journal, 3: 1, 43.

4. Friedlander, S. K., (1967), "Aerosol Filtration by Fibrous Filters," In Biochemical and Biological Engineering Science, Vol. 1, ed. N. Blakebrough, New York: Academic Press.

5. Hull, M. and J. A. Kitchener, (1969), Trans. Faraday Soc., 65, 3093.

6. Marshall, J. K. and J. A. Kitchener, (1966), J. Coll. Interface Sci., 22, 342.

7. Spielman, L. A., and S. L. Goren, (1970), Env. Sci. Tech., 4, 135.

8. Verwey, E. J., and J. Th. G. Overbeek,(1948), "Theory of the Stability of Lyophobic Colloids. Amsterdam: Elsevier Publishing Co.

SEPARATION OF HYDROCARBONS BY SELECTIVE PERMEATION THROUGH

POLYMERIC MEMBRANES

MORT FELS[*] AND NORMAN N. LI[†]

[*]LAKEHEAD UNIVERSITY, THUNDER BAY, ONTARIO

[†]ESSO RESEARCH AND ENGINEERING, LINDEN, NEW JERSEY

INTRODUCTION

Polymer membranes have the unique ability to allow molecules to diffuse through them. Various factors determine the ability of a molecule to permeate through the membrane - both physical and chemical properties. This dependence of permeation rate on the large number of system properties imparts a unique separating ability to polymer membranes compared with other separation methods which depend on differences in relatively few properties.

The separation of compounds by a membrane permeation process is particularly useful whenever conventional separation techniques cannot be used to get reasonable separation. In general, the use of the membrane separation technique is of special benefit for separating mixtures of compounds of similar chemical or physical properties, mixtures of structural or positional isomers, and mixtures containing thermally unstable components.

One important area in membrane permeation is the diffusion and separation of organic compounds. However, unlike gas permeation, the transport of organic compounds is complicated by the fact that the diffusion coefficient is a function of the concentration of the penetrant molecule in the polymer film. The reason for this is because of the interaction between the penetrant molecule and polymer. In general, this interaction causes a swelling to occur, and a consequent increase in the diffusion coefficient. As a result, the separation and permeation of the mixture becomes a complex phenomenon.

The purpose of this paper is to review some of the models proposed for the concentration-dependent diffusion coefficient and discuss their applicability to actual separation data. In addition, some work on ultra-thin films being done at present will be outlined.

THEORY

The starting point for discussion of any permeation process is Fick's Law, here written as a function of both time and of distance through the membrane.

$$\frac{\partial c}{\partial t} = \frac{\partial}{\partial x}\left[D \frac{\partial c}{\partial x}\right]$$

where: c = concentration in the film
           t = time
           x = distance through the film
           D = the diffusion coefficient.

Note that as the diffusion coefficient is not a constant, it must remain inside the first differential.

At steady-state, the change in concentration with time is zero, and one can define a steady-state permeability as equal to the integral of the diffusion coefficient over the concentration range throughout the film.

$$P = \int_0^{C_0} D \, dc$$

where: P = permeability
           C_0 = upstream concentration of penetrant.

In this case, one side of the membrane is kept at 0 concentration, while the other sees a concentration of C_0.

The addition of other components to the permeating species does not change the permeability equation, however, the diffusion

coefficient is now the diffusion coefficient of the i-th species in
the presence of the other components. In general, because of the
swelling effect of the other species on the membrane, the diffusion
coefficient of the i-th species will not be equal to its diffusion
coefficient as measured when only a single component is present.
The concentration of the i-th species in the upstream side of the
membrane can be approximated by the mole fraction times the concentration of the pure component in the membrane.

The problem of calculating the permeability and separation for
a mixture of components can now be resolved into finding the diffusion coefficients of each individual molecular species while in
the presence of others. Let us now examine two models which have
been used to correlate diffusion coefficients with concentration.
The first, which is well known to most is the exponential model.

$$D = D_{c=0}\, e^{AC}$$

In this model, A is a parameter known as the "plasticizing" factor,
and is related to the ability of a penetrant molecule to swell, or
loosen the particular membrane. The larger the value of A, the
greater the interaction between penetrant and polymer. $D_{c=0}$ is
the diffusion coefficient at zero concentration.

One major advantage of this relatively simple model is that
because it is integrable with respect to concentration, an analytical expression for the permeability results. This model has been
used with good results for many liquid-polymer systems. It is
found that $D_{c=0}$ values are in the order of 10^{-9} cm^2/sec and values
of A vary from about 20 to 90 for common organic liquids[1].
Therefore, if one assigns a typical solubility of 0.1 to the liquid
in the polymer, it can be seen that the diffusion coefficient can
be several thousand times as large as $D_{c=0}$ when the polymer is
swollen.

The major disadvantage with a model of this type is that,
although it successfully correlates the performance of single
liquid-polymer systems, it cannot be used to predict or correlate
separation behaviour of multicomponent systems.

The second model is an application of the Free Volume Theory
to diffusion of a penetrant through a polymer film. The free
volume is a measure of the unoccupied volume available for diffusion within the system. For transport purposes, the free volume

is analogous to the "holes" which are opened by thermal fluctuations of the polymer chains and is expressed as a fraction of the
total volume of the polymer.

The final form of the concentration dependence is shown here.

$$\frac{D_v}{D_{v=0}} = EXP\left\{\frac{v}{[f(0,T)]^2/B_d\beta(T) + [f(0,T)/B_d]v}\right\}$$

where: D_v = the diffusion coefficient at volume fraction
          $D_{v=0}$ = the diffusion coefficient at zero volume fraction
                 of the diffusing specie.

The model incorporates three parameters:

(1) The free volume of the polymer which is given the symbol
     f(0,T). The first term in the bracket refers to the
     concentration of the diffusing species, the second is
     the temperature.

(2) The parameter β(T) which relates the increase in free
     volume to volume fraction of the diffusing species.

(3) A constant Bd corresponding to the minimum "hole" required
     for a diffusional displacement.

For a single component, the above expression models the diffusion coefficient; the free volume at a volume fraction v being
given as the free volume of the polymer plus the increase due to
the swelling of the polymer by the diffusing species. This theory
can be extended to a mixture of components. To simplify the argument, let us assume we are dealing with two liquids, A and B. The
basic assumption is that the free volume of a system polymer plus
A plus B is made up of the free volume of the polymer plus the
amount of increase due to volume fractions of each component A and
B being present in the film. In other words, the free volume at
zero concentration of A is equal to the free volume of the polymer
plus the swelling effect of B on the film. Putting this all together, one obtains the final equation shown here, written for
component A in this case.

$$\frac{D_{v_A}}{D_{v_A=0}} = EXP\left\{\frac{B_d(\beta_A v_A + \beta_B v_B)}{f(0,T)\left[\beta_A v_A + \beta_B v_B\right] + \left[f(0,T)\right]^2}\right\}$$

where: subscripts A and B refer to the components A and B
              respectively which are permeating through the membrane.

So to calculate the permeabilities of individual components
in a binary mixture, all one must do is obtain the diffusion coefficients at zero concentration and the three parameters for each
component, and integrate the resulting expression with respect to
concentration (or volume fraction). Unfortunately, obtaining these
coefficients is a rather difficult technique as will be seen shortly.
However, the model does provide some further insight into the mechanism of polymeric separations.

EXPERIMENTAL

In general, the experimental techniques for obtaining the
parameters for the models consist of absorption and/or desorption
experiments. Absorption and desorption data are obtained as a
function of time as the penetrant is either absorbing into or de-
sorbing out of the polymer.

A typical desorption curve is shown in Figure 1 for n-hexane
and low density polyethylene. It can be seen that as time becomes
long, the curve approaches a straight line. The slope of this line
is equal to $-\pi D_{c=0}/\ell^2$. The next step in the procedure is
to find values of the model parameters which result in the calculated desorption curve most closely approximating the experimental
curve obtained. In the case of the exponential model, only the one
parameter A need be calculated, however, with the free volume model
the three parameters must be obtained. Fortunately, in the latter
case, another relationship between free volume of the pure liquid
and its thermal expansion can be used, reducing the problem to a
two-parameter search.

Therefore, values of the parameters are found such that the
discrepancy between a calculated desorption curve and the actual

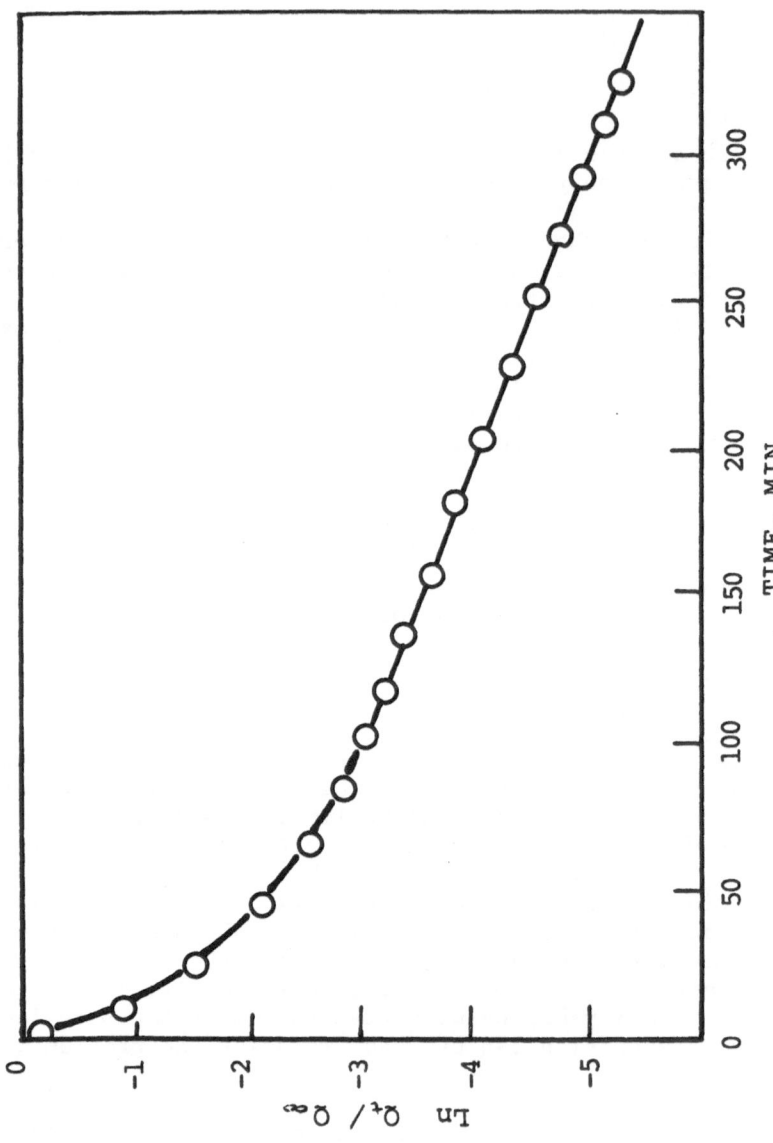

FIGURE 1 TYPICAL DESORPTION CURVE, N-HEXANE IN POLYETHYLENE

curve is minimized. The discrepancy is measured by the square of the deviation of the curve at about 100 points.

The major advantage of the desorption technique is that it is a relatively simple experiment to perform. Unfortunately, however, the computation of the desorption curve for comparison requires the computer solution of the highly non-linear differential equation in the case of the free volume model. The solution requires several minutes of computer time, and even with efficient optimization techniques, much computer time will be used.

The second means available experimentally is the use of absorption data. A typical absorption plot is shown in Figure 2; and it can be seen that the curve tends to become linear at long times. The slope of this line is equal to $\pi^2 D_v / \ell^2$. To perform the experiment, several absorption curves are obtained for corresponding equilibrium volume fractions of the penetrant. The variation in vapour pressure of the experiment provides the variation in volume fraction of the penetrant.

The diffusion coefficient at zero concentration is obtained by extrapolation of the diffusion coefficient data to zero concentration. The data are then plotted as $\ln D_v / D_{v_0}$ versus the reciprocal of the volume fraction. It can be shown that the slope of the resulting line is $[f(0,T)]^2 / \beta(T) B_d$ and the intercept $f(0,T) / B_d$. Coupled with thermal expansion data for the pure liquid, the three parameters may be obtained. A typical plot in this fashion is shown in Figure 3 for n-heptane and low density polyethylene.

Unlike the desorption technique, a minimum of calculations is necessary, however, the method does suffer one large drawback which can be appreciated by looking closely at the figure. The value of the intercept is very sensitive to the slope of the line because the absorption data are obtained far away from $1/v = 0$. Therefore, a relatively large number of experiments should be done in order to ensure the accuracy of the value of the intercept.

A convenient way of obtaining the absorption and desorption data is with an Electrobalance. This balance automatically produces a voltage output proportional to the difference in weight between the two balance arms. In this way, a complete time-absorption history may be recorded. Additional details of this technique can be found in Reference (2).

EXPERIMENTAL RESULTS

The separation behavior of mixtures of organic compounds (which swell the film) through a polymer film is of a fairly

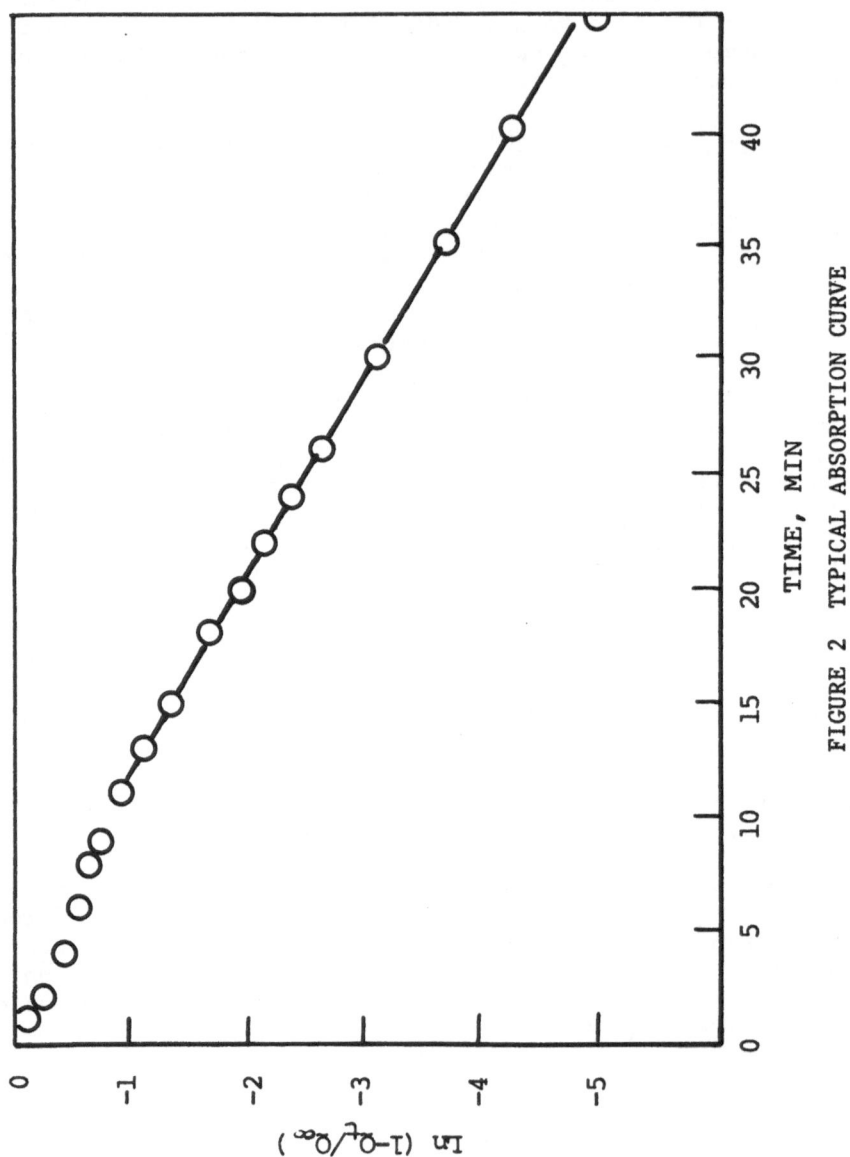

TIME, MIN

FIGURE 2   TYPICAL ABSORPTION CURVE

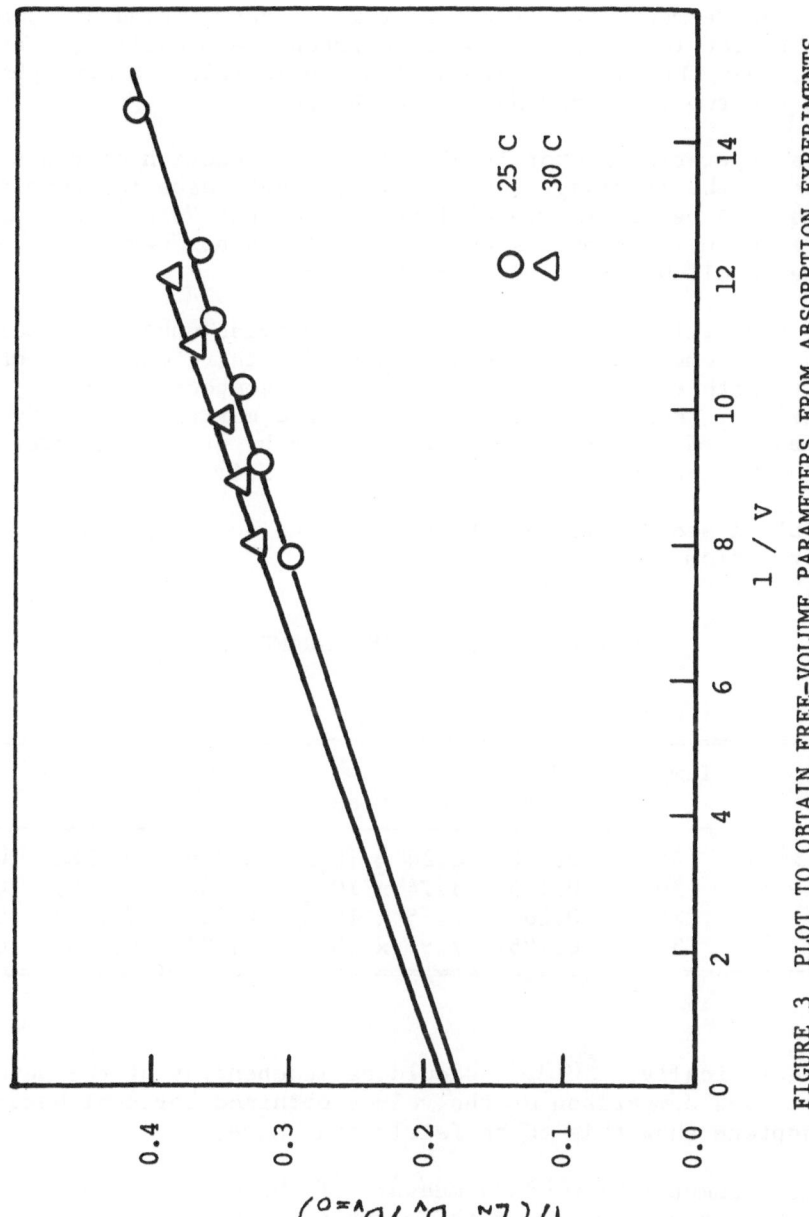

FIGURE 3   PLOT TO OBTAIN FREE–VOLUME PARAMETERS FROM ABSORPTION EXPERIMENTS

complex nature as shown in Figures 4 and 5. Figure 4 shows the steady-state permeability of a mixture of n-heptane and toluene plotted against the weight percent toluene in the mixture, and Figure 5 the functional behavior of the separation factor. The important point to note from the permeability curves is that in no way can the permeability of the mixture be represented by some mole or weight fraction times the pure component permeability. Were this so, then the mixture permeability would follow a straight line between the two pure component permeabilities.

The separation factor is also a strong function of the composition of the liquids. Qualitatively, the reason for this is that liquid A permeating in a mixture, does not "see" the same system as when it permeates by itself. It is now permeating through a polymer swollen by the mixture of A and B.

Li has utilized this phenomenon to provide enhanced separation between two components in the presence of a third component which interacts with the membrane and one of the components to be separated[3]. For example, separation of a mixture of N_2, O_2 and NH_3 is enhanced by a factor of from 2 to 5 by utilizing water vapor in the gas and a hydrophilic polymer film.

Table 1 shows some results of the free volume parameters as calculated from absorption data.

TABLE 1. FREE VOLUME PARAMETERS

Vapor	Temp., C	V_0 cc/cc	$D_{v=0}$ sq cm/sec	B_d	$f(0,T)$	$\beta(T)$
Cyclohexane	25	0.270	2.24×10^{-9}	0.246	0.0382	0.276
Cyclohexane	30	0.275	3.74×10^{-9}	0.246	0.0405	0.278
Heptane	25	0.16	5.25×10^{-9}	0.221	0.0387	0.437
Heptane	30	0.185	7.9×10^{-9}	0.221	0.0405	0.442

Theoretically, $f(0,T)$ should be independent of the penetrant molecule, and comparison of the values obtained for cyclohexane and n-heptane show this to be fairly true here.

The parameter $\beta(T)$ is a measure of the plasticizing effect of a unit volume fraction of the penetrant, and it is seen that n-heptane has a greater tendency for interaction with polyethylene.

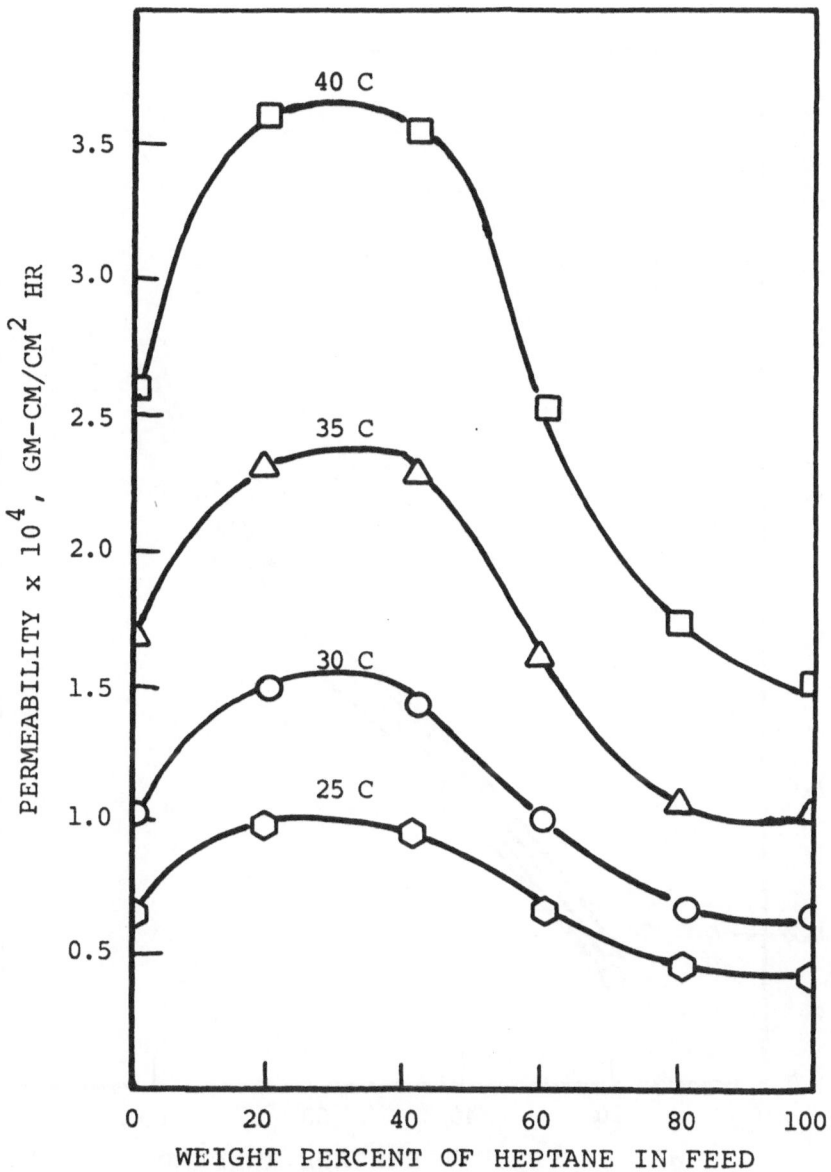

FIGURE 4   EFFECT OF FEED COMPOSITION ON THE PERMEABILITY
OF HEPTANE-TOLUENE MIXTURES

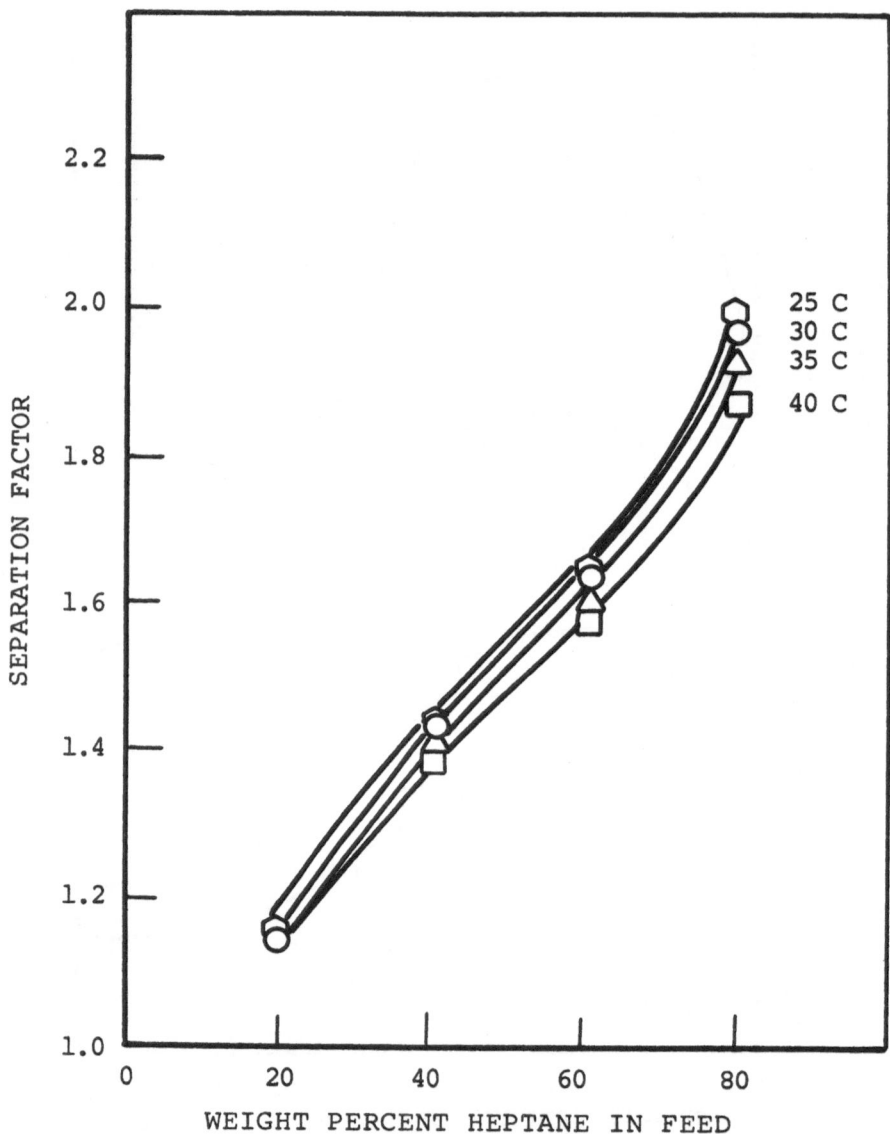

FIGURE 5   EFFECT OF FEED COMPOSITION ON THE SEPARATION
OF HEPTANE-TOLUENE MIXTURES

This is logical in that the heptane molecule is chemically more similar to polyethylene.

The free volume parameters so obtained were then used to calculate permeabilities of the individual species in a mixture of cyclohexane and heptane. Table 2 shows the comparison of the actual permeability as measured under steady-state conditions, with that calculated using the free volume approach.

TABLE 2. COMPARISON OF STEADY AND UNSTEADY STATE PERMEABILITIES FOR MIXTURES OF CYCLOHEXANE AND n-HEPTANE

Temp., C	Feed Composition, weight percent cyclohexane							
	80		60		40		25	
	Cyclo-hexane Ratio	Hep-tane Ratio	Cyclo-hexane Ratio	Hep-tane Ratio	Cyclo-hexane Ratio	Hep-tane Ratio	Cyclo-hexane Ratio	Hep-tane Ratio
25	1.27	1.62	1.15	1.36	0.88	1.32	0.86	0.85
30	1.07	1.65	0.99	1.33	0.80	1.27	0.75	0.80

The ratio shown is the actual permeability divided by the calculated one. Although some of the data show good agreement, for example, cyclohexane at 80 and 60 percent, the general agreement is only fair. The main reason for this is due to inaccuracies in obtaining the calculated diffusion coefficients, and free volume parameters because of the sensitivity of these parameters to the value of the intercepts. Another reason for these discrepancies may be due to conditioning of the membrane by the organic liquid. In the conditioning process, the polymer chains undergo an irreversible rearrangement which is a function of time, temperature and the nature of the liquid. These conditioning effects would be expected to be different for a steady-state and unsteady-state experiment.

Where does this leave us regarding membrane permeation with respect to practical separations? It appears that membrane separations have many advantages and models are available which do a reasonable job of predicting the permeation and separation behavior of polymer films. However, it also appears that it is necessary to increase the rate of permeation by several orders of magnitude in order for the process to compete economically with conventional separation processes. Several methods have been tried and do increase rates, but not by an order of magnitude. These processes include pre-conditioning, irradiation of the film, and grafting of a second component into the membrane.

One means which is presently being investigated is based on the thin-film concept which has proven successful in reverse osmosis. If it were possible to prepare submicron ultra-thin films which were stable, strong and pinhole-free, a large increase in rate would result, and theoretically the good separation could be maintained. One method being attempted at present is to evaporate a thin film onto a suitable substrate. The evaporation takes place in an electron beam welding chamber shown here. A crucible with the polymer is placed in the chamber under high vacuum, and the electron beam is focused on the polymer. The polymer is collected on the surface of the substrate. Figure 6 shows the general layout of the electron beam equipment, and Figure 7 the set-up of the deposition portion.

The films made by this technique were tested for permeation properties and it was found that they had unsatisfactory membrane properties. The reasons for this were deduced by study of scanning electron micrographs and IR measurements. The SEM photographs of the substrate (a Teflon fiber filter) before and after deposition of the polymer (polyethylene) are shown in Figures 8 and 9 respectively. It can be seen that the film had numerous holes in it and did not coat the substrate evenly.

Analysis of IR measurements showed that the polymer deposited was markedly different than the starting material and indicated that substantial depolymerization had taken place. This observation was also confirmed by the fact that deposit on a glass slide had a waxy consistency.

Several modifications of the technique are being tried at present, including ultraviolet irradiation and γ-radiation of the deposited films in an attempt to improve the film properties by polymerization.

CONCLUSIONS

In conclusion, it can be said that membrane processes show definite promise of becoming practical for certain separations. Although the models described here are somewhat impractical to use for predicting separation behavior, nevertheless, an insight has been gained into some of the fundamental processes going on in membrane separation.

FIGURE 6   GENERAL LAYOUT OF ELECTRON BEAM DEPOSITION EQUIPMENT

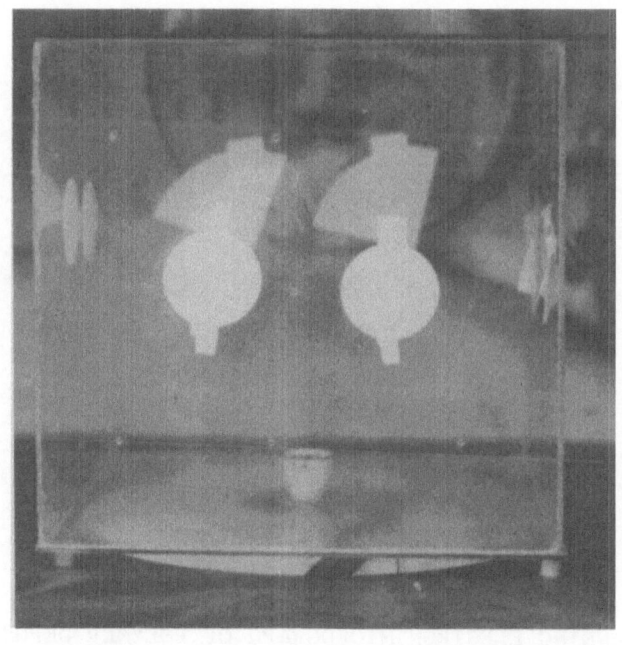

FIGURE 7   DEPOSITION CHAMBER IN ELECTRON BEAM EQUIPMENT

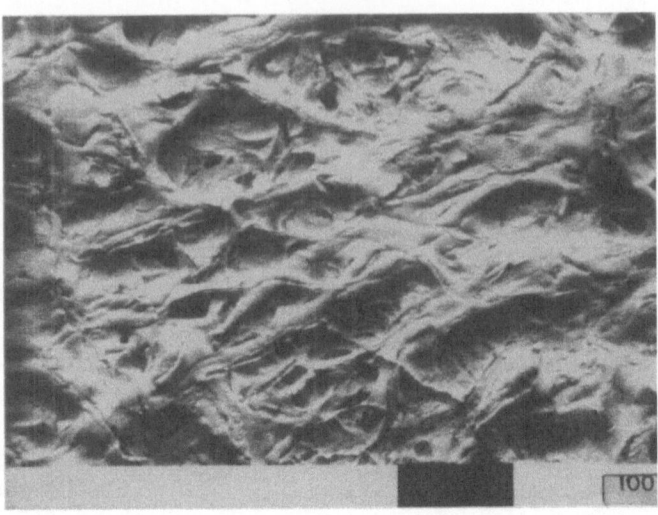

FIGURE 8   SCANNING ELECTRON MICROGRAPH OF SUBSTRATE SURFACE PRIOR TO
POLYMER DEPOSITION

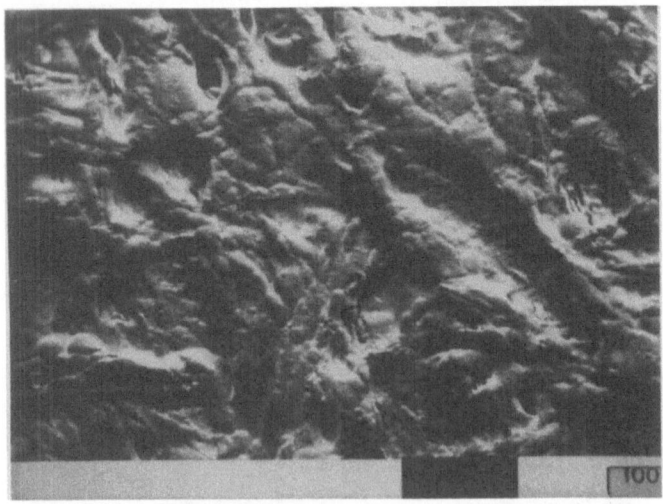

FIGURE 9   SCANNING ELECTRON MICROGRAPH OF POLYMER DEPOSITION BY THE
ELECTRON BEAM

Ultra-thin films show definite promise for providing more practical separations. Their manufacture will require additional experimental efforts.

BIBLIOGRAPHY

1. Rogers, C. E., Fels, M., and Li, N. N., "Separation by Per-
    meation Through Polymer Membranes", Chapter in Recent Develop-
    ments in Separation Science, Vol 11, N. N. Li, ed., CRC Press,
    Cleveland, Ohio, 1972.

2. Recent Advances in Separation Technology, Li, N. N., Fels, M.,
    and Matulevicius, E. S., ed., AIChE Symposium Series 68 (120)
    (1972), "Permeation and Separation Behavior of Binary Organic
    Mixtures in Polyethylene", Fels, M., p 49.

3. Li, N. N., "Membrane Separation", U.S. Pat. 3,566,580
    (March 2, 1971).

THIN-FILM COMPOSITE MEMBRANE PERFORMANCE IN A SPIRAL-WOUND

SINGLE-STAGE REVERSE OSMOSIS SEAWATER PILOT PLANT

R. L. Riley, G. R. Hightower, C. R. Lyons, and M. Tagami

ROGA Division, Universal Oil Products

San Diego, California 92138

INTRODUCTION

Desalination of seawater by reverse osmosis requires a membrane that approaches theoretical semipermeability and is sufficiently thin to provide a rapid transport of water at practical pressures. To be economical, the process must operate in a single stage at a water recovery of 30% and greater. Thus, the membrane must reject about 99.5% of the sodium chloride to produce potable water. Regardless of composition, membranes of this type must essentially be free of imperfections.

The thin-film composite membrane was developed to facilitate single-stage seawater desalination at operating pressures of 68 atm (1000 psi) and less. Unlike asymmetric cellulose diacetate Loeb-Sourirajan brackish water desalination membranes [1], the composite membrane is prepared by forming a thin polymer film directly upon the surface of a finely porous supporting membrane by dipping from a dilute solution [2]. The advantage of this type of construction, shown in Fig. 1, is obvious: the porous substrate provides the necessary strength for the extremely thin semipermeable barrier.

This general method of composite membrane production provides several other advantages, including the following:

1. Independent selection of materials from which to prepare the thin semipermeable barrier and the finely porous supporting membrane.

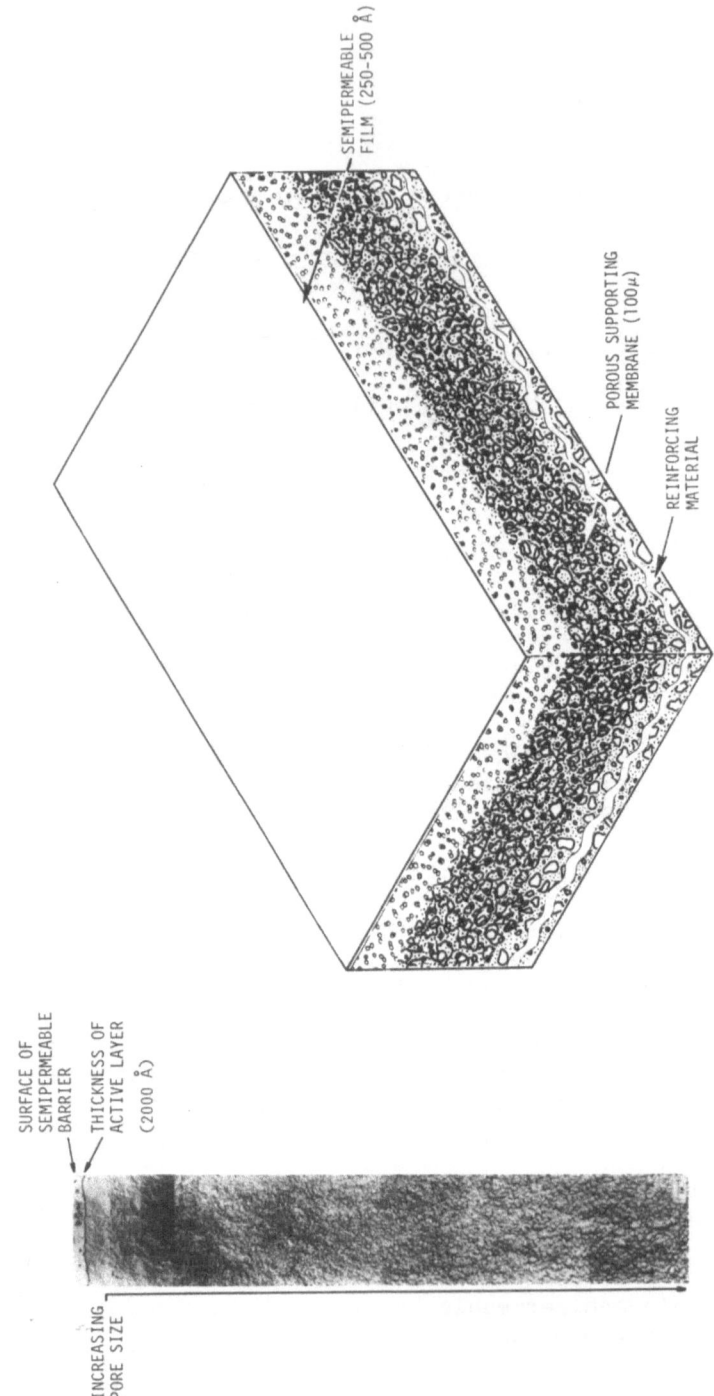

Fig. 1 - Electron Photomicrograph of a 30-μm Cross Section of a Loeb-Sourirajan Modified Membrane (Left) and a Cross-Sectional Drawing of a Thin-Film Composite Membrane (Right)

2. Independent preparation of the thin film and the porous supporting membrane, making it possible to optimize each component for its specific function.

3. Reproducible variation and control over the thickness of the thin semipermeable barrier.

TRANSPORT PHENOMENOLOGY

The movement of water and simple salts through the thin semipermeable film obeys a solution-diffusion model of membrane transport [3]. Thus, the water flux J is given by

$$J_1 = \frac{D_1 c_1 \bar{v}_1 (\Delta P - \Delta \pi)}{RT \Delta x} \equiv A(\Delta P - \Delta \pi) \quad , \qquad (1)$$

where D_1 = the diffusion coefficient of water in the membrane (cm^2/sec),

c_1 = the concentration of water in the membrane (g/cm^3),

$\bar{v}_1$ = the partial molar volume of water in the external phase ($cm^3/mole$),

ΔP = the applied pressure (atm),

$\Delta \pi$ = the osmotic pressure difference across the membrane (atm),

R = the gas constant (cm^3-atm/mole-°K),

T = the absolute temperature (°K), and

Δx = the effective membrane thickness (cm).

The proportionality factor A, referred to as the membrane constant, is a measure of the water flux per unit net pressure (g/cm^2-sec-atm).

The salt flux J_2 through an imperfection-free film is similarly given by

$$J_2 = \frac{D_2 K \Delta \rho_2}{\Delta x} \quad , \qquad (2)$$

where D_2 = the diffusion coefficient of salt in the membrane (cm^2/sec),

K = the distribution coefficient for salt between membrane and solution [(g/cm^3 membrane)/(g/cm^3 solution)], and

$\Delta\rho_2$ = the difference in salt concentration across the membrane (g/cm^3).

The quantities D_1c_1 and D_2K are permeability coefficients.

The salt rejection is defined by

$$S = (\rho_2' - \rho_2'')/\rho_2' \, , \qquad\qquad\qquad (3)$$

where (') and (") refer to feed and product solutions, respectively. In terms of the flux equations, salt rejection is given for dilute solutions by

$$S = 1 - \frac{J_2}{J_1\rho_2'} = \left[1 + \frac{D_2KRT}{D_1c_1\bar{v}_1\,(\Delta p - \Delta\pi)} \right]^{-1} . \qquad (4)$$

The permselectivity increases while the water permeability D_1c_1 decreases and the acetyl content of the membrane increases [4]. The salt rejection is independent of the thickness of the thin film. The water flux, on the other hand, is inversely proportional to the thickness of the thin film [5]. Thus, the water flux through the composite membrane depends on the water permeability of the thin-film material, the thickness of the thin film, and the number and characteristics of the pores on the surface of the supporting membrane [6].

EXPERIMENTAL

Membrane Preparation and Performance

The practicality of producing large quantities of unsupported composite membrane has been demonstrated on a continuous prototype casting machine [2]. The machine, shown schematically in Fig. 2, processes the porous cellulose nitrate - cellulose acetate supporting membrane and the thin semipermeable barrier in a single operation. The latter is formed directly upon the finely porous surface of the support membrane by dipping or by wicking from a dilute solution of cellulose triacetate in chloroform. The thickness of the thin film is controlled by the concentration of the polymer in the dilute solution and by the withdrawal rate according to the Levich-Deryaguin theory [7]. The thin film can be formed either on the prototype casting machine or,

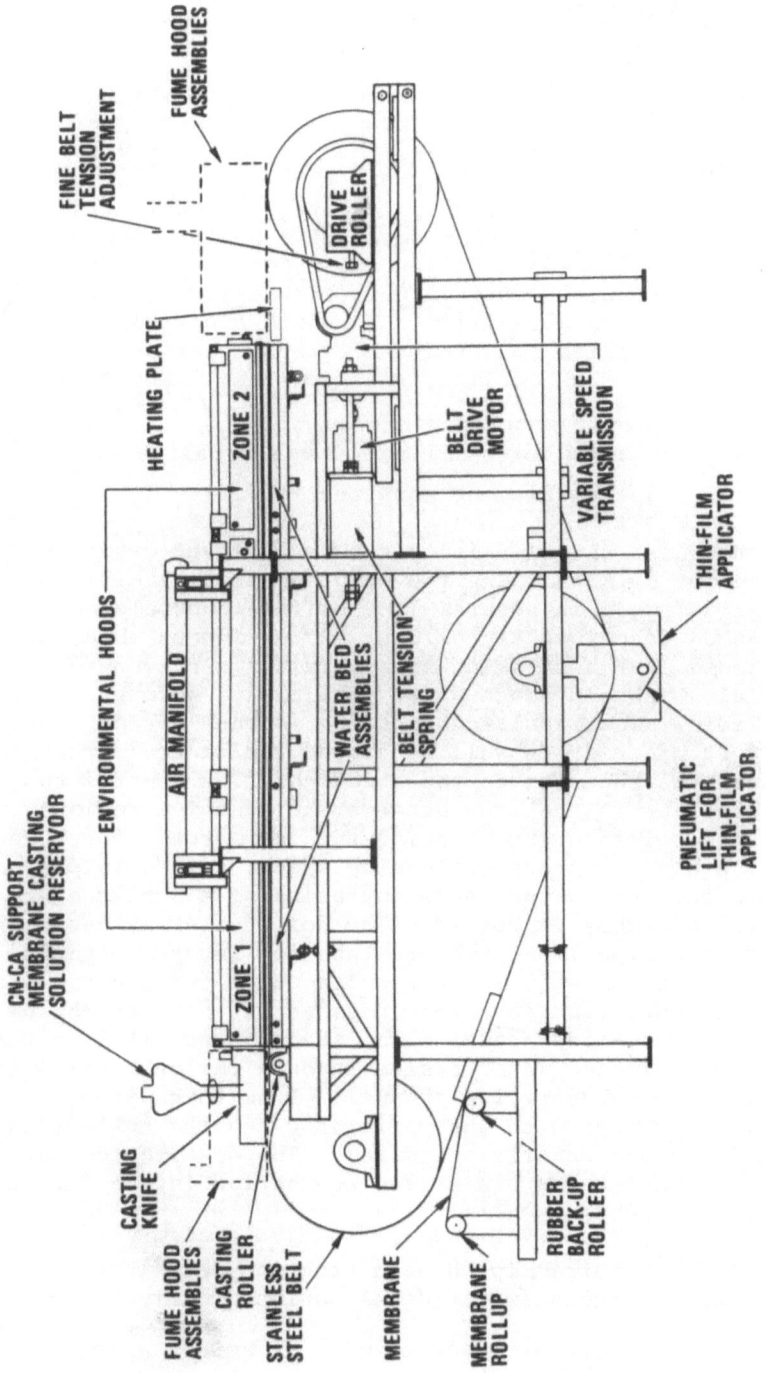

Fig. 2 – Schematic Drawing of Composite Membrane Casting Machine

alternatively, on a device removed from the prototype machine [8]. Thus, membrane is produced under conditions that represent a meaningful reduction to practice on a scale sufficient to permit valid estimates of performance, costs, and equipment configuration.

The estimated thickness of the cellulose triacetate thin film of the composite membrane is 250 Å. Membranes of this type exhibit water fluxes of 30 gal/ft^2-day when tested in reverse osmosis at an applied pressure of 102 atm (1500 psi) using a 3.5 wt % sodium chloride brine; the sodium chloride rejection is generally in excess of 99.5%. The comparative performance of composite and modified Loeb-Sourirajan membranes is presented in Fig. 3. Also of particular importance is the reproducibility of membrane performance, the durability and hydrolysis resistance of the membrane in unacidified seawater for extended periods of time, membrane resistance to flux decline, and membrane insensitivity to chlorine.

Spiral-Element Construction

Plant compactness and simplicity can be achieved with the spiral-wound design by maximizing the membrane area within a given pressure vessel volume. In general, the spiral-wound element concept [3] can be described as follows. A continuous sheet of membrane is folded over a sandwich of several product-water-side backing materials with a product water tube placed between the two layers at the fold. The resulting sandwich is then sealed along both sides and at one end, as well as around the central product water tube where the tube leaves the sandwich. The central tube wall has perforations in the area between the adhesive regions. This unit is then rolled up by wrapping the leaf around the tube. Before rollup, a highly porous brine-side spacer screen is placed on one membrane surface so that adjacent brine-side surfaces are separated as the assembly is rolled. The coiled element assembly is then placed in a pipe-like pressure tube for reverse osmosis testing. The element is sealed between its outer surface and the lining of the pressure tube to direct passage of the total brine flow through the brine-side spacer in a direction parallel to the axis of roll. Desalinated water passes through the membrane into the porous product-water-carrying channel. From this point, the water flows spirally inward to the perforated central collecting tube at essentially atmospheric pressure. The desalinated water subsequently flows out through a sealed connection in the capped end of the pressure tube.

The perfection required in the construction of spiral-wound elements for single-stage seawater desalination greatly exceeds

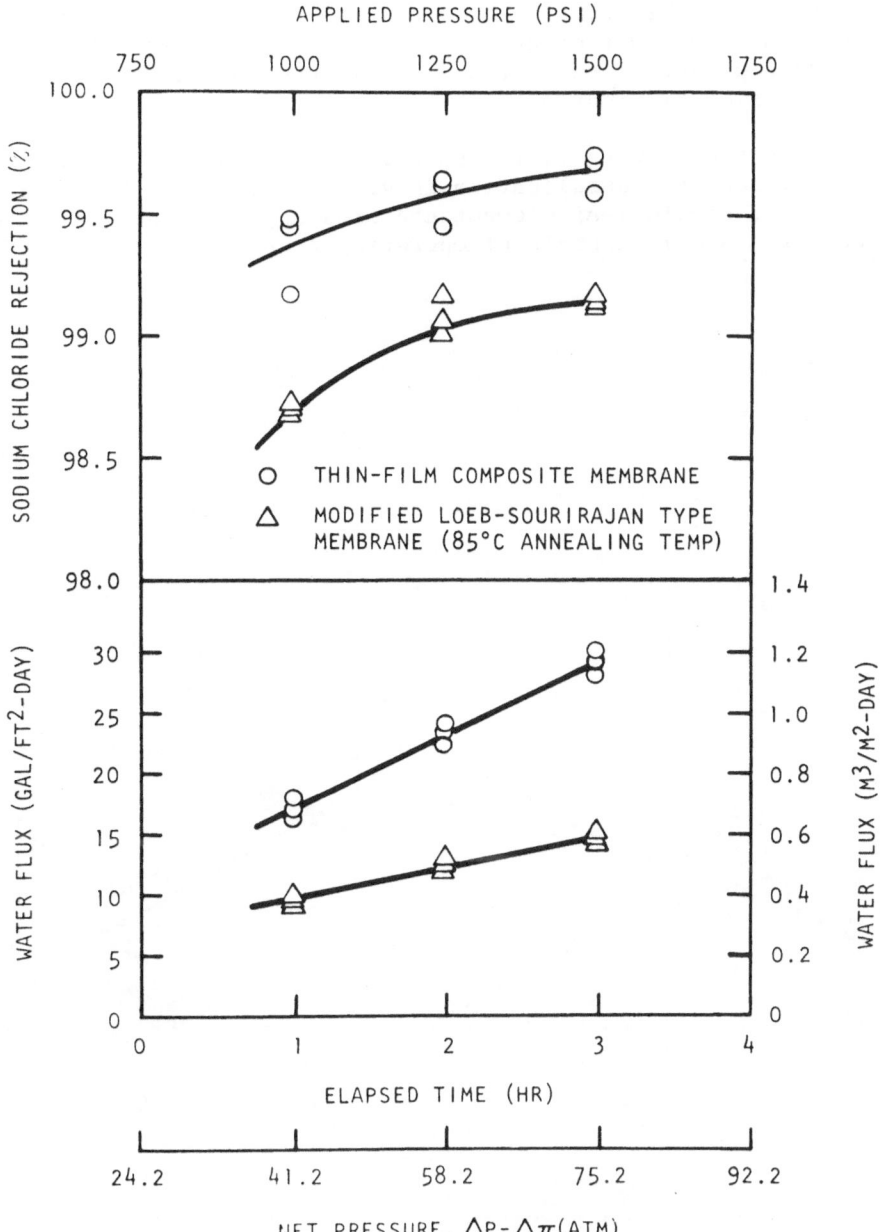

Fig. 3 — Reverse Osmosis Performance for Three Cellulose Triacetate
Thin-Film Composite Membranes and Three Modified Loeb-
Sourirajan Type Membranes as a Function of Applied Pressure
With 3.5 wt % Sodium Chloride (Simulated Seawater)

that required for brackish water elements. A single imperfection
in either the construction and/or materials used in the element
represents a potential salt leak. It is fundamental that
imperfections be eliminated.

A schematic drawing of the spiral-wound composite membrane
element for seawater desalination is shown in Fig. 4. The 2 in.
by 10-3/4 in. single-leaf element contains approximately 4 ft^2 of
membrane area and is capable of operating at pressures up to 102
atm (1500 psi).

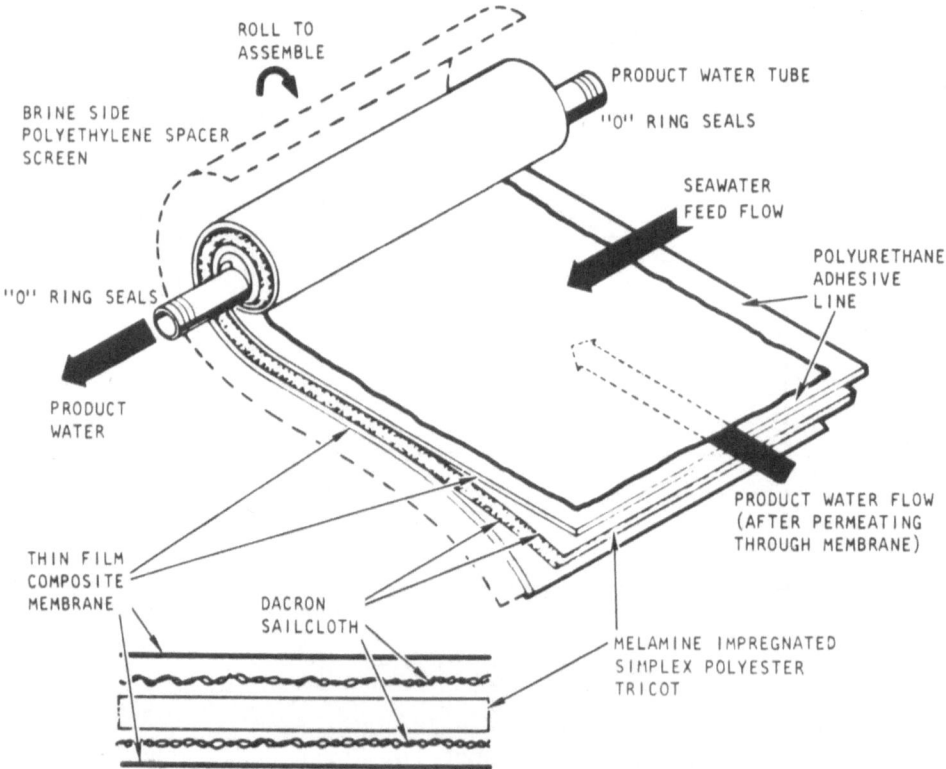

Fig. 4 - Schematic Drawing of Spiral-Wound Thin-Film Composite
Membrane Element Construction

PILOT PLANT OPERATION AND PERFORMANCE

A single-stage seawater pilot plant, shown in Fig. 5, was designed and constructed to evaluate the composite membrane [9,10]. The plant, located at Seaworld on Mission Bay in San Diego, California, operates with twenty-four 2 in. by 10-3/4 in. spiral-wound composite membrane elements. The purpose of the test plant, at least during the initial phase of element development, was to collect data. As a result, a large amount of instrumentation was included as an integral part of the plant.

Operation of the pilot plant is shown schematically in Fig. 6. Seawater is taken from Mission Bay at a depth of about 10 ft, chlorinated with 2 ppm chlorine, and filtered through the sand filtration beds shown in Fig. 7. The pretreated seawater, which contains 0.5 ppm chlorine, is passed through a 5-μm cartridge filter to remove sand before entering the high-pressure pump.

Fig. 5 – Single-Stage Seawater Pilot Plant Operated at Seaworld on Mission Bay in San Diego, California

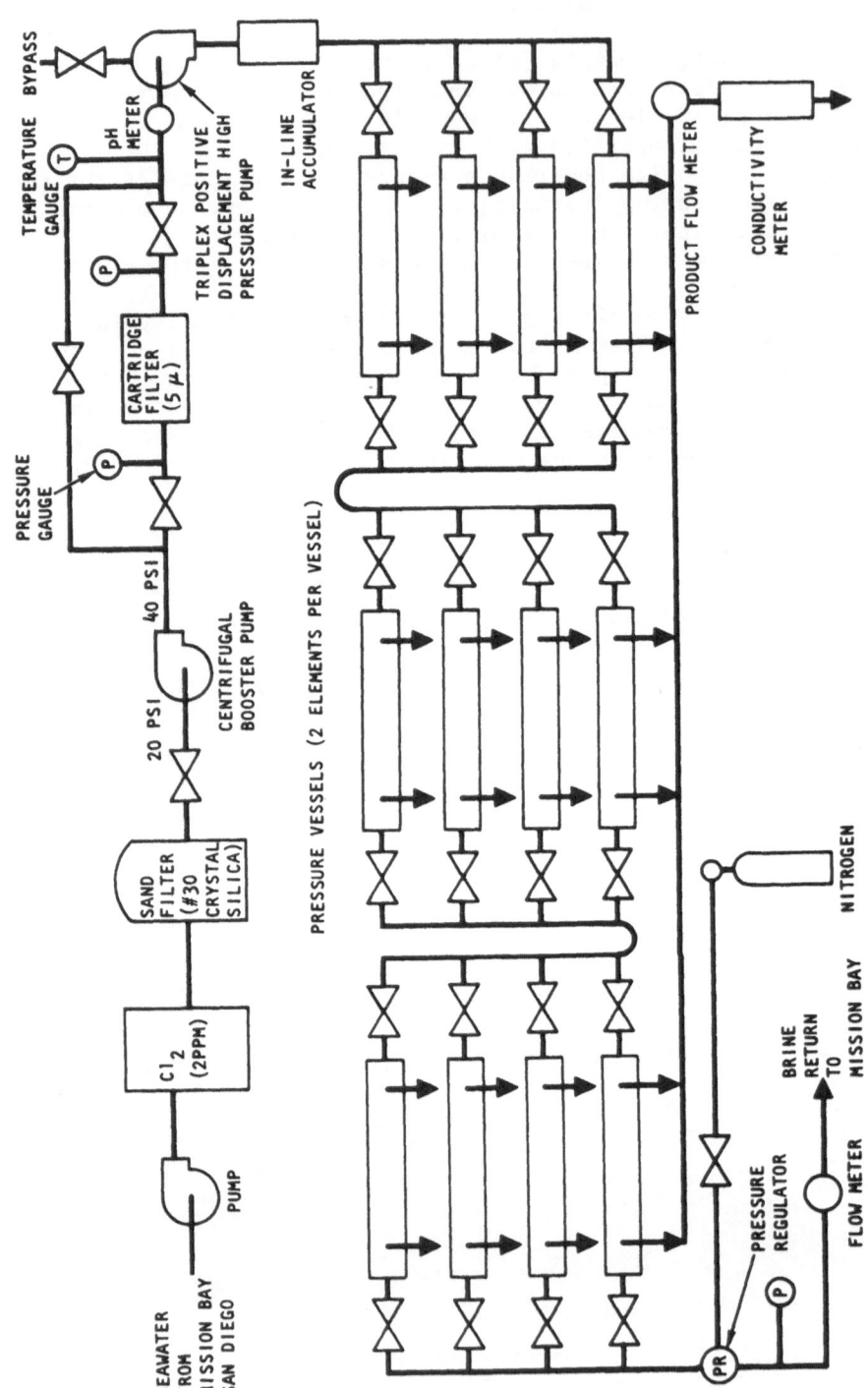

Fig. 6 – Operational Schematic of Single-Stage Seawater Pilot Plant

Fig. 7 - Sand Filtration Beds for Single-Stage Seawater Pilot
        Plant

    The plant operates on unacidified seawater (pH = 8) at an
applied pressure of 68 atm (1000 psi). The temperature of the
seawater ranges from 13° to 20°C. The brine flow across the
elements is maintained at 2 gal/min; the water recovery is
approximately 7%. The concentration of the seawater is
essentially equal to that of the open Pacific Ocean.

    The performance and reliability of the plant have been
gratifying. The total plant performance, of twenty-four spiral-
wound elements corrected to 25°C, is shown in Fig. 8.

    During the first 2200 hr, the plant produced an average of
700 gal of potable water per day. The membrane flux averaged
about 8 gal/ft^2-day. The flux decline was abnormally high based

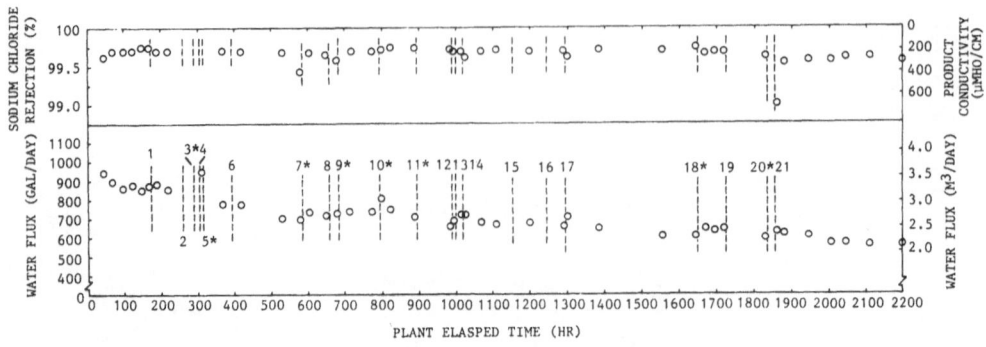

1,2,4,6,8,14,15,16,19,21: SHUTDOWN

3,7,13,17,18: SHUTDOWN; SYSTEM CLEANED WITH ALCOZYME DETERGENT

5,9,10,11,20: PRESSURE REDUCED TEMPORARILY TO 200 PSI

12: SYSTEM CLEANED WITH CITRIC ACID

* ELEMENT CHANGE

Fig. 8 - Total Plant Performance of Twenty-Four Spiral-Wound Thin-
Film Composite Membrane Elements

on the properties of the composite membrane. This is attributed
to membrane fouling, which developed as a result of the minimal
pretreatment available at Seaworld. A periodic cleaning cycle was
established with an enzyme detergent but was not particularly
effective.

The quality of the product water was excellent. The
conductivity of the product water averaged 200 μmho/cm. The
sodium chloride rejection exceeded 99.7%.

At 40.8 atm (600 psi), the plant produced 300 gal of potable
water per day having a conductivity of 500 μmho/cm. The sodium
chloride rejection was 99.2%. It is apparent, therefore, that the
composite membrane is capable of desalinating seawater in a single
stage at relatively low operating pressures.

SUMMARY

A general method of preparing semipermeable membranes has
been developed whereby a 250-Å thin film of a polymer is formed
directly upon the finely porous surface of a support membrane by
dipping from a dilute solution. The thickness of the thin film
can be reproducibly varied and controlled. The membrane, capable
of desalinating seawater in a single stage, combines high flux and
stability with excellent selectivity. The membrane is processed
in large quantity on a continuous prototype casting machine.

A single-stage seawater pilot plant has been in operation for 1 year using the cellulose triacetate thin-film composite membrane in the spiral-wound configuration. The plant, located at Seaworld on Mission Bay in San Diego, California, operates on unacidified seawater at an applied pressure of 68 atm (1000 psi) with twenty-four 2 in. by 10-3/4 in. spiral-wound elements, each containing approximately 4 ft^2 of membrane. The performance and reliability of the system have been excellent. To date, more than 200,000 hr of element testing have been completed.

ACKNOWLEDGMENTS

The authors would like to acknowledge Frank Powell and John Rognlie of Seaworld Marine Park in San Diego for their cooperation and for providing the site to operate the seawater test plant. This work is supported by the Office of Saline Water, U.S. Department of the Interior, under Contract 14-30-3016. We are grateful to Drs. W. S. Gillam, J. L. Leiserson, and L. M. Kindley of that office for their continued support and encouragement.

REFERENCES

1. S. Loeb and S. Sourirajan, Advan. Chem. Ser., 38, 117 (1962).
2. R. L. Riley, C. E. Milstead, W. J. Wrasidlo, G. R. Hightower, C. R. Lyons, and M. Tagami, "Research and Development on a Spiral-Wound Membrane System for Single-Stage Seawater Desalination," Office of Saline Water Research and Development Program Report, U.S. Government Printing Office, Washington, D.C., to be published.
3. U. Merten, (Ed.), Desalination by Reverse Osmosis, M.I.T. Press, Cambridge, Massachusetts, 1966.
4. H. K. Lonsdale, U. Merten, and R. L. Riley, J. Appl. Polymer Sci., 9, 1341 (1965).
5. R. L. Riley, H. K. Lonsdale, and C. R. Lyons, in Proceedings of the 3rd International Symposium on Fresh Water from the Sea, Dubrovnik, Yugoslavia, September 13-16, 1970, Vol. 2, p. 551, 1970.
6. H. K. Lonsdale, R. L. Riley, C. R. Lyons, and D. P. Carosella, Jr., Membrane Processes in Industry and Biomedicine, Plenum Press, New York, 1971, p. 102.
7. V. G. Levich, Physiochemical Hydrodynamics, Prentice Hall, Englewood Cliffs, New Jersey, 1962.
8. R. L. Riley, G. R. Hightower, and C. R. Lyons, Thin Film Composite Membrane for Single-Stage Seawater Desalination by Reverse Osmosis, No. 22, p. 255, John Wiley and Sons, Inc., 1973.

9. R. L. Riley, G. R. Hightower, C. R. Lyons, and M. Tagami, in _Proceedings of the 4th International Symposium on Fresh Water from the Sea_, Heidelberg, Germany, September 9-14, 1973, Vol. 4, p. 333, 1973.

10. R. L. Riley, C. R. Lyons, G. R. Hightower, M. Tagami, and F. K. Lesan, "Research and Development of Composite Membrane Technology," Office of Saline Water Research and Development Progress Report, U.S. Government Printing Office, Washington, D.C., in press.

THE EVOLUTION OF PHASE INVERSION MEMBRANES

R. E. Kesting

Chemical Systems Incorporated

1852 McGaw Avenue, Irvine, California 92705

INTRODUCTION

The general technique for the preparation of thin polymeric structures by the controlled desolvation of (usually multicomponent) solutions is known as <u>phase inversion</u>. It is one of the oldest and most versatile fabrication processes and has been applied to the manufacture of synthetic leathers, fibers, films and membranes. The present review seeks to trace the evolution of the latter at four different levels: Configuration, physical structure, chemical nature, and manufacturing process.

CONFIGURATION

The oldest and simplest membrane configuration is the flat sheet (fig. 1). This configuration was employed by *Fick* for collodion (nitrocellulose) membranes in his 1855 studies on diffusion [1]. Flat sheet membranes possess a number of very real competitive advantages relative to their alternatives. In the first place they exhibit a higher degree of perfection. For a given permeability, flat sheet membranes exhibit the highest degree of permselectivity; conversely, for a given degree of permselectivity, they have the highest permeability. In other words for a given effective area per volume of membrane element, flat sheet membranes will be the most efficient. The reasons for this are not known in any quantitative sense; qualitatively, however, the maintenance of the casting solution in the plane horizontal to the casting surface imposes restraints on the consolidation and compaction of the sol and gel precursors to ultimate membrane gel structure. Shrinkage occurs primarily in one dimension, i.e., in the plane vertical to the casting surface. Another factor which must play a role to an as yet undetermined extent is the greater ease of maintaining specific envir-

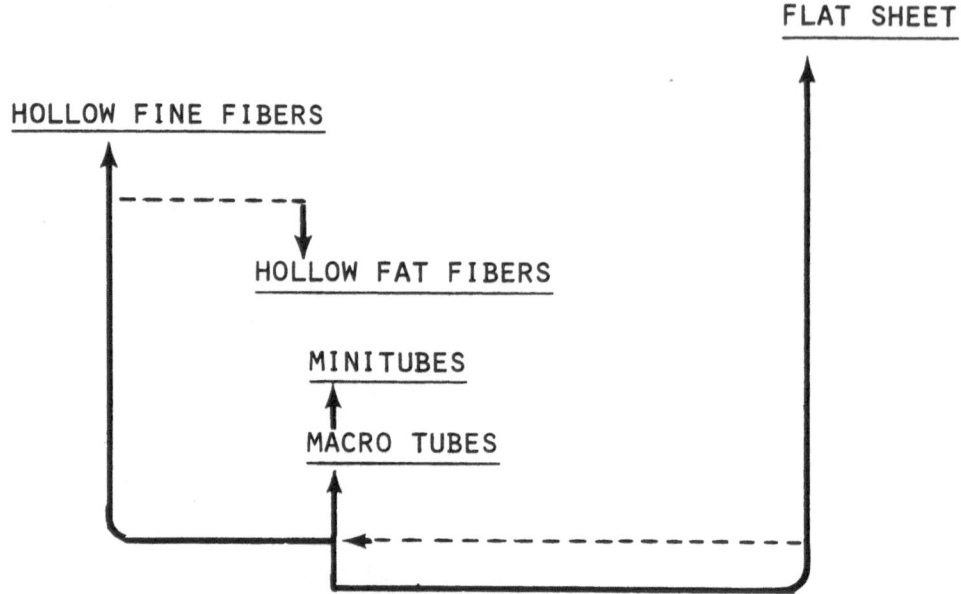

Figure 1. Evolution of Membrane Configurations

onmental conditions during the fabrication process. The greater
perfection of flat sheet relative to, e.g., macrotubular (> 5 mm
in diameter) membranes, has prompted some proponents of the latter
to actually prepare membranes in flat sheet form for ultimate use
as tubes. Thus American Standard once utilized a ribbon of flat
sheet Wet Process membrane which it wound around a mandrel for insertion into a tube. In this case the active surface of the membrane was on the inside of the tube. The alternate procedure, that
of wrapping a ribbon of membrane on the outside of the tube is currently being utilized by Universal Water Systems. In both cases a
net gain in performance characteristics was achieved by the prior
preparation of the membrane in the flat sheet configuration.

     Today flat sheet membranes are employed both as flat sheet
disks (primarily for microfiltration applications) and as membrane
envelopes sealed along three sides and wound up in such a manner as
to permit the product to spiral towards a perforated product tube
into which the permeate enters from the unsealed edge of the membrane
envelope. This so-called spiral configuration incorporates a large
effective membrane area per unit volume of element.

What of macrotubular membranes cast as tubes? The history of macrotubular membranes is almost as long as that of flat sheets. *Schumacher* in 1860 dipped test tubes into collodion solutions to produce nitrocellulose membranes which remained popular through the turn of the present century [2]. Macrotubular membranes are currently produced by UOP (formerly Calgon-Havens), Abcor, Raypak, Aquachem and by a number of manufacturers in England and Germany. The first two of these utilize fiber glass tubes with the membrane on the inside in the design pioneered by Havens. Raypak utilizes ceramic tubes with the membrane on the outside. Philco Ford once utilized membranes which were cast on the inside of metal tubes, removed, and supported in a separate operation by a braided polyester tube. Although macrotubes are not considered economical for water purification and desalination applications, they are of interest where higher costs can be tolerated and where high solids contents are encountered. Membranes cast on the exterior surface of tubes tend to be highly permselective with a modest permeability. These characteristics are a consequence of the shrinkage onto an unyielding tube. Membranes cast on the inside of the tubes tend to be more permeable because of the high concentration of solvent vapors which acts to retard the formation of a skin.

The advantages of the tubular configuration are well-defined fluid dynamics, low tendency to foul, and capability for working with solutions or suspensions with a high solids content. The principal disadvantage of the tubular configuration is an extremely low effective membrane area per volume element. This makes tubular configurations excessively large and costly. Because the high intrinsic cost of macrotubes is attributable to their size, substantial efforts at size reduction have been made. These attempts fall into two categories: Minitubes and Hollow Fibers. The former (2-4 mm in diameter) are truly a miniaturization of macrotubes. An interesting variation has been the spaghetti tubing of Patterson-Candy, an English firm. Spaghetti-sized polypropylene tubing with two indentations in the cylinder is extruded, covered with a polyester sleeve, and then coated with a cellulose acetate membrane. The product permeates the membrane and polyester sleeve and is carried along the clefts in the cylinder to the ends of the tubing for collection.

Hollow fibers represent a more fundamental departure from macrotubes than do the minitubes. In this regime the effective membrane area per unit volume can be very high indeed. The first hollow fibers to be developed were the hollow fine HF (50-300 μm in diameter) fibers. The skin side of HF fibers is most commonly encountered on the outside. Because of its strength under compression, this type is able to serve as its own pressure vessel in which product flows down the interior of the fiber to the end of a tube sheet for collection.

The principal advantage of HF fibers is their high packing density which is an absolute necessity inasmuch as their performance characteristics on an area basis are not competitive with those of flat sheet membranes. Their principal disadvantage is their pronounced tendency to be fouled by colloidal solutes. This feature is so marked as to necessitate a very high degree of pretreatment. Furthermore, once fouled, HF fibers are cleaned only with the greatest difficulty. This tendency to foul is not a function of the chemical nature of the membrane polymer since fouling occurs on both cellulosic and polyamide fibers.

Partially to minimize the HF fiber fouling problem and partially to accommodate those applications for which somewhat larger fiber diameters are preferred, attention is currently being focused on hollow "Fat" fibers. Fat (400-1000 μm in diameter) fibers can be made with external and/or internal skins. They are less subject to fouling than the HF type. Fat fibers appear certain to find application in such areas as hemodialyzers for patients with chronic uremia.

At the present time, commercial interest appears to be about equally divided between flat sheet (spiral elements) and hollow fibers, with macrotubes and minitubes restricted to specialty applications. For the most general case wherein only a moderate amount of pretreatment can be tolerated, the flat sheet configuration appears to be gaining favor. Where pretreatment cost is no object, hollow fine fibers may be in the ascendancy.

PHYSICAL STRUCTURE

Unskinned Membranes

Although phase inversion membranes have been in use approximately 120 years - *Fick* [1] utilized cellulose nitrate membranes in 1855 for his studies of diffusion - serious attempts to elucidate membrane structure and to correlate structure and function date back only two decades. In 1953 both *Helmcke* [3] and *Maier* [4] were concerned with the structure of unskinned, microporous membranes, since the asymmetric, or skinned, types had not yet been developed. They found that microporous phase inversion membranes exhibited an open-celled foam type of structure (fig. 2), rather than the idealized cylindrical pores which had sometimes been assumed. The surface of the membrane which had been exposed to air during its fabrication was found to be less porous than both the internal structure and the bottom surface.

In 1960, *Maier* and *Scheuermann* [5] (MS) published a paper on the formation of (unskinned) phase inversion membranes. Their orig-

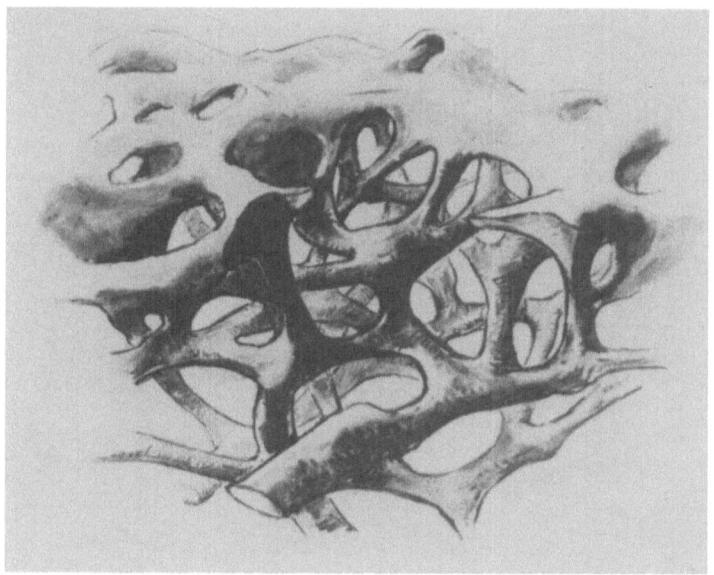

Figure 2. Schematic Representation of the Surface and Contiguous
Substructures of a Microporous Membrane (*Helmcke* [*16*])

inal hypothesis has recently received experimental verification and
has been expanded by the present author to cover the important class
of asymmetric membranes [*6*]. Their scheme envisioned the following
sequence of events: A solution containing a polar polymer, sol-
vent(s), and swelling agent(s) or nonsolvent(s) is allowed to desol-
vate. As the loss of the volatile solvent progresses, the solvent
power of the remaining solution declines and a separation into two
interdispersed liquid phases occurs. The dispersed phase consists
of micelles of the swelling agent and nonsolvent components of solution covered with a coating of polymer molecules. (Because of
their surface activity, the polymer molecules act in concert with
the decreased solvent power of the solution to effect the separation.) As additional solvent is lost, the micelles become smaller
and their polymer coating thickens (fig. 3a). They approach one
another (fig. 3b); make contact (fig. 3c); allow the polymer chains
in their shells to diffuse into those of their neighbors (fig. 3d);
deform into polyhedra (fig. 3e); and suffer a rupturing of their
walls to yield the final open-celled structure (fig. 3f).

At about the same time as the MS mechanism for the formation
of unskinned microporous membranes was postulated, *Reid* [*7*] discovered the high permselectivity of cellulose acetate CA and *Loeb*
and *Sourirajan* (LS) [*8*] developed the first <u>highly</u> asymmetric phase
inversion membranes of this material. The LS membrane possessed a

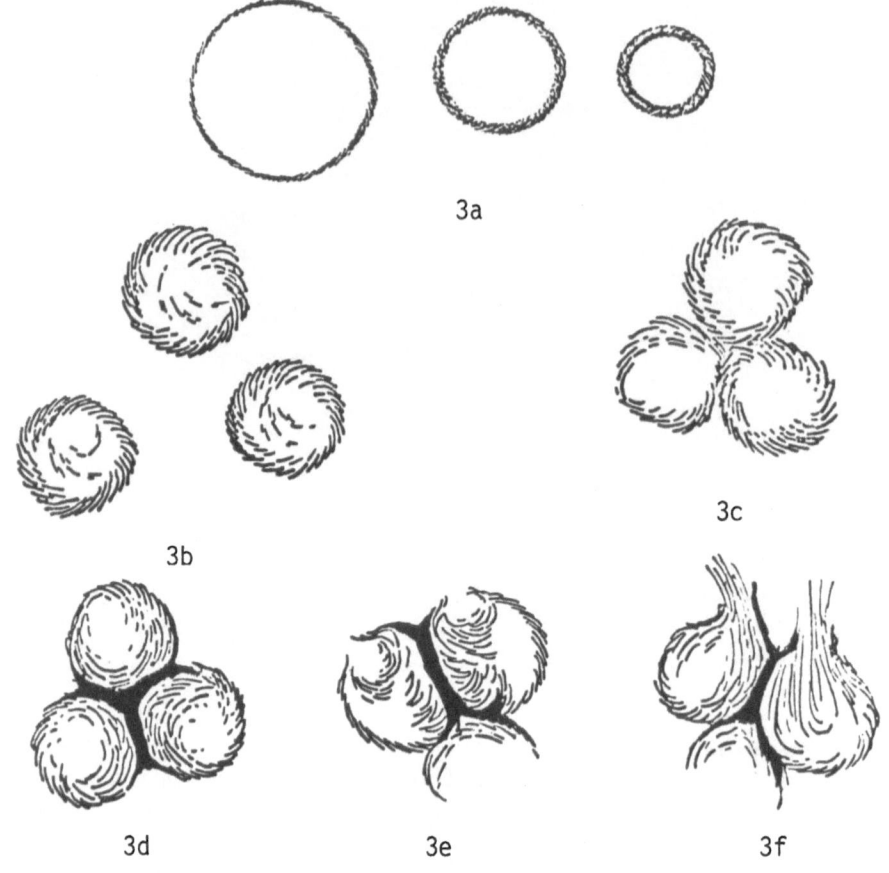

Figure 3.   The Membrane Formation Process
(*Maier and Scheuermann* [5])

thin, dense skin supported by a thicker, porous, substructural layer.
Because the skin layer was not only permselective, but also very
thin and hence permeable, this development made the related pressure-
driven processes of reverse osmosis RO and ultrafiltration UF technically feasible. With this vindication of its early support of the
RO process, the Office of Saline Water (USDI) was then able to expand its sponsorship of membrane research with the result that our
current understanding of the structure of asymmetric membranes has
reached its present level.

In the remainder of this section the important class of asymmetric membranes will be considered first in its nascent stages and
then in a stepwise description of the final membrane beginning with
the skin and proceeding through the transition, substructure and

bottom surface layers.

Nascent Membranes

By employing the technique of freeze quenching and lyophiliza-
tion of membranes in their formative stages and obtaining SEM photomicrographs of the "ghosts of the nascent membranes," the validity
of the MS hypothesis for the formation of unskinned microporous
membranes has been verified and shown to be applicable to asymmetric
membranes as well [6]. The principal structural features of an
asymmetric membrane are found to exist in the sol prior to gelation.
The skin of an asymmetric membrane is formed in the same manner as
the substructure: Polymer-coated micelles make contact and their
walls inter-diffuse, with the quantitative difference that micellar
compaction and wall inter-diffusion occur to a greater extent than
in the substructure. The result is a higher polymer density in the
skin layer and a disappearance in EM photomicrographs of its micellar
origin (owing to the inter-diffusion of micellar walls).

Skin

The skin thickness of the LS membrane was determined by *Riley*
[9] and co-workers to be 0.25 μm out of a total membrane thickness
of between 100 and 150 μm. However, this value is not constant but
can vary with such factors as drying time and concentration of swelling agent in the casting solution [6]. The structure of the skin
was more difficult to establish. An early approach was to consider
that all dense films were structureless or at least equivalent in
structure. This was implicit in the phenomenological approach of
Lonsdale [10] who employed a dense film of unspecified structure to
determine salt diffusion coefficients from which "theoretical" salt
rejections for asymmetric membranes of the same chemical composition
as the dense film were calculated. The difficulty with this concept
lies in reconciling order of magnitude differences between salt diffusion coefficients from variously prepared dense films of the same
polymer. Indeed, the fact that such diffusion coefficients differ
so widely is an indication that the physical structure of dense
films can vary in such a way as to affect salt diffusion.

The first evidence for the existence of fine structure in the
skin of asymmetric LS membranes was obtained by *Schultz* and *Asunmaa*
[11] (SA). They employed an ion etching technique which apparently
modified the interfaces between the micellar walls where inter-
diffusion had obscured the micellar origins to earlier EM investigators. SA were not concerned with the origins of the 188A spherical structures which they observed (fig. 4). (They believed that
they represented lamellar microcrystallites.) SA felt that the

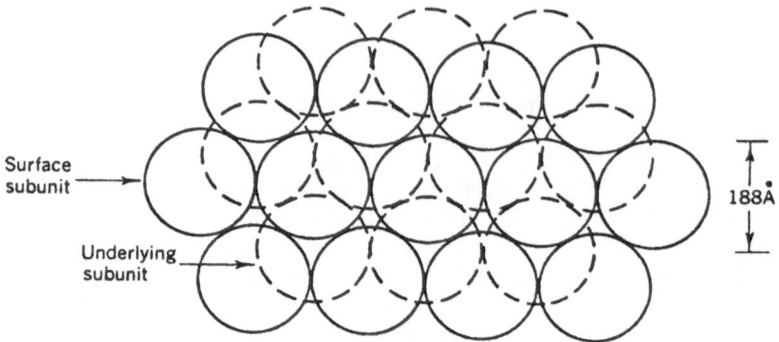

Figure 4.   Active Desalination Layer of Cellulose Acetate Membrane
            Idealized as Assembly of Close-Packed 188-A-Diameter
            Spheres (*Schultz and Asunmaa* [11])

~ 20A interstices between the spheres represented the channels where
permeation occurred. It now appears probable that the precise diameter of these channels will vary with the hydrophilicity of the
polymer since this property will affect intermicellar swelling. Such
an expectation is consistent with the higher permselectivity and
lower permeability of the hydrophobic cellulose triacetate compared
to the relatively hydrophilic diacetate.

    Water within the interstitial spaces was held to be ordered with
a decreased capacity for solvating salt ions. Although there is
little doubt that the intermicellar channels represent the path of
least resistance, the question as to the differences in density between the micelles proper and the domains where they meet has been
raised [12]. It does not appear necessary to postulate channels
completely free of polymer chains bridging the intermicellar gaps.
Micellar structures with dimensions virtually identical to those
found by SA have been discovered in ion-etched Dry-RO[tm] membranes
[6]. The fact that the same skin structure is present in both Wet
and Dry Process membranes indicates that water plays no essential
role in the formation of the skin.

    The structural interpretations presented here are not limited
to membranes of CA or even other cellulose derivatives since *Panar,
et al* have found that they are applicable to polyamide membranes as
well [13].

                        Transition Layer

    It has been found that for those membranes prepared from cast-
ing solutions containing a low concentration of swelling agent,

closed-, rather than open-cells, predominate in the substructure
[14]. Closed cells are characterized by a polymer density between
that of a dense film and that of open cells. It is not too surprising, therefore, that *Gittens, et al* [15] observed that a transition
layer of closed cells was sometimes found beneath the dense skin.
What is surprising is that they are not found in every case. In solutions which contain a concentration of swelling agent high enough
to be close to the swelling limit, the transition layer appears to
be absent. The transition layer is more commonly found in the case
of Wet- than in the case of Dry-Process membranes apparently because
of the higher concentration of polymer in casting solutions of the
former. The effect of the transition layer upon membrane performance characteristics has not yet been unequivocally demonstrated.
On *a priori* grounds, however, since closed cells are denser than
open cells, the transition layer can be expected to decrease permeability by: (1) offering higher resistance to transport *per se*
and (2) undergoing compaction into a dense film thereby increasing
the effective thickness of the overlying skin.

Substructure

As stated above, the porous substructure consists of open-
celled voids formed from the micelles in the two phase solution
which results after solvent loss has effected a phase inversion.
The difference between this layer and the less porous transition and
skin layers is one of degree rather than kind. Owing to their more
gradual desolvation, micelles in the former do not compact to the
same extent as do those in the layers which are closer to the air/
solution interface. A considerable variation in the size of substructural voids is possible, depending upon which of the two [the
LS Wet-(*Loeb-Sourirajan* [8])] or the KD Dry-(*Kesting* [6])] Processes
for phase inversion is employed in the fabrication of asymmetric
membranes. In the former small (0.1-0.5 μm) voids are the rule
whereas in the latter large (1-2 μm) voids predominate. See Section
entitled "Membrane Manufacturing Processes."

Bottom Surface

In most cases the bottom surface of asymmetric membranes is
simply an extension of the structure found in the interior [16].
Since void diameter increases as the distance from the top surface
increases, the largest voids are usually found at the bottom surface. Occasionally, however, atypical structures appear: large
sinkhole cavities at one extreme, and skins at the other [6]. The
large cavities, called microbubbles or macrovoids, are two orders
of magnitude larger than the normal matrix voids and when present,
are located throughout the membrane as well as at the bottom surface.

Because the interior surface of these macrovoids is lined with a
highly porous skin, they probably originate from pockets of solvent
vapor which have been trapped after the top surface skin has formed,
and not as others [17] have suggested, from aqueous intrusion. The
presence of macrovoids is detrimental to permselectivity since applied pressure may rupture the overlying skin.

The presence of a bottom surface skin was first observed by
Frommer and *Matz* [18]. Such skins are almost invariably nonintegral,
i.e., they possess a number of widely-spaced pores. The simultaneous
existence of both top and bottom surface skins occurs when a membrane
is cast on a porous substrate. At other times, however, a skin can
be made to appear at either surface, simply by making a small change
in casting solution formulation. Changes in the relative rates of
desolvation at top and bottom surfaces owing to a surface effect
appear to be involved in this as yet obscure phenomenon.

Trends in the Evolution of Physical Structures

The evolution of the different physical structural types of
phase inversion membranes is depicted in fig. 5. Thick dense membranes (solvent-cast films) have been employed commercially for about
50 years as packaging materials where their barrier properties are of
interest. Thick films of viscose and cuprammonium regenerated cellulose are used in their water swollen condition as membranes in dialysis. *Reid* [7] employed thick dense films in 1959 for his pioneering study to find polymeric materials suitable for use as RO membranes (fig. 5a).

The first class of nondense membranes, i.e., structures with a
high void volume, were unskinned membranes designed for microfiltration (fig. 5b). *Bechhold* in 1907 produced the first graded series,
i.e., varying pore size, nitrocellulose membranes [20]. *Brown* and
Elford in England produced such series employing cellulose acetate
CA [21] and nitrocellulose [22], respectively.

The present intense interest in membrane separation processes
is due not to unskinned membranes but rather to the development of
LS of their Wet Process for the fabrication of asymmetric CA membranes (fig. 5c). This giant step in the evolution of membranes can
be likened to the biological evolution of the fishes from their in-
vertebrate ancestors. In common with the fish, LS or Wet membranes
must remain in water throughout their lifetime. The LS membrane is
of such great interest because it permitted for the first time membrane separations involving the smallest of solutes, viz., inorganic
ions. Moreover, because the LS Wet membranes consisted of a thin
dense permselective skin surmounting a porous low resistance substructural support layer, permeability and permselectivity were very

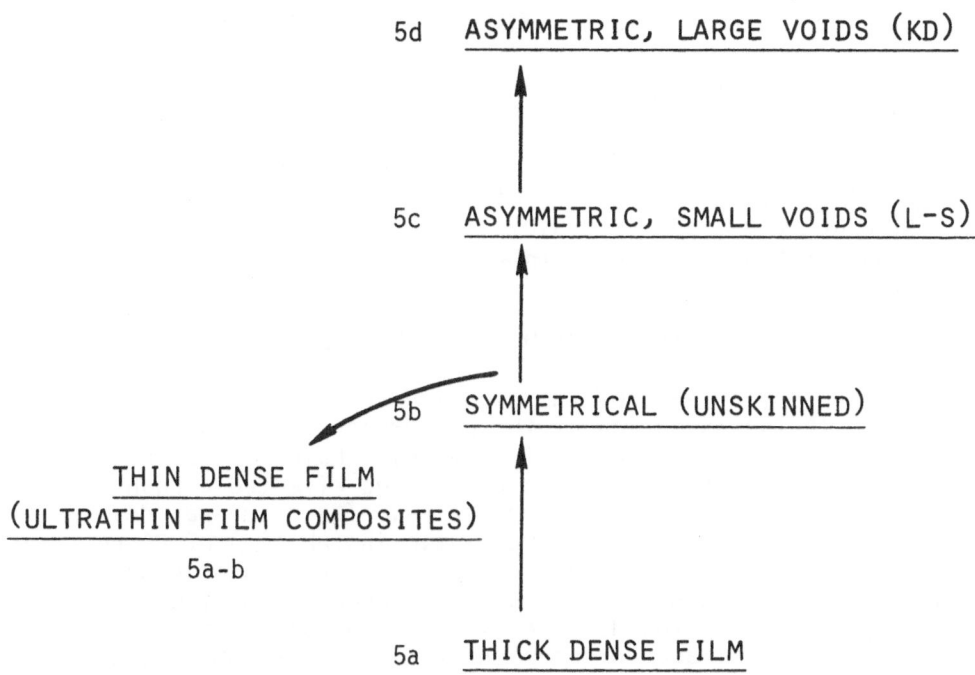

Figure 5.   Physical Evolution of Membrane Structure

high. Thus the most difficult separations could be made at an economically attractive rate. The substructure of the LS or Wet
membrane consists of small (0.1-0.5 μm) voids. Although this portion
of the membrane serves merely as a physical support for the thin skin,
its microstructure is of practical interest in that the small void
size and its consequent high internal surface area results in capillary forces which cause the membrane to densify irreversibly upon
drying. This fishlike necessity for being maintained wet at all
times has adverse affects upon the handling, shipping, and, especially, storage.

Evolution occurs as the result of mutations, most of which are
detrimental to the evolving organism. Because of this most mutations
do not persist but result in the demise of the mutant organism. To
the present author the class of ultrathin film composite membranes
(i.e., those membranes made by the separate preparation of the porous substrate and subsequent coating with a thin dense film) appears
to fall into the category of an organism with a lethal mutation.
The reasons for this are its increased complexity from the manufacturing point of view and its fragility. The ultrathin film composite,
therefore, is seen as a return to dense film technology, but coupled

with that of symmetrical microporous membranes (fig. 5a-b).

The latest development in the evolution of the physical structure of phase inversion membranes is the appearance of asymmetric
membranes of the KD type [6], i.e., skinned membranes with large
(1-2 μm) substructural voids (fig. 5d). Known as Dry-RO^{tm}, these
membranes have all the advantages of membranes of the LS type, and
in addition are wet $\rightleftharpoons$ dry reversible so that they can be handled,
shipped, and stored in the dry state.

CHEMICAL EVOLUTION OF MEMBRANE POLYMERS

The discovery by *Reid* of the high permselectivity of secondary
CA coupled with the invention of the LS membrane necessarily focused
early attention upon this material (fig. 6). Later the development
by duPont of HF fibers of aromatic polyamides effected a rush to develop noncellulosic polymers to replace CA. The principal driving
force for this movement was the hydrolytic instability of CA relative
to certain noncellulosic polymers. However, noncellulosic polymers
have been shown to exhibit weaknesses not found in CA itself. The
polyamides, e.g., were found to be extremely sensitive to oxidative
degradation by even traces of free chlorine, a very common constit-
uent in most aqueous feeds. Thus an extra carbon pretreatment step
is necessary for demineralization when polyamide membranes are utilized. Furthermore, the vaunted hydrolytic superiority of polyamide

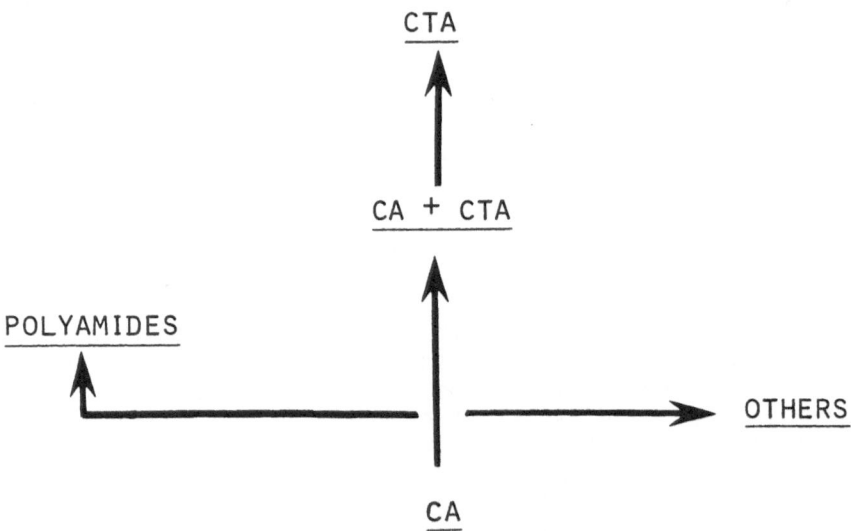

Figure 6. Chemical Evolution of Membrane Polymers

membranes was not usually realizable in practice because the plating out of insoluble salts at high pH necessitated either pH control (in the same range in which CA is quite stable) or the utilization of an ion exchange column upstream from the RO element.

Perhaps of even greater significance is the fact that development of the highly versatile CA polymers themselves has been very intense. The first important post-LS development was the cellulose triacetate CTA-CA blend membrane of Cannon [23]. The blend membrane exhibits higher flux and rejection than the original LS membranes of straight CA. Much recent research has centered on the utilization of CTA by itself. In retrospect the considerable amount of CA research can be viewed as a striving toward CTA, in which the blend was the halfway point. Membranes consisting entirely of CTA have hydrolytic stability and resistance to biodegradation far in excess of that of CA or the CTA-CA blends. They also have outstanding resistance to free chlorine, the nemesis of the polyamides. It is felt that the CTAs offer such considerable advantages over other membrane polymers, including the noncellulosics, that they will ultimately constitute the workhorse of the membrane polymer stable. The reason why they have not yet entered the commercial arena is related to their processing, rather than to their end use, characteristics. A similar situation was encountered in the use of CTA fibers in the textile field. CA fibers were utilized for years before CTA fibers were introduced because the former were so much more soluble than the latter. Now, however, after years of research, the solubility behavior of CTA is better understood and laboratory results from a number of membrane manufacturers indicate that a number of previously impossible separations such as the single pass desalination of sea water is technically feasible with CTA membranes. The present author believes that CTA·Dry-RO^{tm} which is now undergoing development represents the evolutionary apex in both physical and chemical planes.

Events of the past year related to the worldwide petroleum shortage have introduced still another element into the question of membrane polymers, viz., their long term availability. Because it is derived from regenerable cotton and wood, cellulose will never be unavailable as a raw material for membranes. This, of course, is not the case for petroleum derived products such as the noncellulosic polymers.

The future of other noncellulosic RO polymers such as the sulfonated polysulfones and polyphenyleneoxides and polyureas is seen as very bleak. The polyureas originated in the wool industry as the Wurlan Process to shrink proof wool and were developed as ultrathin composite membranes by North Star Research Institute. They have two weaknesses: chlorine sensitivity and fragility. The sulfonated polymers are difficult to prepare and incapable of very high rejec-

jection — no serious attempt to commercialize these polymers seems
likely. In summary, the cellulosic polymers, particularly the CAs
and CTAs appear perfectly capable of maintaining and even extending
their preeminent position as the membrane polymers of choice.

MEMBRANE MANUFACTURING PROCESSES

A discussion of membrane manufacturing processes is an important part of any discussion of the evolution of phase inversion
membranes. The choice of process will affect reproducibility, membrane performance characteristics, physical properties, wet $\rightleftharpoons$ dry
reversibility (or lack thereof), lifetime, ease of fabrication into
the end use configuration, etc. In this section the potentially
more important class of asymmetric membranes, rather than its similar but unskinned counterpart, has been chosen for detailed consideration. To achieve the proper perspective, a description of the
generalized phase inversion technique precedes a discussion of the
Wet- and Dry-Process subdivisions.

The Generalized Phase Inversion Technique

Solutions from which phase inversion membranes are cast consist
of polymer(s) and a solvent system. The solvent system includes all
the casting solution components with the exception of the polymer.
It is in most cases prepared from a minimum of two compounds located
in the Polymer-Solvent Interaction Spectrum, i.e., reagents with an
affinity for the polymer which can vary from that of a strong solvent
at one extreme to a strong nonsolvent at the other, with weak solvents,
swelling agents and weak nonsolvents in the middle. The swelling
agent and nonsolvents comprise what can be termed the Swelling Agent
Subsystem, and are primarily responsible for the formation of voids
in the membrane substructure. The true solvent should have excess
solvation capacity so that it can accommodate in a compatible casting solution not only the polymer itself, but a high concentration
of the Swelling Agent Subsystem. Porosity (void volume) varies directly, and skin thickness inversely, with increasing concentration
of the swelling agent subsystem. Since the flux of product water
increases with increasing porosity and decreasing skin thickness, it
is understandable that swelling agents have played a crucial role in
membrane development.

The generalized case of the phase inversion process will now be
considered. A casting solution consisting of a polymer, a solvent,
and a swelling agent is cast onto a suitable substrate. The op-
tically clear solution, initially homogenous on the colloidal level,
starts to lose solvent owing to evaporation. After sufficient solvent has been lost, the solution becomes turbid — evidence that it
is no longer structureless on the colloidal level. Phase inversion

has occurred and the membrane gels shortly thereafter to assume its primary gel configuration. The primary gel membrane can subsequently be modified by heat or pressure to form a more dense secondary gel structure. Deductions have been made as to the sequence of events on the colloidal level which accounts for this structure.

Immediately after the solution is cast, solvent starts to evaporate. Since solvent is lost more rapidly from the casting solution/air interface than from the interior of the solution, solvent is rapidly depleted at this interface and polymer concentration increases — rapidly rising to a level where the remaining solvent is insufficient to maintain the polymer in solution. (The concentration of polymer at the air/solution interface is also higher than in the bulk solution owing to the surface activity of the polymer.) The polymer then comes out of the solution and forms the skin layer. This skin layer originates in essentially the same manner as does the porous substructure. It consists of micelles which move closer together and compact as drying continues. <u>This concept is of importance to an understanding</u> of the differences between the primary skin structures of Wet and Dry membranes.

After the skin has been formed, desolvation is slowed considerably, so that the rate of gelation in the substructure can be considerably slower than it was in the skin. Eventually, however, remaining solvent power becomes insufficient to maintain all the components in a homogeneous true solution. Since the solvent is usually more volatile than the swelling agent, a large amount of the latter remains even after much of the solvent has been lost. The swelling agent now separates from the continuous phase as tiny droplets of a dispersed phase. The polymer molecules aggregate about these droplets because of surface activity and because insufficient solvent is available to keep them in the continuous phase. As more solvent is lost, the polymer-coated droplets of swelling agent approach one another more closely and deform into polyhedra. Eventually, as both solvent and swelling agent depart, the polymer coating of the droplets is stretched too thin and ruptures, leaving behind an open-celled structure. It should be recalled that all this happens in the substructure <u>after the dense skin consisting of spherical micelles has been formed</u>.

The Loeb-Sourirajan (LS) Wet Process

The original LS Process casting solution consisted of cellulose acetate, acetone, water and magnesium perchlorate. The Loeb-Manjikian modification of this solution consisted of cellulose acetate, acetone, and formamide. The only significant difference between the two was the swelling agent, consisting of water and magnesium perchlorate in the one case and formamide in the other. The Wet Process employed a concentrated solution containing about 20% of polymer. The following steps were found necessary:

(1) The solution was cast at -11°C (somewhat higher temperatures are permissible with the formamide-containing solution).

(2) The solution is allowed to dry for one to three minutes.

(3) The partially dry solution is immersed in water at 0°C and washed for one hour to yield the primary gel membrane.

(4) The primary gel membrane is heated to 70-90°C to produce a secondary gel membrane which is ready for desalination. Because the solution is concentrated, very little excess solvation capacity is left after the polymer has been added to the solvent. For this reason, accommodation of high concentrations of pore-producing constituents in the casting solution is only possible if they are highly compatible with the polymer. (High concentrations of swelling agents are essential if an asymmetric membrane with a thin skin is to be formed.) By definition, however, a high degree of compatibility ensures that no phase inversion will occur by itself, and so extensive drying will result in a dense film rather than the desired asymmetric membrane. Therefore, in the case of the Wet Process, phase inversion must be evoked by an <u>external</u> source, viz., the nonsolvent gelation bath. After the skinned nascent Wet membrane has been immersed in the bath, diffusion of solvent out, and water in, occurs through the skin. Owing to the high viscosity of the casting solution and the high compatibility of the pore-producing constituents, very small micelles are formed in Wet Process solutions when phase inversion takes place. These small micelles ultimately are converted into small (0.1-0.5 μm) voids in the finished membrane. Because of this small void size, a large internal surface area is present and the capillary forces upon drying are sufficient to cause void collapse. For this reason Wet Process membranes must be kept wet at all times.

It will be recalled that the skin consists of micelles which move more closely together in the second step of the Wet process with continued solvent evaporation, thereby decreasing channel diameter. However, before the micelles have made their closest possible approach, the entire membrane is immersed in ice water. This quenching has a pronounced effect on the nascent (forming) skin layer. It both freezes the micelles in place and inhibits their compaction before the skin has achieved its maximum density. Intermicellar diameter is large enough to accommodate some ordinary ion-solvating water and salt rejection is, therefore, very low or altogether absent. However, this situation cannot be overcome by allowing the wet nascent membrane to dry completely because the concentrated Wet Process solution will lose its porosity. The only alternative, therefore, is to add a wet annealing step as a result of which the skin layer is densified, i.e., the micelles are driven closer together by the introduction of thermal energy. Annealing to progressively higher temperatures permits the membranes to become progressively denser so that intermicellar spacing decreases and permselectivity increases.

The LS Process is simple, but it requires careful control of several separate fabrication steps. Nevertheless, because wet membranes have been under intensive development for ten years, their functional performance characteristics vary from good to excellent. Their chief failing is the absence of wet $\rightleftharpoons$ dry reversibility, a property which is inherent in its structure and method of manufacture. This lack of wet $\rightleftharpoons$ dry reversibility is a severe disadvantage in that wet membranes are subject to hydrolysis, microbiological degradation and freezing during storage.

The Kesting Dry KD Process

The KD Process is a complete evaporation process consisting of one step: A solution is simply cast and allowed to evaporate completely. Dilute polymer solutions are employed. The resultant excess solvation capacity after the addition of the polymer to the solvents permits the incorporation of less compatible pore-producing constituents within the casting solution. The importance of this is that the potential incompatibility sufficient *per se* to effect phase inversion (after the requisite solvent loss has occurred) is an internal characteristic of the system. Thus complete evaporation results not only in phase inversion, but in the maintenance of an asymmetric structure in the absence of water. Dry Process (Dry-RO^{tm}) membranes are not only more reproducible (owing to lesser number of fabrication steps), but are produced in the dry form and may be wet and dried repeatedly without undergoing densification. This property of wet $\rightleftharpoons$ dry reversibility may be ascribed to the immediate effect of large (1-2 μm) substructural voids for which the appearance of large micelles during phase inversion is ultimately responsible. Low polymer concentration and relatively incompatible swelling agents are the reasons for the large size of micelles in nascent Dry-RO^{tm} membranes.

Because the Dry-RO^{tm} membrane is wet $\rightleftharpoons$ dry reversible it has a virtually infinite shelf life. Membranes can be made, fabricated into modules and stored indefinitely without any fear of microbiological degradation. Once used, elements containing these membranes can be dried completely, stored indefinitely, and reused without suffering any loss in either permeability or permselectivity.

REFERENCES

1. A. Fick, Ann. Physik Chem., 94, 59 (1855).
2. W. Schumacher, Ann. Physik Chem., 110, 337 (1860).
3. J. G. Helmcke, Optik, 10, 147 (1953).
4. K. H. Maier and H. Beutelspacher, Naturwiss, 23, 605 (1953).
5. K. H. Maier and E. A. Scheuermann, Kolloid Z., 171, 122 (1960).
6. R. E. Kesting, J. Appl. Polymer Sci., 17, 1771 (1973).
7. C. E. Reid and E. J. Breton, J. Appl. Polymer Sci., 1, 133 (1959).
8. S. Loeb and S. Sourirajan, UCLA Rept. 60-60, 1960.
9. R. L. Riley, J. O. Gardner, U. Merten, Science, 143, 801 (1964).
10. H. K. Lonsdale, Chap. 4 in *Desalination by Reverse Osmosis*, U. Merten, ed., M.I.T. Press, Cambridge, Mass. (1966).
11. R. Schultz and S. Asunmaa, Recent Progr. Surface Sci., 3, 291 (1970).
12. O. Kedem, discussion at NSF-sponsored membrane conference, Cleveland, Ohio, May, 1973.
13. M. Panar, H. Hoehn, and R. Hebert, Macromolecules, 6, 777 (1973).
14. R. E. Kesting, *Synthetic Polymeric Membranes*, McGraw-Hill, New York, 1971.
15. G. J. Gittens, P. A. Hitchcock, D. C. Sammon and G. E. Wakely, Desalination, 8, 369 (1970).
16. J. G. Helmcke, Kolloid Z., 135, 29 (1954).
17. G. J. Gittens, P. A. Hitchcock and G. E. Wakley, Desalination, 2, 315 (1973).
18. M. A. Frommer and M. Matz, Second OSW Conference on Reverse Osmosis, Miami, Fla., April, 1969.
19. G. Belfort, Paper Presented at the 4th International Symposium on Fresh Water from the Sea, Heidelberg, Germany, Sept. 1973.
20. H. Bechhold, Z. Physik, Chem., 60, 277 (1907).
21. W. Brown, Biochem. J., 9, 591 (1915); 11, 40 (1917).
22. W. Elford, Trans. Faraday Soc., 33, 1094 (1939).
23. C. Cannon, U. S. Patent 3,497,072, Feb. 24, 1970.

III. Membrane-Moderated
Biomedical Devices

THERAPEUTIC SYSTEMS FOR CONTROLLED ADMINISTRATION OF DRUGS:

A NEW APPLICATION OF MEMBRANE SCIENCE

Alan S. Michaels, Sc.D.

ALZA Corporation

950 Page Mill Road, Palo Alto, California   94304

In recent years, increasing attention has been given to the medical and health problems associated with the administration of pharmacologically active substances (drugs and biologicals) to humans and animals for the treatment of acute and chronic diseases, in the face of growing awareness that these substances are frequently toxic if present in too high concentration, or for too long a time-period in body tissues, and often therapeutically ineffective if present at too low a level or too short a time in the affected body site. It is both paradoxical and apprehending that, despite the phenomenal progress made by the pharmaceutical industry over the past three decades in development of novel and powerful compounds for the treatment and care of a host of diseases that man is heir to, the methods of administering these products to the patient - principally by injection, or by oral ingestion, as pills, capsules, or elixirs - remain virtually unchanged from those practiced a century ago. It has only been with the recent evolution of the disciplines of pharmacology, pharmacokinetics, and bioengineering (whose origins lie in the sciences of biology, physical chemistry, mass transport, fluid mechanics, and system dynamics) that the inadequacies of traditional modes of drug-administration have been adequately recognized, and the immense safety- and efficacy-benefits of controlled drug-administration fully appreciated.

Most drugs which are administered by mouth or parenteral injection are absorbed more or less rapidly into the gastrointestinal mucosa or the tissues surrounding the injection-site, and ultimately transferred into the blood stream, where they are disseminated widely throughout all body tissues. Typical plasma concentration-time profiles following administration of a single

409

intravenous or oral bolus of drug are shown in Figure 1. Ideally, however, a chemotherapeutic agent should be localized within the specific body organ, or specific type of tissue, requiring treatment. Obviously, the absorption/circulatory dissemination process which follows conventional oral or parenteral administration presents to the diseased organ an exceedingly small fraction of the administered dose, and to all other body tissues, a substance which is certainly unnecessary, and frequently unwelcome. Since, further, a beneficial response of a diseased organ or tissue to a specific drug ensues only if the concentration of drug in that tissue exceeds a certain threshold value, and since other body tissues may be adversely affected by drug-concentrations not far different from that needed for therapeutic efficacy, the physician is faced with an often unresolvable dilemma: he must choose between a dose large enough to insure beneficial pharmacologic activity on the diseased organ with probable toxic side effects elsewhere in the body, or small enough to avoid side effects with risk of inadequate treatment of the disease.

Routine methods of drug administration are further hobbled and confounded by both the necessary periodicity of drug delivery to the patient, and the vagaries of the natural biological processes of drug absorption, metabolism, and elimination from the body. This is reflected by corresponding periodic fluctuations in blood levels, as shown in Figure 2. Conventional drug dosage forms are designed largely to suit the convenience and comfort of patient and physician as regards administration schedule; we consume pills on a "X times a day" regimen, to accommodate our waking hours, and to facilitate our remembering when we should take them, and we receive injections on schedules which are governed by hospital rounds or when we can get to the doctor's office. This kind of periodicity of drug administration is totally unrelated to the therapeutic requirements for maximum drug safety and efficacy. Usually, it is desirable to maintain, with reasonable constancy for an extended time-period, a therapeutically effective level of drug within the body; periodic administration presents the body with a sudden surge of drug and transient high tissue concentration, followed by an often rather rapid exponential decline in concentration by metabolic degradation and/or excretion. Clearly, if minimum therapeutically effective drug levels are to be sustained between medication intervals, then following each dosage, inordinately high (and often toxic) levels will be present for significant periods; whereas if the dosage is reduced in magnitude to prevent unduly high drug levels, the concentration in the body during much of the treatment-period will be below the threshold of efficacy. In either event, the fraction of the total administered dose which is beneficially utilized by the patient is depressingly small.

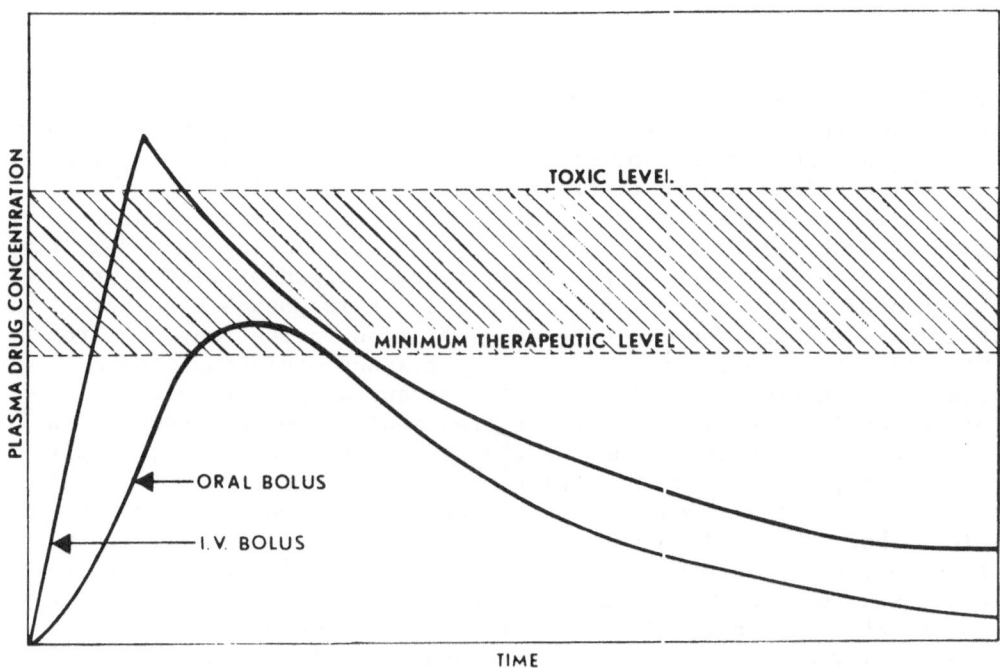

Figure 1.   Dynamics of Drug Administration - Elimination (IV/Oral
            Bolus)

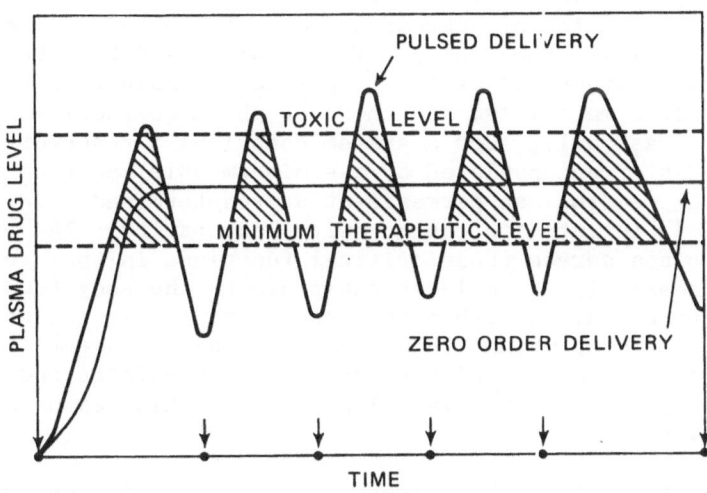

Figure 2.   Pulsed vs. Zero Order Delivery - Concentration/Time
            Curves

The deficiencies and limitations of conventional drug-administration techniques can in large measure be circumvented by (1)
confining the drug within a delivery system which contains a supply
of drug sufficient to provide therapy for an extended time-period,
and which releases drug to the body at a predetermined, precisely
controlled rate adequate to maintain the optimal tissue-concentra-
tion of drug during that time-period, and (2) placing such delivery
system in contact with or proximity to that organ or tissue which
is in need of treatment. By this means, it is possible both to
eliminate fluctuations in drug-concentration in the body, and to
minimize exposure to the drug of body tissues other than those
requiring therapy. Localization of therapy by proper system-
placement is generally limited to treatment of those organs which
are readily accessible, such as the eye, uterus, GI tract, respira-
tory tract, vagina, or skin; placement of systems within internal
organs by surgical implantation methods is, of course, also possible, but usually involves risks to the patient which are justi-
fied only for treatment of life-threatening conditions. In those
disease-situations where localization of therapy is impossible or
unwarranted, the systemic administration of drug via delivery
systems which control both the rate and time of drug supply to the
body is invariably therapeutically superior to periodic administration by traditional means.

Developments, during recent years, in the fields of permselective membrane technology, and biomaterials for surgical and
medical uses, have provided the impetus for the development of
devices and systems which allow both temporal and spatial control
over the administration of pharmaceuticals to the body. The
principles underlying the concept of the Controlled Drug Delivery
System involve a synthesis of the principles of molecular transport in polymeric materials, and those of pharmacokinetics and
biodynamics. Basically, such a system comprises a reservoir of
drug surrounded or encapsulated by a semipermeable membrane
through which the drug can permeate at a predetermined rate, when
the membrane is in intimate contact with body cavity. The encapsulating membrane serves these critical functions in the drug-
delivery process; (1) it isolates and protects the body from the
source of drug contained within the reservoir; (2) it protects
and isolates the drug within the reservoir from the deteriorative
environment of the body chemistry; and (3) it regulates and controls the rate at which drug is released to the tissues or fluids
in contact with it.

In order to perform these protective and kinetic control
functions, the encapsulating membrane of the system must be selectively permeable to the active drug-substance, and substantially
impermeable to the components of body fluids. In addition, of
course, the membrane material must be biochemically compatible
with the surrounding biological tissues, must remain unattacked

or unaltered by the biological fluids to which it is exposed, and the entire system must possess mechanical properties and be of a configuration which provides minimum distortion or disturbance of the surrounding tissues, in order to avoid pain or discomfort to the patient. These requirements, needless to say, place severe constraints upon the selection of materials for such delivery systems, and limit the number of polymers which can be employed successfully in their construction.

Polymeric membranes which are permeable to drugs by virtue of the dissolution and diffusion of drug molecules in the polymer are attractive candidates for Delivery Systems, since the rate of transport of drug through such a membrane can be controlled by adjusting its area and thickness, and by controlling the drug activity- or chemical potential-difference across the membrane. Such a system is shown schematically in Figure 3. Since the body constitutes, in most instances, an "infinite sink" of negligible drug activity, the drug delivery rate from a membrane-moderated device is uniquely determined by the drug activity within the reservoir; if this activity can be maintained constant, then the delivery rate from such a system will also remain constant. Maintenance of constant reservoir activity despite gradual depletion of drug from the reservoir can be accomplished by dispersing solid, crystalline drug in a reservoir vehicle within which the drug is sparingly soluble, but freely diffusible. Such a system is exemplified by ALZA's "PROGESTASERT"^{TM} - an intrauterine progesterone delivery system for contraceptive use, which delivers progesterone to the uterine lumen at an essentially constant rate of 65 micrograms/day for a period of more than one year. A cross sectional view of this system, showing the solid-drug containing reservoir and its surrounding membrane, is illustrated in Figure 4; its placement in the uterus, in Figure 5; and its release rate-time profile in vivo, in Figure 6.

Certain drugs - for example, the miotic pilocarpine - are exceedingly soluble in water, and exist as liquids rather than solids at ambient temperature. Confinement of these drugs within a membrane-encapsulated reservoir presents special problems, since it is virtually impossible to maintain constant drug activity with changing drug content; moreover such drugs, when present in high concentration, tend osmotically to absorb water from ambient body tissues or fluids. If the encapsulating membrane is even moderately permeable to water as well as drug, reverse transport of water into the reservoir not only dilutes the drug, but also causes swelling of the system. By selecting a membrane-polymer of suitably low water permeability whose intrinsic drug-permeability increases with decreasing drug-concentration, and using a reservoir vehicle in which drug activity varies but little with drug concentration over a wide range, it is possible to produce a system displaying a nearly constant delivery rate during

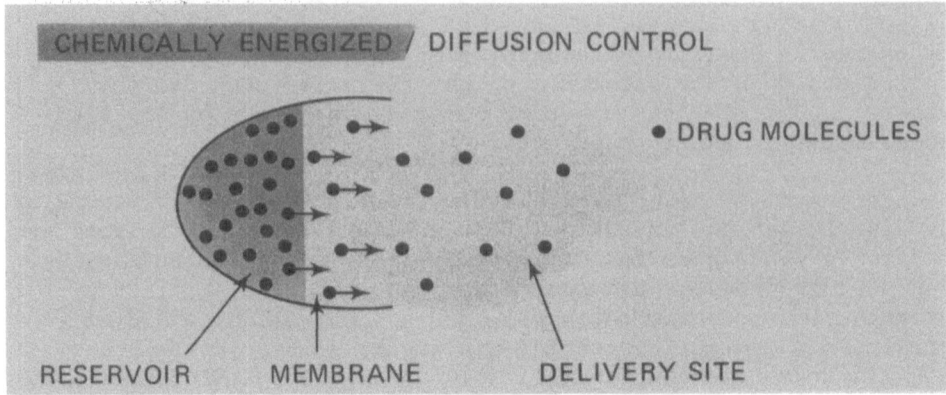

Figure 3.   Diffusional Membrane System

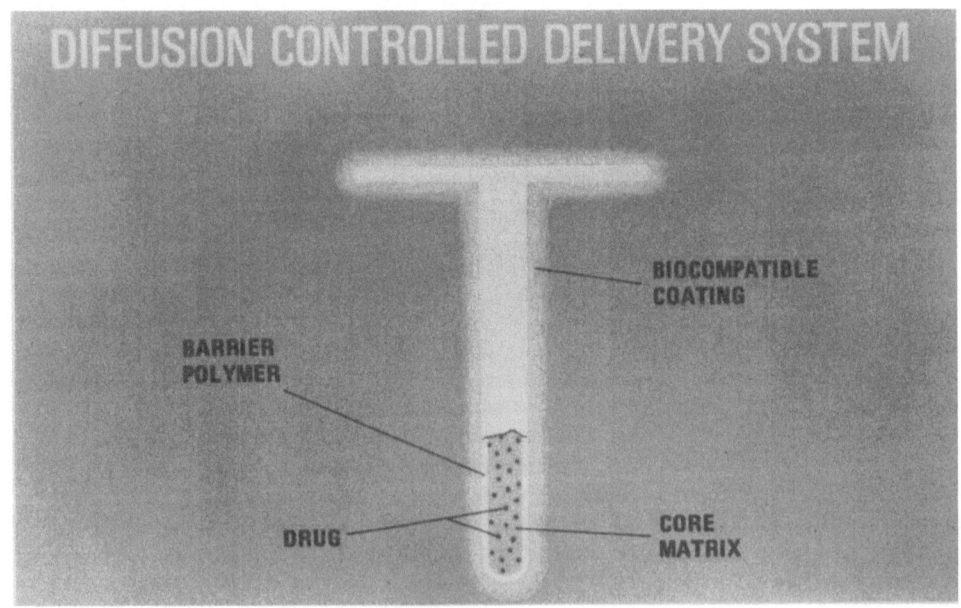

Figure 4.   PROGESTASERT^{TM} Cross Section

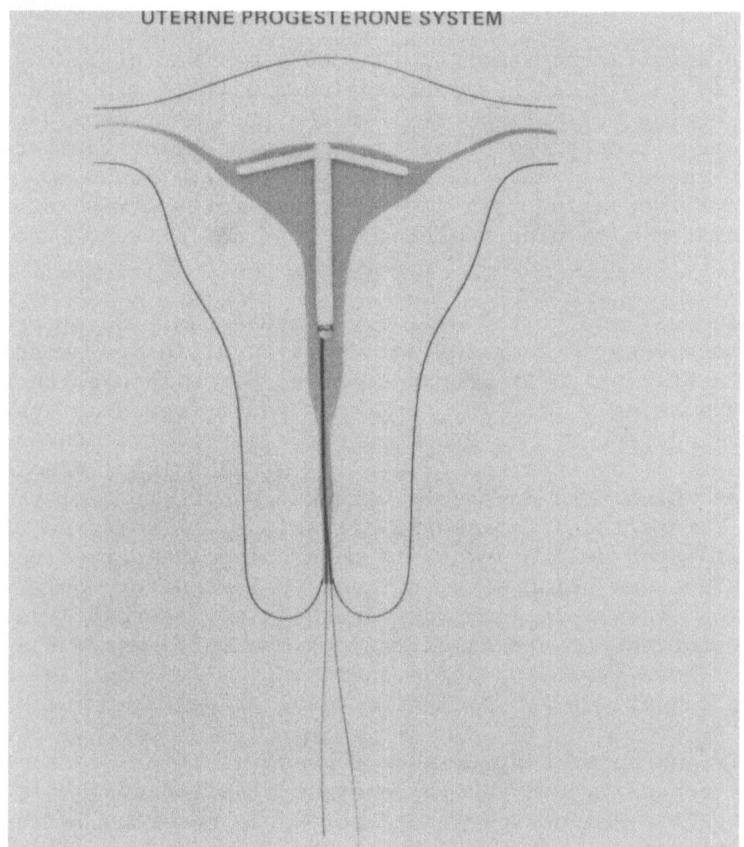

Figure 5. PROGESTASERT[TM] In Utero

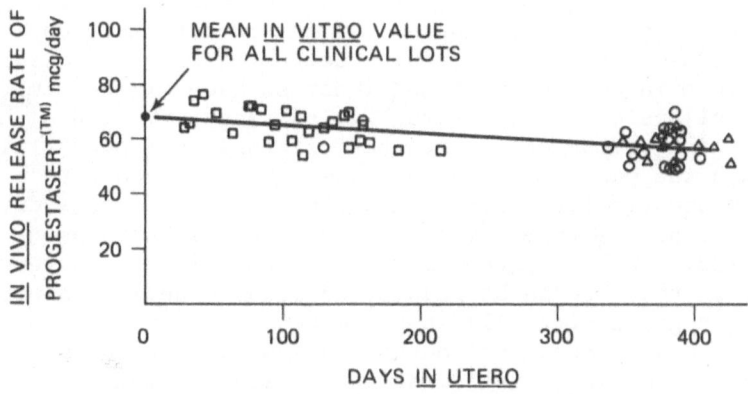

Figure 6. PROGESTASERT[TM] Release Rate/Time Curves In Vivo

release of a significant fraction (say 60%) of the total drug content. This result has been accomplished with ALZA's "OCUSERT"® Pilocarpine Ocular Delivery System for the treatment of glaucoma, which delivers pilocarpine to the eye from a membrane-laminate placed under the eyelid at a nearly constant rate (20 or 40 micrograms per hour, depending upon the membrane) for a period of one week. The system, its placement in the eye, and a typical release-rate-time profile, are shown in Figures 7, 8 and 9.

Diffusionally controlled Delivery Systems can also be produced by homogenously dispersing solid, particulate drug within a polymeric matrix which is drug-permeable, but within which the drug is only sparingly soluble. Delivery rates from such systems are not time-independent, as are membrane-encapsulated constant-activity-reservoir systems, but (in film- or slab-form) display release rates which vary inversely with the square root of time of operation. For certain therapeutic situations, this degree of variability in drug delivery rate is tolerable; laminar structures of this type are now being developed and evaluated for ocular administration of anti-inflammatory drugs (e.g., steroids) and antibiotics, for the treatment of various eye inflammations and infections. These systems provide therapeutically effective treatment of such conditions for periods of days or weeks.

There is one major body-component whose activity or chemical potential is virtually everywhere constant at all times; this is water. Thus, if a selectively water-permeable membrane is contacted with tissue or body fluid while the opposite face of the membrane is in contact with a crystalline, water-soluble solute (e.g., a simple salt), water will be transported through the membrane to the solute-containing compartment at constant flux under a constant osmotic gradient, so long as solid solute is present. This principle (illustrated schematically in Figure 10) can be used to produce a constant-rate, "zero-order" drug delivery system. Such a system is comprised of (a) an external, encapsulating rigid osmotic membrane, (b) a compacted mass of solid osmotic solute in contact with the inner surface of the membrane, and (c) a collapsible, liquid-drug filled reservoir within the enclosure which communicates with the external environment via a port through the encapsulating membrane. When placed within the body, water osmotically imbibed into the system displaces drug from the collapsible reservoir into the external environment at constant rate. Such a system, it can be appreciated, is very versatile, in that it can be readily adapted to deliver a wide variety of drugs at virtually any selected rate. Devices of this type have been developed for controlled administration of drugs to the gastrointestinal tract, and as surgically implantable delivery systems for the local treatment of tumors.

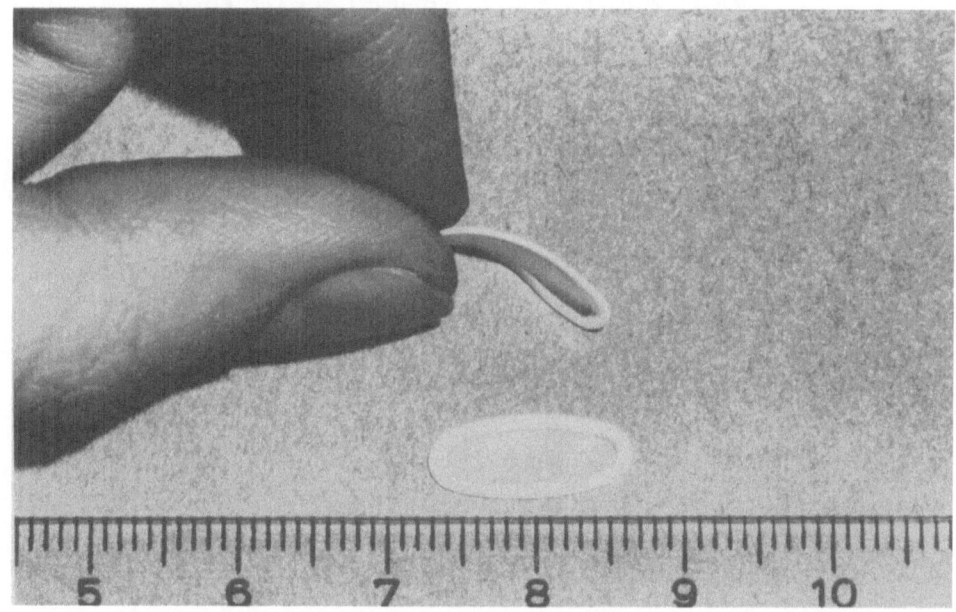

Figure 7. OCUSERT ® System

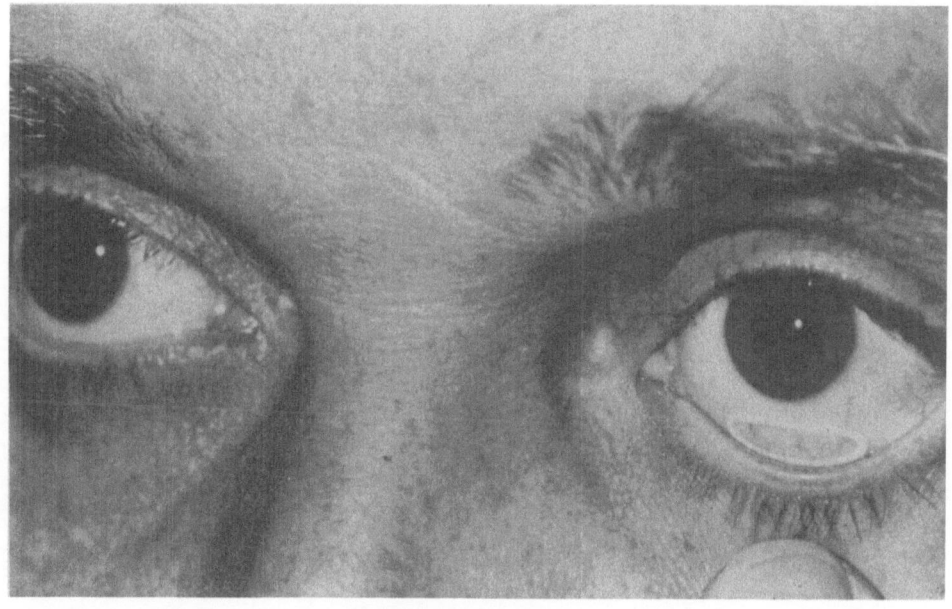

Figure 8. OCUSERT ® System - Placement in Eye

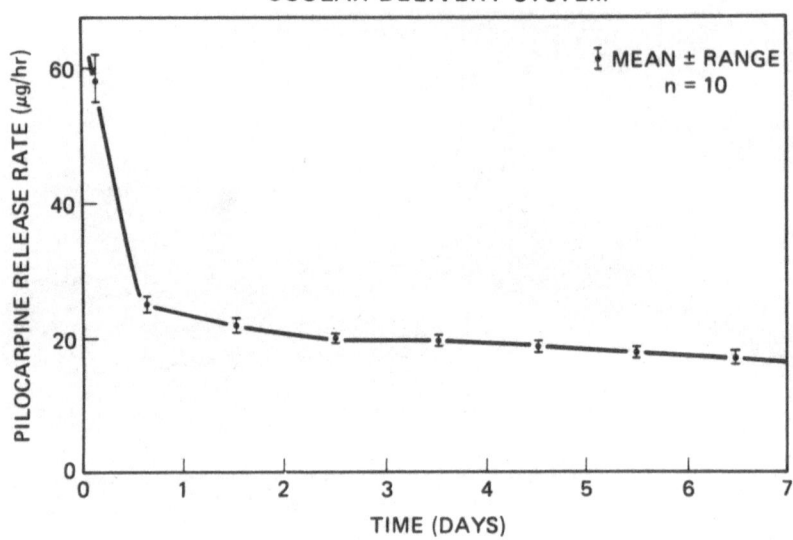

Figure 9. OCUSERT ® System - Release Rate/Time Profile

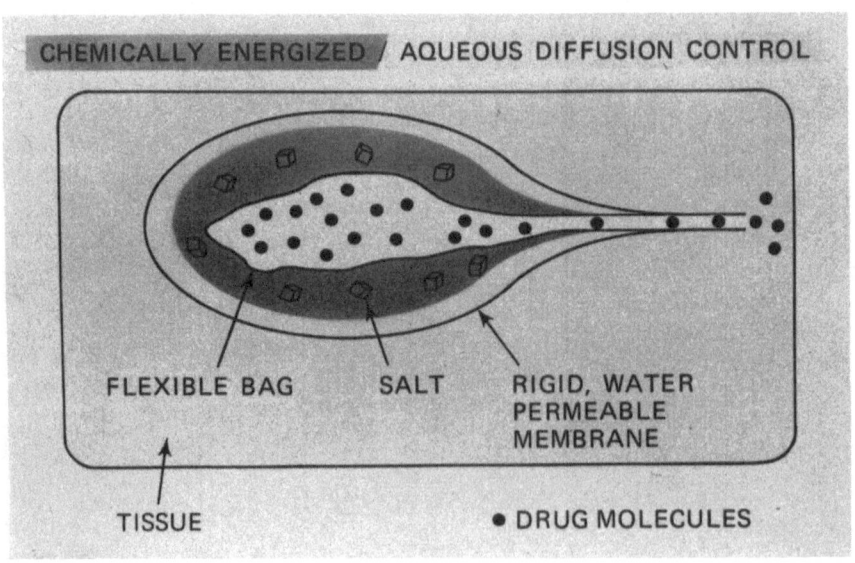

Figure 10. Osmotic System

The administration of drugs via topically-applied delivery
systems by controlled permeation through the (intact) skin is
particularly appealing for systemic therapy, since such a delivery
route can supply directly to the circulation drugs which are
destroyed or poorly absorbed in the GI tract, which cause GI dis-
turbances, or which must presently be administered by injection
to be therapeutically effective. Moreover, this route of administration has the advantage of great patient convenience and comfort,
and maximum safety, since medication can if necessary be promptly
terminated by removal of the topically-applied delivery system.
The transdermal delivery of drugs is beset with unique problems,
since the skin - which is obviously designed to protect us from
our hostile environment - is a membrane of quite low permeability
to most substances. Obviously, if one aims to develop a Transdermal Delivery System which will administer drug at a controlled
and predictable rate to the systemic circulation, one must render
the resistance to permeation of the drug offered by the skin a
small fraction of the resistance offered by the Delivery System.

By studying the mechanism and kinetics of transport of various drugs across intact skin in vitro, we have found certain classes
of drugs whose skin-permeability is sufficiently high to permit
their administration at therapeutically useful rates. Moreover,
we have found certain non-toxic compounds which temporarily elevate
the skin permeability of many drugs when present in quite small
amounts. With this information, it has been possible to design a
membrane-modulated delivery system resembling a self-adhering
adhesive patch which, when applied to the skin surface, delivers
drug to (and through) the skin at a predictable and constant rate
for a period of 24 - 36 hours, independent of the locus of application and with minimal variability from subject to subject. A
Transdermal Delivery System embodying an anti-nauseant drug is now
in clinical trial. This device (schematically shown in Figure 11)
is a multilayer laminate comprising (1) an external, moisture- and
drug-impermeable film, (2) a thin, solid layer of active drug dispersed as a separate phase within a highly drug-permeable matrix,
(3) a drug-permeable membrane of carefully controlled intrinsic
permeability, (4) a pressure-sensitive, compliant adhesive coating
which provides both adherence to and intimate contact with the
skin surface, and whose resistance to drug-transport is very low,
and (5) a protective strippable film applied to the adhesive layer,
which is removed by the patient before application. A prototype
system, and its application to the skin surface, are shown in
Figures 12 and 13. In vitro and In vivo drug release rates from
this device are found to be virtually identical, confirming that
drug-administration kinetics are indeed governed by the delivery
system and not the patient.

Central to the development of controlled drug delivery systems is a thorough understanding of the physicochemical and

SCHEMATIC DIAGRAM OF ALZA'S TRANSDERMAL THERAPEUTIC SYSTEM

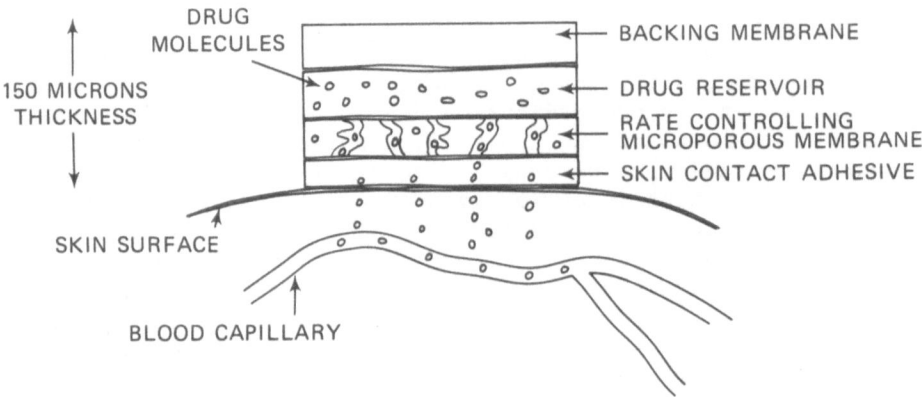

Figure 11. Transdermal Delivery System - Schematic Diagram

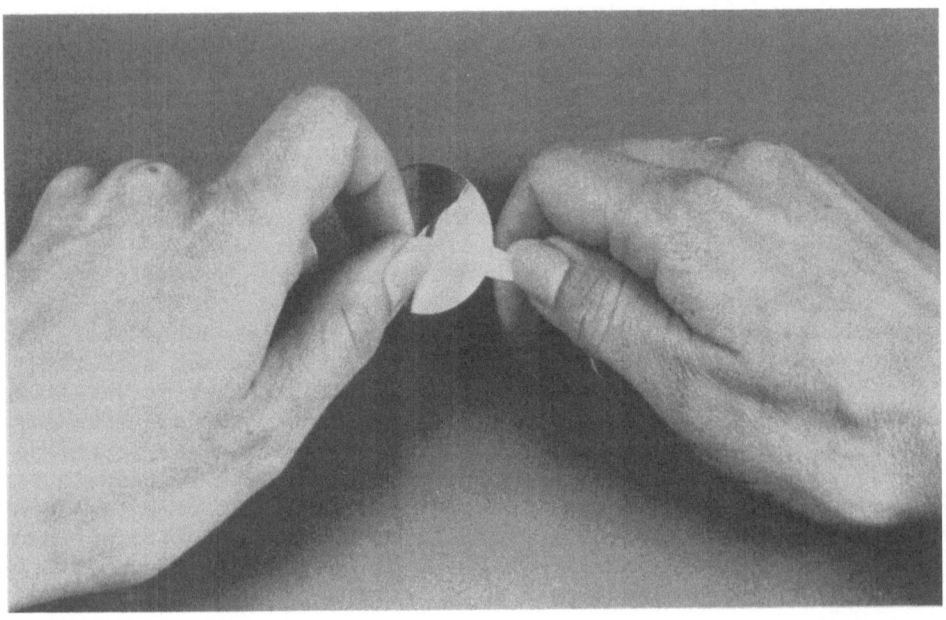

Figure 12. Transdermal Delivery System - Pre-Application

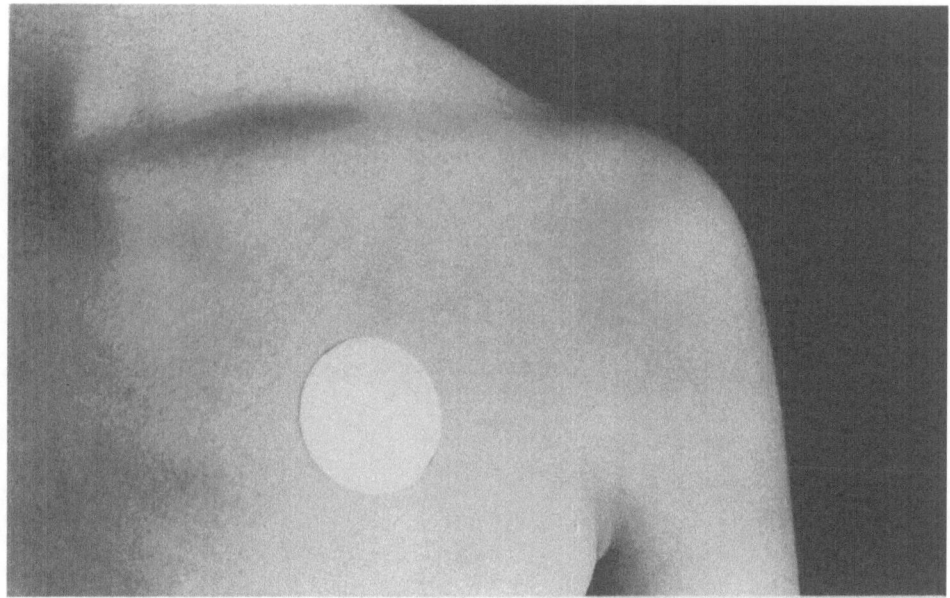

Figure 13.   Transdermal Delivery System - Applied to Skin Surface

molecular kinetic factors influencing the transport of drugs and
other small molecules through polymers, and techniques for "tail-
oring" polymers to display the proper permeation properties, along
with mechanical and chemical characteristics needed to assure
biocompatibility. Few areas of research provide a more challenging
creative opportunity for the innovative polymer- and membrane-
technologist.

PERMEABILITY STUDIES WITH HEMODIALYSIS MEMBRANES

Elias   Klein, J. K. Smith and F. F. Holland

Gulf South Research Institute

P.O. Box 26500, New Orleans, Louisiana,   70186

INTRODUCTION:
    The clinical demonstration of hemodialysis as a treatment
methodology (1,2) and the development of permanently implantable
blood access cannulae (3) have evolved into a broadly based
dialysis technology. In addition to the generally known use of
hemodialysis in the amelioration of the effects of uremia, clinical dialysis has been used in the treatment of acute poisoning
episodes, the reduction of serum barbiturate levels, and as a
means of achieving rapid serum electrolyte balance (4).

    The increased use of hemodialysis as a treatment methodology,
coupled with an appreciation of the potential benefits to be obtained from a multidisciplinary approach in solving problems
associated with the treatment of chronic uremia led to the establishment of the Artificial Kidney and Chronic Uremia Program within the National Institute of Arthritis and Metabolic Diseases.
The publications of this program's conferences, the reports of
progress in the Transactions of The American Society for Artificial
Internal Organs, and the summaries of conferences addressing particular aspects of the hemodialysis problem constitute an impressive
record of the progress, and as yet unsolved problems in the field.

    A number of studies delineating the relative importance of the
mass transfer devices utilized in hemodialysis have been reported
(5). There is a clear interdependence of dialysis membranes and
dialyzers, and the improvements in one dictate that improvements
in the other must be sought. Flat sheet dialyzers with multiple
supports which have been developed in recent years provide good
fluid geometry conditions, given the restraints that whole blood

imposes on the fluid dynamics that can be employed safely.

Under the auspices of the Chronic Uremia Program, a research effort has been undertaken by several laboratories to develop new membrane materials which hopefully will provide improved performance in combination with well designed dialyzers. This paper reports on the transport properties of three such experimental membranes tested both under laboratory conditions which minimize boundary resistance, and in clinical devices such as are commonly used today.

Current clinical practice utilizes three dialyzer types: coiled tube dialyzers using large diameter tubing as the membrane material; hollow fiber dialyzers; and flat sheet dialyzers.

With these dialyzers the membrane generally used is a cellulosic film; the uncoated cellulosic membrane called Cuprophan, manufactured by Enka Glanzstoff AG, is the most commonly used material, and comparisons of experimental membranes are generally drawn against this material. Its properties have been reported upon extensively (6,7,8).

EXPERIMENTAL
Solute permeation rates were measured in a rotating cell characterized previously (9). All measurements were made at 37°C, using radioactively labeled organic solutes in isotonic saline (0.15 M NaCl). The solutes and their concentrations are listed in Table I, together with their molecular weights, diffusion coefficients in saline at 37°C and characteristic molecular radii, as estimated by Farrell, et al., (6). The data of Farrell reveal a good correlation between the radius calculated from solute molal volumes at normal boiling points and molecular weight. For convenience molecular weight is employed in the analyses presented here, since molecular radii estimates of the high molecular weight solutes are speculative.

Equilibrium water contents of the membranes were measured by blotting excess water from the surfaces of membrane samples. From the measurement of the wet weight and the weight following drying at 100°C, a weight fraction of water was calculated. This datum, together with a knowledge of the bulk polymer density, allowed calculation of the void volume fraction in the membrane occupied by water.

Hydraulic permeability of the membranes was measured by applying a known hydraulic pressure to a precisely measured area of the membrane supported on a porous metal disc, and measuring the resulting flow rate through the membrane. Well filtered saline solutions were used in these measurements to avoid filter effects by the membrane; only steady state values were taken.

Ultrafiltration rates were also determined in a clinical dialyzer (Model D-3 Mini Kiil) using either saline solution or heparinized outdated whole blood as the test fluid. The Kiil dialyzer was operated at 200 cc/min blood flow rate, and 500 cc/min dialysate flow rate; the dialysate was prepared and continuously thermostated in a Cobe Century home dialysis monitor. Fluid losses from the whole blood were replenished with saline solution to maintain a constant hematocrit throughout the measurements.

The four membranes examined in this investigation are:

Cuprophan — a cellulosic membrane regenerated commercially from a cuprammonium solution of cellulose; it contains approximately 15 w/w% humectant and plasticizer after drying. Prior to testing, these agents are thoroughly washed out.

Cellulose Acetate -A — an asymmetric membrane prepared by casting a solution of secondary cellulose acetate in acetone containing other non-solvents onto a release paper, evaporating a portion of the solvent, and then quenching the membrane in cold water. The membrane cannot be dried and was tested in the as received form after thorough washing.

TABLE I

Experimental Test Solutes

Solute	Molecular Weight	Solute Concentration % (wt/vol)	Diffusion* Coefficient $\times 10^5$ (cm^2/sec)	Molecular Radius from MVNBP** Å
Urea	60.1	0.5	1.81	2.84
Creatinine	113.1	0.1	1.29	3.58
Uric Acid	168.1	0.005	1.16	3.87
Glucose	180.2	0.5	0.934	4.30
Sucrose	342.3	0.5	0.697	5.23
Raffinose	504.5	0.2	0.534	5.87
Vitamin B-12	1355.4	0.01	0.379	8.35
Inulin	5200	0.01	0.215	Nonspherical
Dextran	16000	0.1	0.166	Nonspherical

* Taken from Colton, et al., (7), or estimated from Wilke-Chang (11).

** Taken from Farrell and Babb (6), estimated by method of LeBas.

Cellulose   - a membrane similar to Cellulose Acetate A, but pre-
Acetate -B   pared from a different solvent-non-solvent combina-
               tion.

Polycar-   - a membrane prepared by solvent casting a solution of
bonate   a block copolymer of bis-phenol-A and polyethylene
               glycols from an organic solvent, evaporating a por-
               tion of the solvent, and then coagulating the nascent
               membrane in water. The membrane cannot be dried with-
               out catastrophic loss in permeability and thus was
               tested in the as received state, after thorough
               rinsing.

DISCUSSION

In water filled, porous membranes it is anticipated that solute transport occurs only through the water phase, and then only if the water is contained in contiguous void spaces whose dimensions are adequate to allow passage of the hydrated solute. Since the permeability coefficient relates observed transport in terms of the external solution concentrations, it can be expected that this coefficient will be a function of the partition coefficient of the solute between the external solution and the solution in the membrane void volume, the magnitude of the void volume, the probability that a given void space is contiguous and adequate to allow transfer, and the diffusion coefficient of the solute in the pore held solution.

The distribution coefficients for a series of representative solutes between saline and the solution contained in Cuprophan have been reported by Colton (7), who found that until the molecular dimension approaches a radius of 16 A, the value is approximately 1.0. Thus in the absence of Donnan effects resulting from charged sites, the only constraint to equal distribution is the size of the membrane surface void.

The magnitude of the void volume (without differentiation of whether the voids are contiguous) can be estimated from the water held in the pore structures. The probability that the void volumes are contiguous, and have the requisite dimension must be lumped into the membrane diffusion coefficient. Thus one can stipulate the following relationships:

$$J_s = \bar{P}\Delta C = \frac{D_m K_d \Delta C}{l} \tag{1}$$

where J_s is solute flux, $\bar{P}$ is the measured permeability coefficient for solute transfer in the absence of bulk flow, D_m is the diffusion

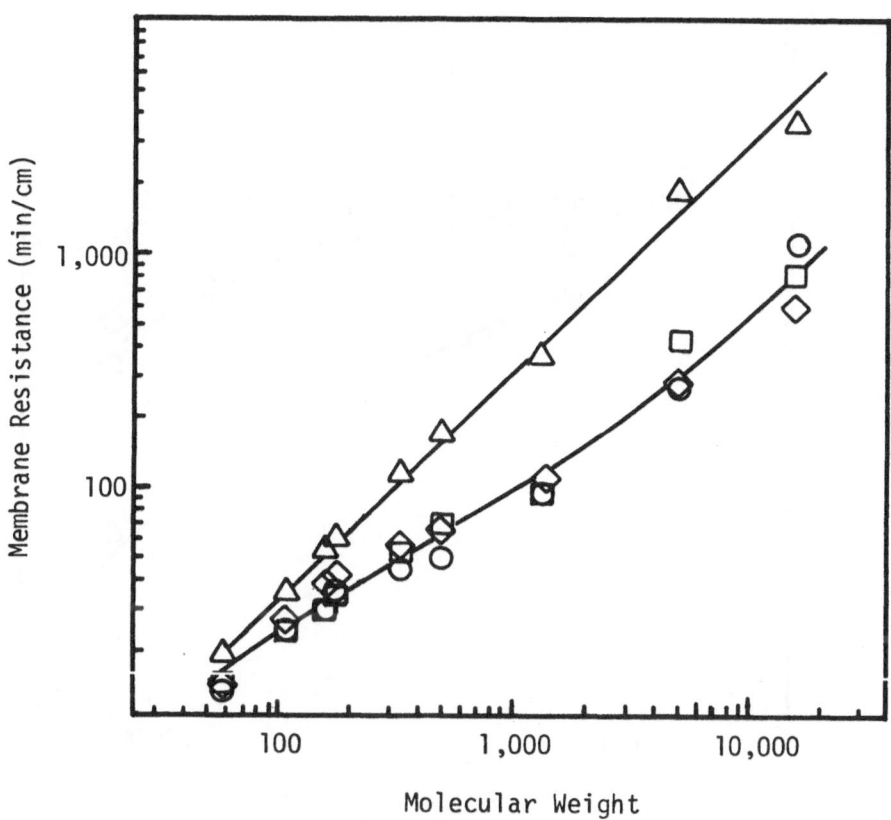

Figure 1.   Plot of membrane resistance as a function of
molecular weight.

△ Cuprophan PT-150,   □ Polycarbonate,

◇ Cellulose Acetate A, ○ Cellulose Acetate B.

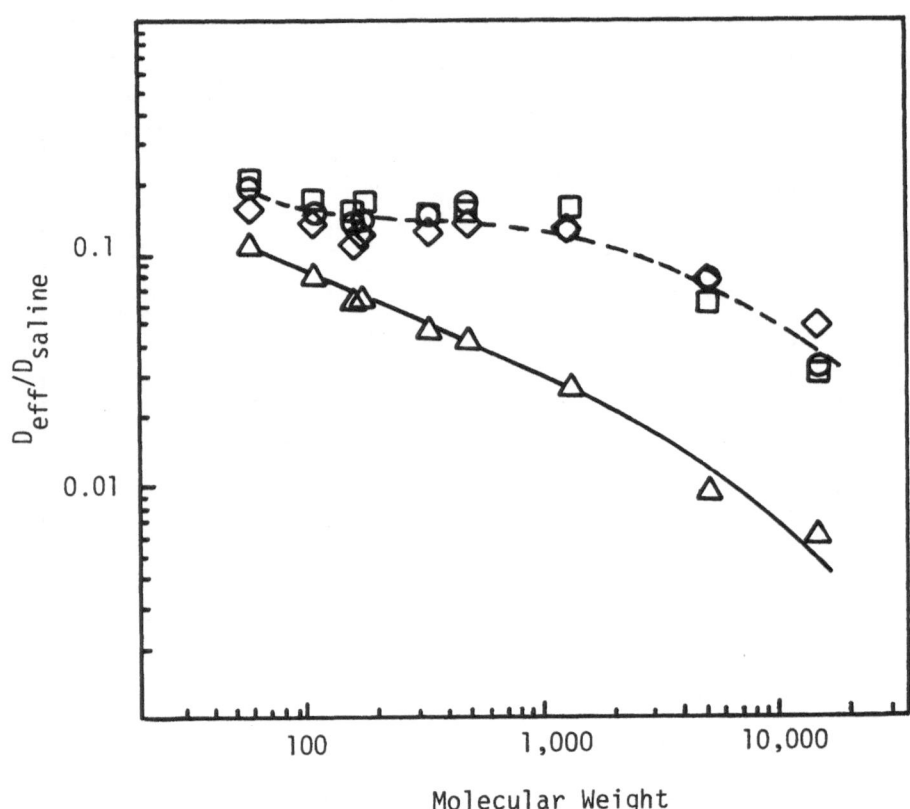

Figure 2.   Plot of the solute diffusion ratio as a
            Function of Molecular Weight.
            △ Cuprophan,   □ Polycarbonate,
            ◇ Cellulose Acetate A, ○ Cellulose Acetate
            B.

coefficient of the solute in the membrane, ΔC is the concentration gradient measured in the solution at the membrane surface, K_d is the distribution coefficient of the solute and 1 is the membrane thickness. In such an expression D_m is expected to be a function of the solute diffusivity in water, the void volume in the membrane, and the nature of the void volume; i.e., the pore sizes, contiguous or not, size distribution, etc. The ratio of D_m/D_w, where D_w is the diffusion coefficient of the solute in water would be expected to be a function of the characteristic membrane void volume, and the solute characteristic dimension. In the absence of data for K_d, the product $D_m K_d$, which is a practical diffusion coefficient, was used. These data are shown in Table II.

In Figure 1 is shown the log (1/P) as a function of molecular weight. The cellulosic membrane data fall on a straight line as reported by Farrell (6) and by Colton, et al., (7); however, the three new membranes exhibit a different relationship, with all membranes quite similar in response to molecular weight. The membrane resistance (1/P) reflects the sum total of all of the forces governing transport through the membrane; consequently it yields very little information about the distinguishing features which might govern the transport.

In Figure 2 a more refined analysis is attempted. Here correction is made for gross membrane thickness, and the resulting effective diffusion coefficient is normalized by the solute diffusion coefficient in saline. Two membrane variables are still unaccounted for: the void volume and the distribution of this volume among various pore sizes and types. Nevertheless, a clear distinction begins to take shape. When diffusivity of the solute in saline is considered, it appears that the response of transport to molecular weight is flat up to a value of 1,000. The implication is drawn that for solute species of this mass and volume there is no selective impedance to diffusion through the membrane, but at higher molecular weights, the sieving effects and/or adsorption phenomena begin to retard transport.

By contrast, the cellulosic membrane is impeding transport for even the low molecular weight species on a size selective basis. The void volume corrections are no guide in correlating relative diffusion coefficients for these membranes. The most rapid transport is exhibited by the polycarbonate and the Cellulose Acetate B membranes, followed by the Cellulose Acetate A and the cellulose membranes. However, the void volume fractions are not in this same order, nor are the hydraulic permeabilities found to be in this order. Since the cellulose acetate membranes are known to exhibit asymmetric structures, it is likely that the relative thicknesses of the layers, together with their characteristic impedances override the analysis based on a homogeneous porous structure.

TABLE II

Properties of Membranes Tested

	Cuprophan	Polycarbonate	Cell Acetate A	Cell Acetate B
Hydraulic Permeability cm/sec atm X 10^5	3.02	8.1	29.7	21.3
Void Volume	.646	.590	.684	.645
Thickness (microns)	22.2	34.0	29.2	27.0
Solute Diffusivity cm^2/sec X 10^6				
Urea	1.96	3.94	2.79	3.45
Creatinine	1.03	2.39	1.77	1.94
Uric Acid	.70	1.91	1.27	1.55
Glucose	.60	1.61	1.17	1.27
Sucrose	.32	1.05	.86	1.03
Raffinose	.23	.80	.74	0.91
Vitamin B-12	.10	.61	.48	0.48
Inulin	.020	.13	.17	0.17
Dextran	.010	.050	.082	0.054

In Figure 3 are shown the hydraulic ultrafiltration rates using both blood and saline solution as the test fluid. Because blood is a polydisperse fluid, a separation of blood serum and formed elements can take place during the passage of blood through the dialyzer. The resulting phase separation, plus possible protein absorption on the membrane pore surfaces leads to non-linear responses not found when membranes are tested only with saline solution. The degree of this non-linearity appears to be a function of ultrafiltration rate, with the more permeable membranes showing the greater non-linearity.

APPLICATIONS TO CLINICAL PRACTICE

The inherent mass transfer capability of each membrane is reduced in clinical devices because of several factors. The blood and dialysate channels impose fluid boundary layers which diminish the effective concentration gradients; there are limitations on the degree of shear to which the blood can be subjected so that laminar flow is imposed; and the geometry of some dialyzers reduces the solute concentration gradient more than others as the blood traverses its length. The effective overall mass transfer that the membrane/dialyzer combination permits is expressed in a convenient form by the clearance K_D, defined as:

$$K_D = \frac{Q_a C_a - Q_v C_v}{C_a} = \frac{1}{C_a}\frac{dM}{dt} \qquad (2)$$

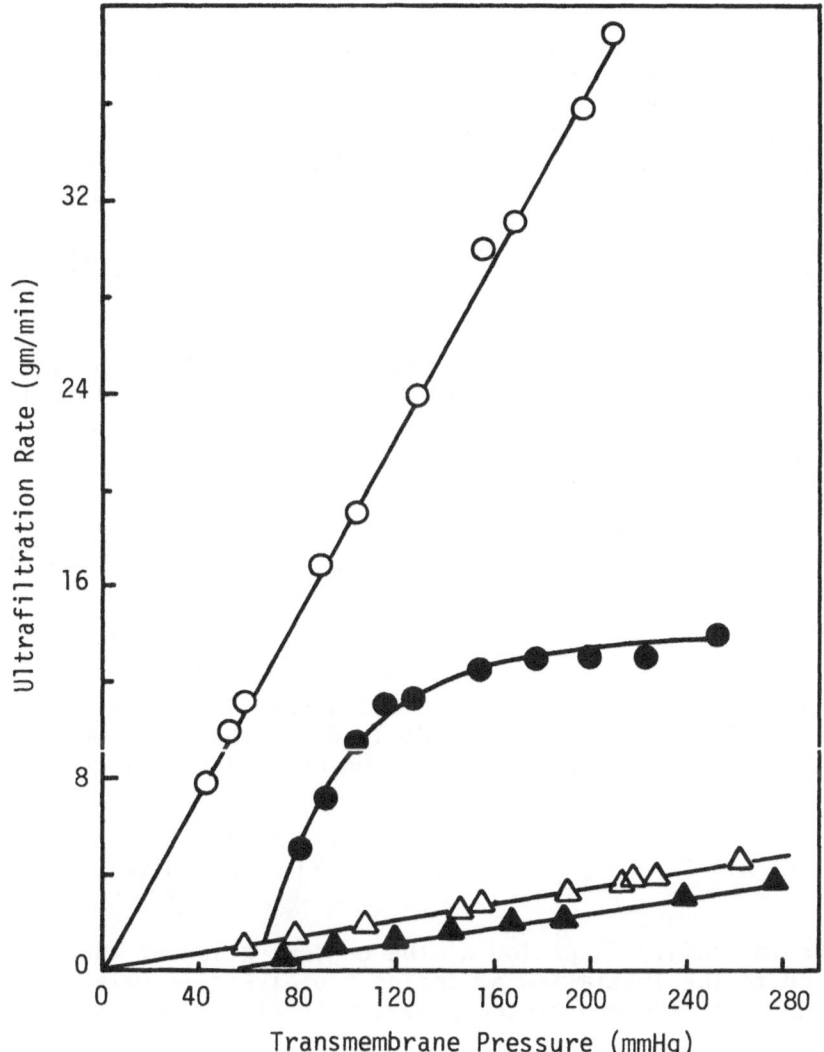

Figure 3. Plot of ultrafiltration rate as a function of
transmembrane Pressure. Cuprophan PT-150
△ Dialysate vs. Dialysate, ▲ Blood vs.
Dialysate; Cellulose Acetate ○ Dialysate vs.
Dialysate, ● Blood vs. Dialysate.

Where Q and C are the blood flow rate and solute concentration respectively; subscripts a and v denote arterial and venous
connections. The product of clearance times arterial concentration gives the mass rate of metabolite removal from the body.

The removal of metabolite by the dialyzer is only one of
several processes that governs the ambient concentration of the
solute in blood serum. In addition to the dialyzer clearance,
the individual may have residual kidney function. In a stable
patient the combined dialyzer clearance (K_d) and renal clearance
(K_r) processes are in balance with the endogenous generation rate
of metabolite. A number of investigators have described the
inter-relationships between these variables, and the time cycle of
treatment (12-18). The defining equations arise from the following
considerations.

When the patient is between treatment periods (in waiting
period of length W), the change in serum level of the solute is
given by:

$$V(dC/dw) \; = \; G - K_r C \tag{3}$$

When the patient is undergoing treatment, for a period of duration,
T, the concentration change in the serum is given by:

$$V(dC/dt) \; = \; G - (K_d + K_r) \, C \tag{4}$$

Integration of these equations with the assumptions that G, V, and
K_r are constant through both periods leads to the relationship between the serum concentration at the end of dialysis C_T to that at
the beginning of dialysis C_0 given by:

$$C_T \; = \; C_0 e^{-\alpha} + \frac{G}{K_r + K_d} (1 - e^{-\alpha}); \; \alpha = \left(\frac{K_r + K_d}{V}\right) T \tag{5}$$

After the interdialytic period W, the concentration of the serum
just prior to the initiation of the next dialysis is related to the
concentration at the end of the previous treatment period by:

$$C_{o1} \; = \; C_T e^{-\beta} + \frac{G}{K_r} (1 - e^{-\beta}) \qquad \beta = \frac{K_r}{V} W \tag{6}$$

From a knowledge of the membrane permeability, and an estimate
of the conversion of this value to a mass transfer value in a given
dialyzer, one can calculate the effective clearance that is to be
expected under a specified treatment protocol. Together with a
measure of residual renal function in a patient, this should allow

an estimate of the maximal serum concentrations that the patient
should experience, using the preceding equations. The simplifying
assumptions are extensive, and should not be ignored. Nevertheless,
the derivations provide a valuable insight into the results that a
variety of treatment schedules can achieve.

For predicting the concentration of solutes that are rapidly
and uniformly distributed throughout the entire body water, the
equations should be quite accurate. However, when the rate of
solute distribution between the various intracellular volumes and
extracellular volumes is not rapid, the equations will ignore a
significant kinetic factor. To illustrate the kinds of informa- (19)
tion which can be deduced, we show in Figure 4 the calculated pre-
and post-dialysis blood levels of urea to be expected in a patient
having a total water volume of 37.0 1, dialyzed on two different
machines: one has a clearance of 90 cc/min, and the other has a
clearance of 135 cc/min. For the purposes of the calculation
values of residual function of 0.5 cc/min were assumed, together
with a urea generation rate of 6 mgm/min.

Figure 4 illustrates the effect of both the dialyzer,
clearance together with the effect of treatment schedules. Four
hours of daily treatment will provide approximately the same mean
levels of serum concentration, but will reduce the maxima significantly in comparison to the same total treatment hours distributed
over alternate days.

In Figure 5 the effect of the dialyzer clearance is shown more
clearly. The pre-dialysis levels decrease with increasing clearance, with very little effect on the difference between the pre-
and post-dialysis levels. The schedule selected for this calculation is a 6-hour treatment on alternate days.

In Figure 6 are shown the effects on the serum concentration
that can be expected with high molecular weight metabolites having
a clearance value of 10 cc/min, a generation rate of 0.6 mgm/min,
and still the same renal clearance as used in the previous examples. The most striking difference is the "leveling" effect that
the residual clearance exerts in reducing the difference between
pre- and post-dialysis serum concentrations. The serum concentration can still reach significant levels, but the cycling of concentration is only a fraction of the median concentration.

In Figure 7 the effect of the residual clearance is shown by
calculating the pre- and post-dialysis serum levels for a fixed
dialyzer clearance, generation rate, and schedule, but varying the
residual clearance from 0.1 cc/min to 5.0 cc/min.

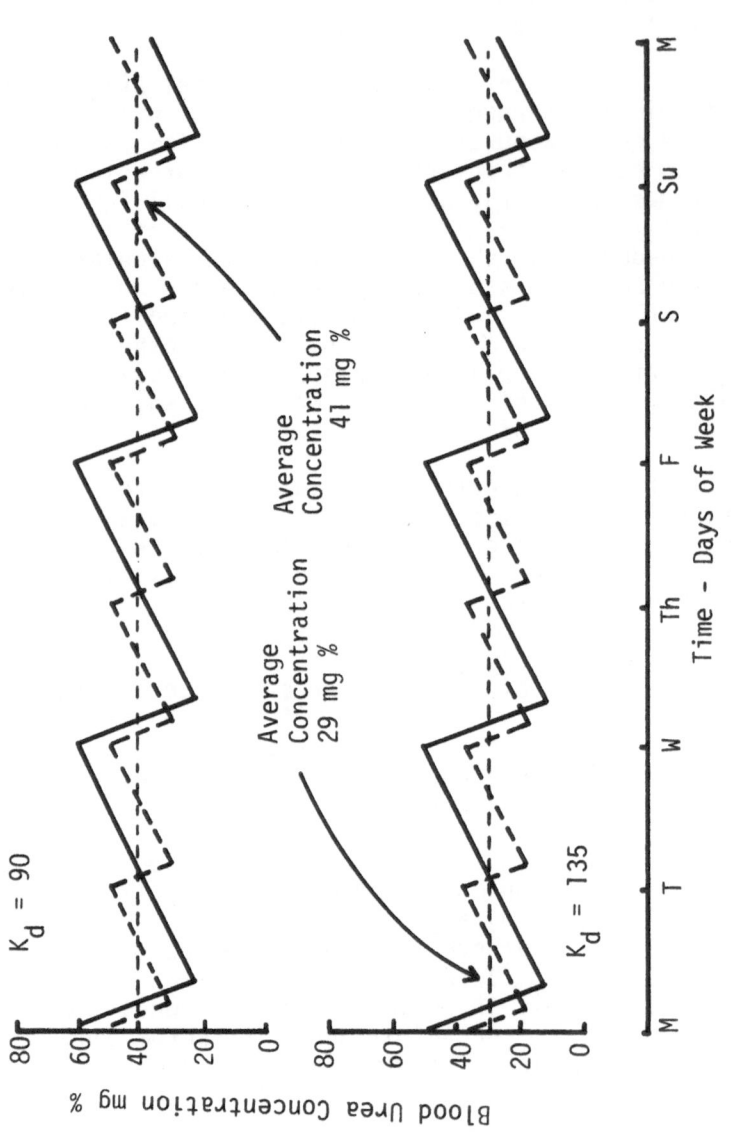

Figure 4.   Comparison of blood urea concentration as a function of dialysis schedule and of dialyzer clearance. V = 37 liters, G = 6 mg/min, K_r = 0.5 ml/min. ——— 8-hour dialysis, alternate days; ------ 4-hour dialysis, daily.

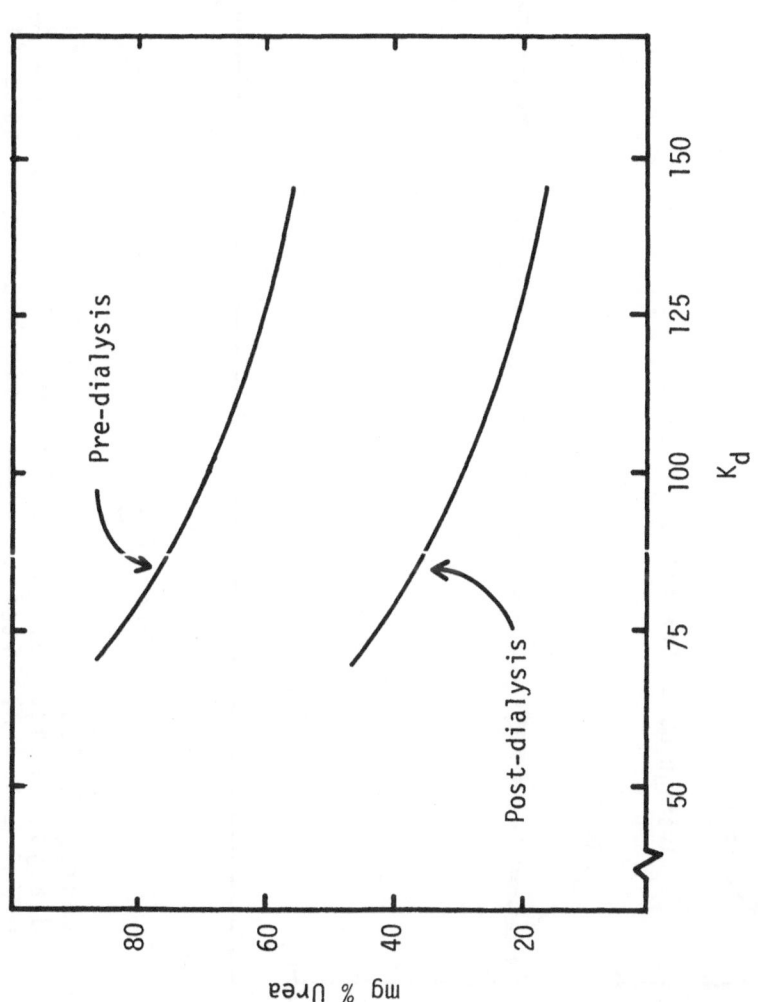

Figure 5.   Blood urea as a function of dialyzer clearance K_d.   Alternate day dialysis.   Six-hour schedule.

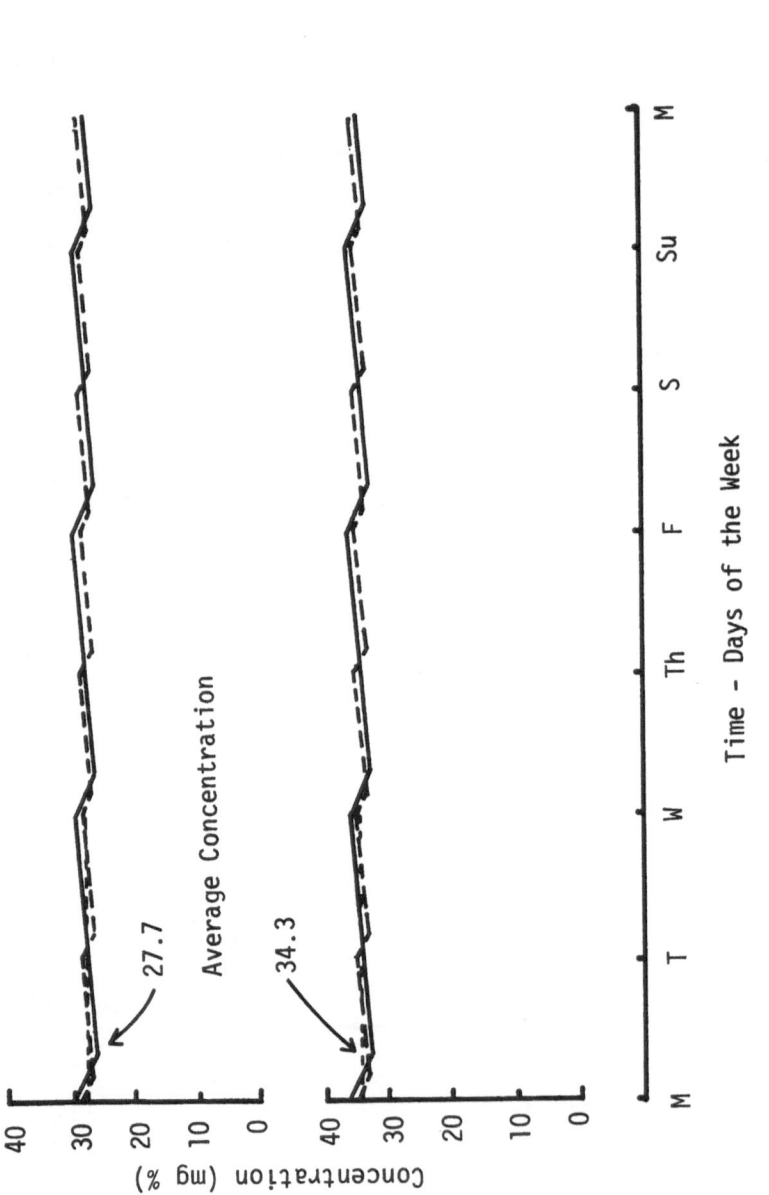

Figure 6. Concentration of 1350 molecular weight solute as a function of dialysis schedule. $G = 0.6$ mg/min, $K_d = 10$ ml/min, $K_r = 0.5$ ml/min, $V = 37$ liters. ——— six-hour alternate day dialysis, ------ three-hour daily dialysis.

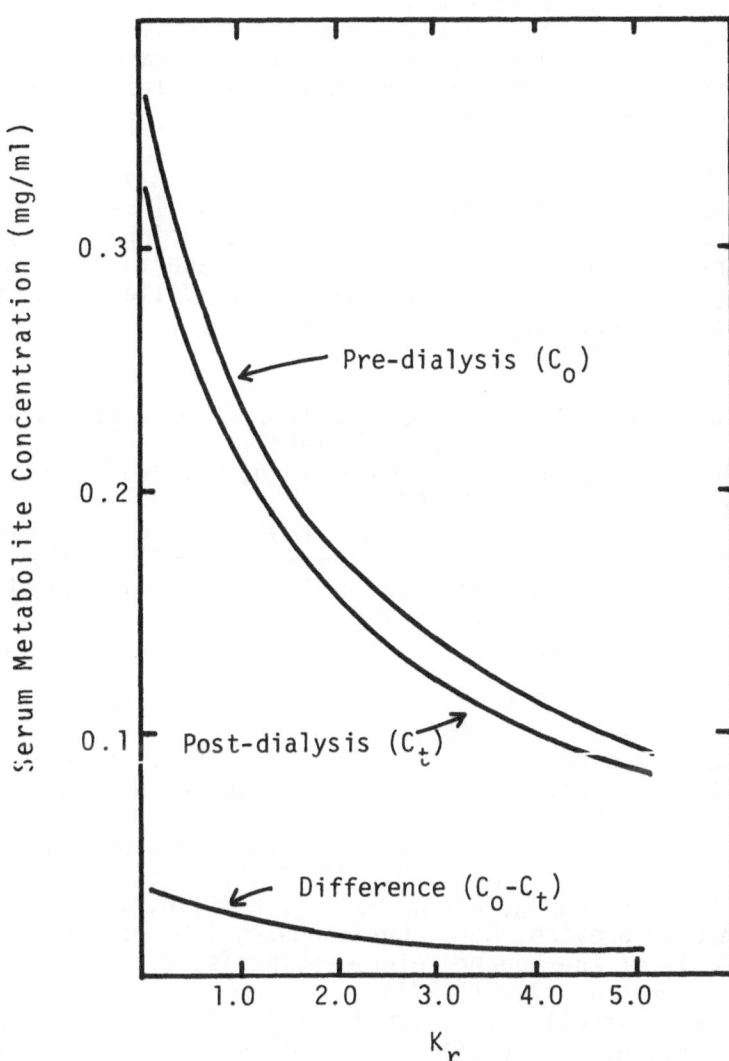

Figure 7.   Effect of Residual Renal Clearance on
            Serum Pre- and Post-dialysis high
            molecular weight metabolite concentration.
            V = 37 liters, G = 0.6 mg/min, K_D = 10
            ml/min; dialysis schedule 8 hours on
            alternate days.

SUMMARY
    The permeability of hemodialyzer membranes is a significant
variable in the treatment of renal failure. However, other design
and physiological limitations must be considered together with the
membrane properties. These factors become increasingly more significant when the permeability of the membrane decreases with molecular weight to the point that the residual renal clearance becomes
a major fraction of total solute removal rate.

ACKNOWLEDGMENT
    This work has been supported by the National Institute of
Arthritis, Metabolism and Digestive Diseases, National Institutes
of Health through Contract NIH NIAMDD-72-2221 and the assistance
and many suggestions of Doctors R. Wineman, F. Villarroel, and
E. Swilley are appreciated.

REFERENCES
1. Kolff, W.J. and Berk, H.T.J., *Acta Fed. Scand.*, <u>117</u>, 121(1944).
2. Kolff, W.J., *Circulation*, <u>15</u>, 285(1957).
3. Scribner, R.H., Quinton, W. and Dillard, D., *Trans. Am. Soc.
    Art. Int. Organs*, <u>6</u>, 104(1960).
4. Knepshield, J.H., Schreiner, G.E., Lowenthal, D.T. and Gelfand,
    M.C., *Trans. Am. Soc. Art. Int. Organs*, <u>19</u>, 590(1973).
5. Chemical Engineering Symposium Series, "The Artificial Kidney",
    Am. Inst. Chem. Eng., 84, (1968).
6. Farrell, P.C. and Babb, A.L., *J. Biomed. Mater. Res.*, <u>7</u>, 275
    (1973).
7. Colton, D.K., Smith, K.A., Merrill, E.W. and Farrell, P.C., *J.
    Biomed. Mater. Res.*, <u>5</u>, 459(1971).
8. Craig, L.C. and Koenigsberg, W.J., *Phys. Chem.*, <u>65</u>, 166(1961).
9. Wendt, R.P., Toups, R.J., Smith, J.K., Leger, N. and Klein, E.,
    I&EC Fundamentals, <u>10</u>, 406(1971).
10. Fisher, B.S., Higley, W.S., Cantor, P.A. and Stone, W., *Trans.
    Am. Soc. Art. Int. Organs*, <u>19</u>, 429(1973).
11. Wilke, C.R. and Chang, P., *AIChEJ*, <u>1</u>, 264(1965).
12. Babb, A.L., Popovich, R.P., Christopher, T.G., Scribner, B.H.,
    The Genesis of the Square-Meter Hypothesis, *Trans. Am. Soc.
    Art. Int. Organs*, <u>17</u>, 81(1971).
13. Babb, A.L., Farrell, P.C., Uvelli, D.A., Scribner, B.H., Hemo-
    dialyzer Evaluation by Examination of Solute Molecular Spectra,
    Trans. Am. Soc. Art. Int. Organs, <u>18</u>, 98(1972).
14. Babb, A.L., Farrell, P.C., Strand, M.J., Uvelli, D.A.,
    Milutinovic, J., Scribner, B.H., Residual Renal Function and
    Chronic Hemodialysis Therapy, *Proc. of CD&T Forum*, <u>11</u>, 142
    (1972).
15. Kjellstrand, C.M., Evans, R.L., Petersen, R.J., Rust, L.W.,
    Shideman, J., Buselmeier, T.F. and Tozelle, L.T., Considerations
    of the Middle Molecule Hypothesis, *Proc. of CD&T Forum*, <u>11</u>,
    127(1972).

16. Edson, H., Keen, M., and Gotch, F., Comparative Solute Transport and Therapeutic Effectiveness of Multiple Point Support and Standard Kiil Hemodialyzers, *Trans. Am. Soc. Art. Int. Organs*, <u>18</u>, 113(1972).
17. Sargent, J.A., Gotch, F.A., Caskey, T.L., Keen, M., Landau, J.I., Look, A.T., and Seid, M.A., Studies on the Molecular Etiology of Uremia using High Flux Cellulose Acetate Dialyzers, Third Annual Progress Report, Dow Chemical Co., submitted to Artificial Kidney Program, NIAMDD, 1973.
18. Gotch, F.A., Sargent, J.A., Keen, M.L., Seid, M. and Foster, R., *Proc. CD&T Forum*, 1973 (in press).
19. Cavanaugh, K.L., "Physiological Transport Parameters in the Patient-Artificial Kidney System" Thesis, U. of Texas, May, 1973.

PREPARATION AND APPLICATION OF RADIATION-GRAFTED HYDROGELS

AS BIOMATERIALS

A. S. Hoffman, G. Schmer, T. A. Horbett, B. D. Ratner,
L. N. Teng, C. Harris, W. G. Kraft, B. N. L. Khaw,
T. T. Ling, and T. P. Mate
Dept. of Chemical Engineering and Center for Bioengineering
University of Washington, Seattle, Washington

Hydrogels refer to a class of water swollen, lightly cross-linked polymeric gels which have been mentioned often as useful biomaterials (e.g., 1,2). Although they may be used in bulk forms, as in the soft contact lens, there are many applications where the mechanical demands are too great for typical hydrogels, and then such materials are best used as coatings on stronger supports. Although some polymeric supports are readily coated by solution dip coating techniques (3), there are many polymers of bio-medical interest which are not so easily coated. The technique of radiation grafting (e.g., 4) is especially suitable for depositing coatings of a wide variety of hydrophilic polymer compositions and thicknesses onto and within almost any shape or composition of support polymer (e.g., 5-14). Over the past three years* our group has been applying this method, using cobalt-60 radiation, to produce directly or precursors of a whole family of new biomaterials (6-8, 10-12). This paper describes some of the highlights of our studies; most details of materials and procedures may be found in these references.

The process of radiation grafting hydrophilic polymers onto "inert" polymeric supports is shown in Figs. 1, 2. An additional process step is possible in which biologically active molecules (as enzymes) are chemically bonded either directly onto the grafted polymer -OH, $-CO_2H$, etc. groups or via a short molecular "arm." (Fig. 1) There are a great variety of potential applications for such materials (Table I).

Not all hydrogels are equivalent in composition, chemical, physical, mechanical and/or biochemical properties (2, 6) and the

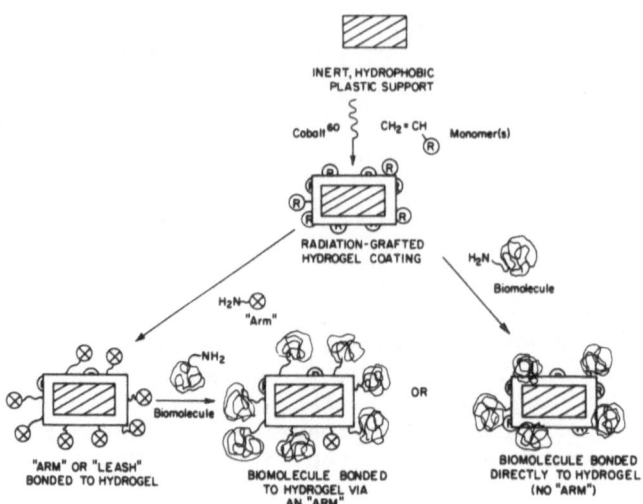

Fig. 1   Schematic process for direct radiation grafting of hydrophilic polymers onto other polymeric supports followed by
immobilization of biological molecules onto these grafted
surfaces (6).

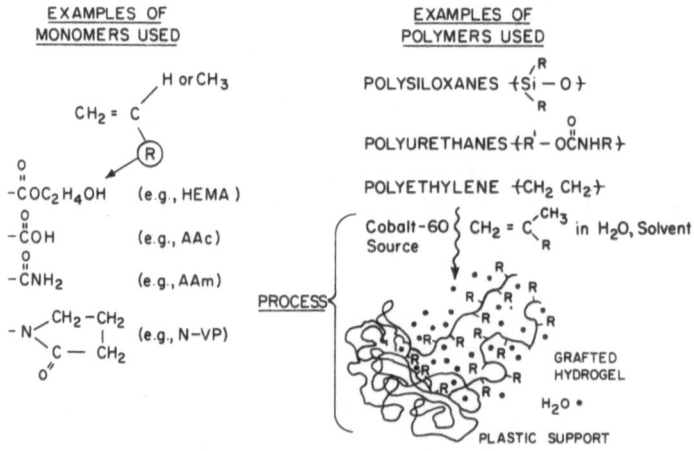

Fig. 2   Some examples of hydrophilic monomers and polymeric
support materials which may be used in the radiation
grafting step.

Table I

USEFUL APPLICATIONS OF IMMOBILIZED
BIOMOLECULE/HYDROGEL SURFACES

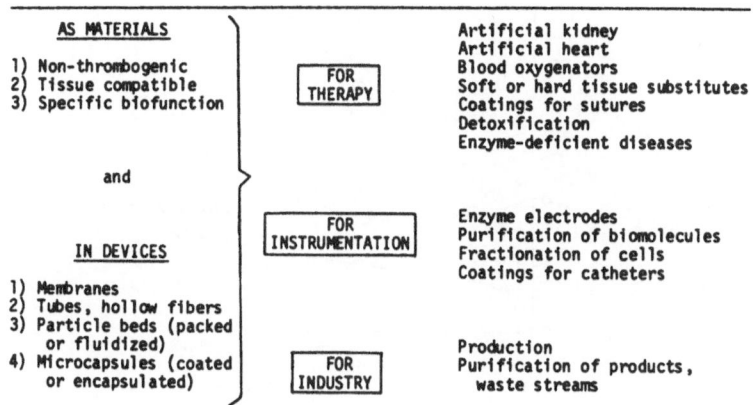

AS MATERIALS		
1) Non-thrombogenic	FOR THERAPY	Artificial kidney
2) Tissue compatible		Artificial heart
3) Specific biofunction		Blood oxygenators
		Soft or hard tissue substitutes
		Coatings for sutures
		Detoxification
		Enzyme-deficient diseases
and		
	FOR INSTRUMENTATION	Enzyme electrodes
IN DEVICES		Purification of biomolecules
		Fractionation of cells
1) Membranes		Coatings for catheters
2) Tubes, hollow fibers		
3) Particle beds (packed or fluidized)		
4) Microcapsules (coated or encapsulated)	FOR INDUSTRY	Production
		Purification of products, waste streams

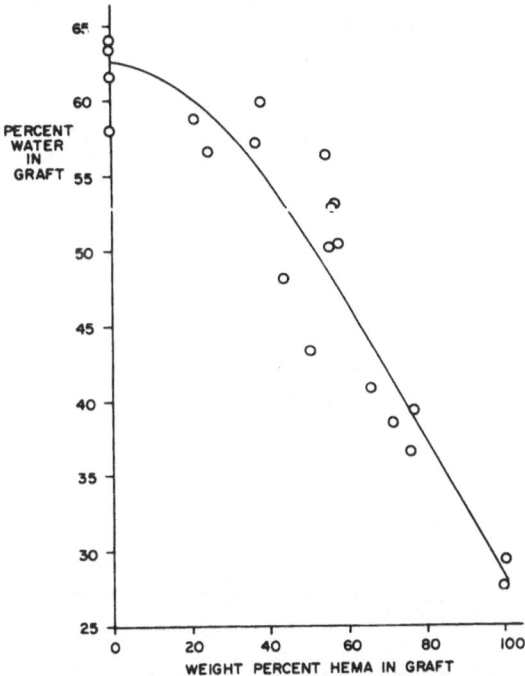

Fig. 3. Water content of copoly- (HEMA/N-VP) on Silastic films
as a function of the observed HEMA content of the
grafted copolymer. (12)

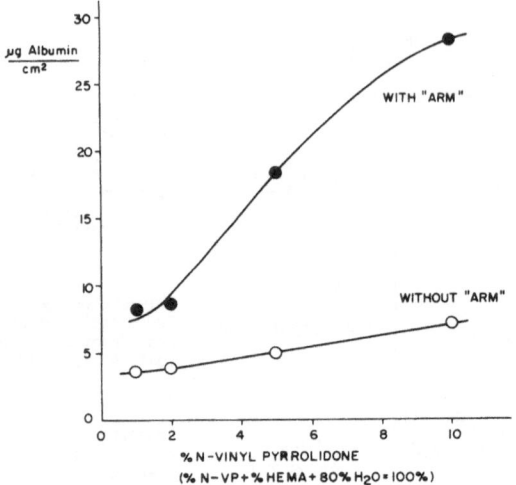

Fig. 4.   Effect of the composition of the HEMA/N-VP grafting
          solution on the amount of albumin immobilized directly
          or via an ε-amino caproic acid arm; the support is
          Silastic rubber. (6)

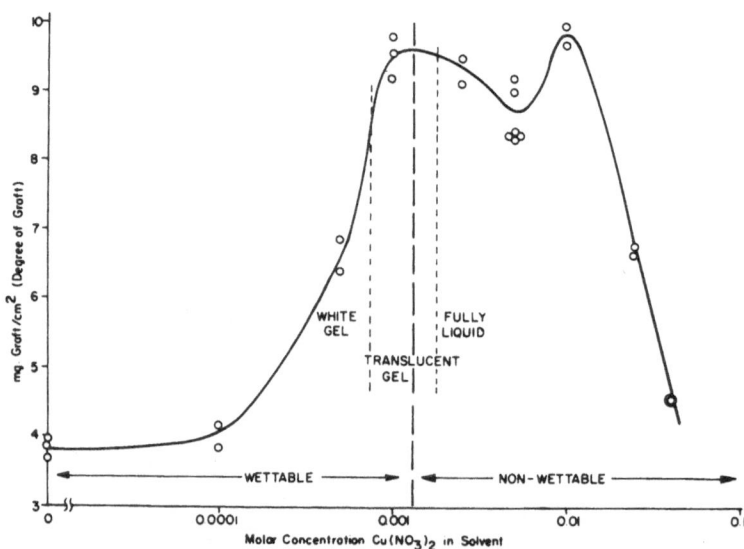

Fig. 5.   Effect of cupric ion on the extent of grafting copolymers
          of HEMA and N-VP to Silastic film supports. Grafting
          solution composition is 10 HEMA, 10 N-VP and 80 H$_2$O
          (+Cu(NO$_3$)$_2$), all parts by volume.

          Dose = 0.25 Mrad.

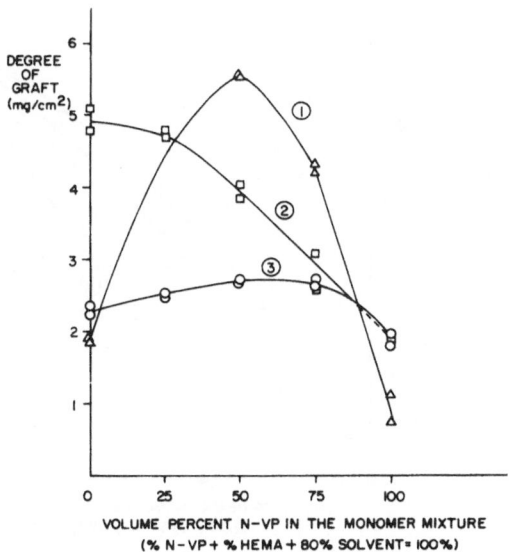

Fig. 6. Effect of methanol and monomer contents in the grafting
         solution on the extent of grafting to Silastic film
         supports. Solvent compositions are:

       1:H_2O

       2:MeOH/H_2O = 1/1 (vol.)

       3:MeOH/H_2O = 3/1 (vol.)

       Dose = 0.25Mrad.(<u>12</u>)

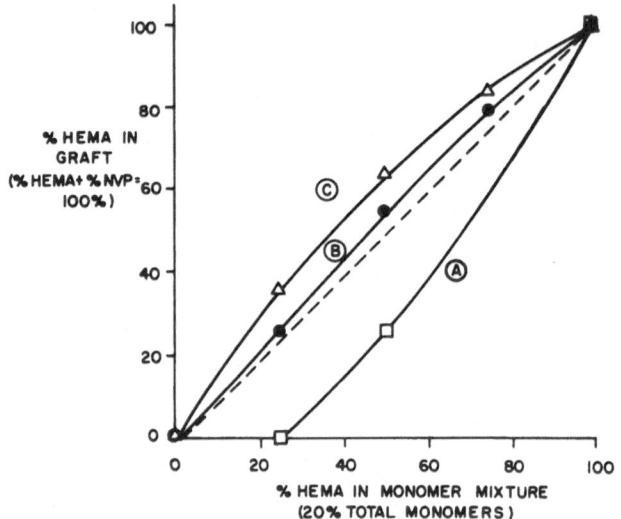

Fig. 7. Effect of grafting solution composition on composition of
         the grafted copolymer. Solvent (80% by vol. of total
         solution) compositions are:

         A:H_2O

         B:$MeOH/H_2O$ = 1/1 (vol.)

         C:$MeOH/H_2O$ = 3/1 (vol.)

         Dose = 0.25Mrad. (12)

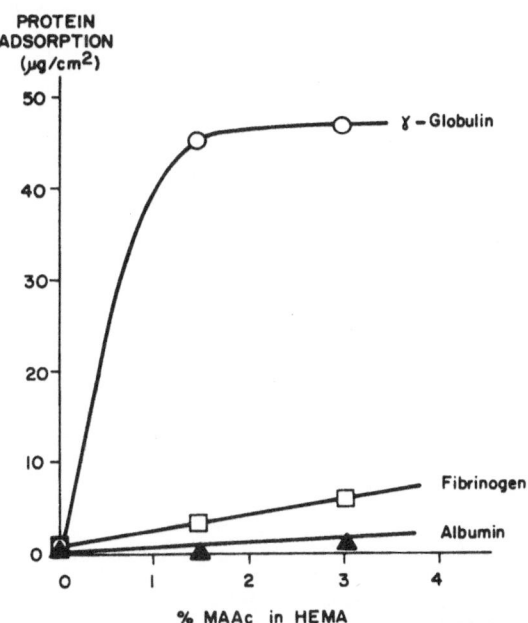

Fig. 8.   Adsorption of plasma proteins onto methacrylic acid/HEMA
copolymer grafts on Silastic supports.   Dose = 0.25 Mrad.   Protein concentrations = 0.5mg./ml., pH=7.4, no added salt, T=37°C.

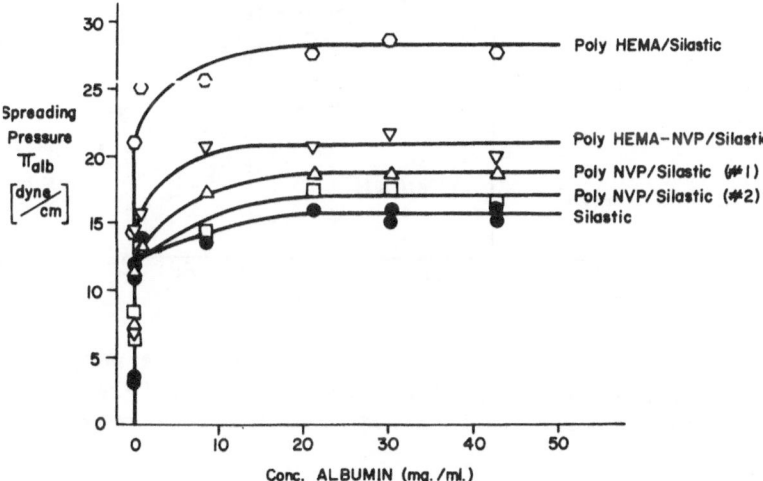

Fig. 9.   Spreading pressure of albumin on Silastic and HEMA and/or
N-VP grafted Silastic as a function of albumin concentration.
Data are based on contact angle measurements.   The two NVP grafts
were prepared with 60/20(#1) and 40/40 (#2) MeOH/H$_2$O ratios (vol.)
in the grafting solutions.   Protein solutions all contain 0.15M
NaCl; pH = 7.4; T = 22°C (10,11).

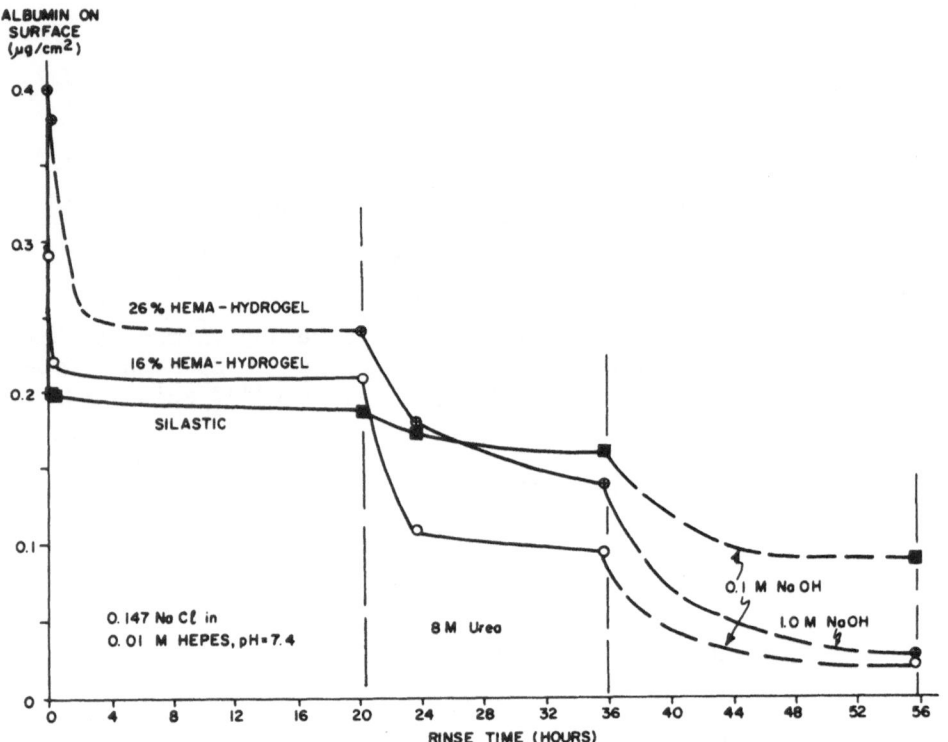

Fig 10.   Desorption of albumin from two different polyHEMA/Silastic
          hydrogels and from Silastic. Three different elution
          solvents were used in sequence, each at room temperature.

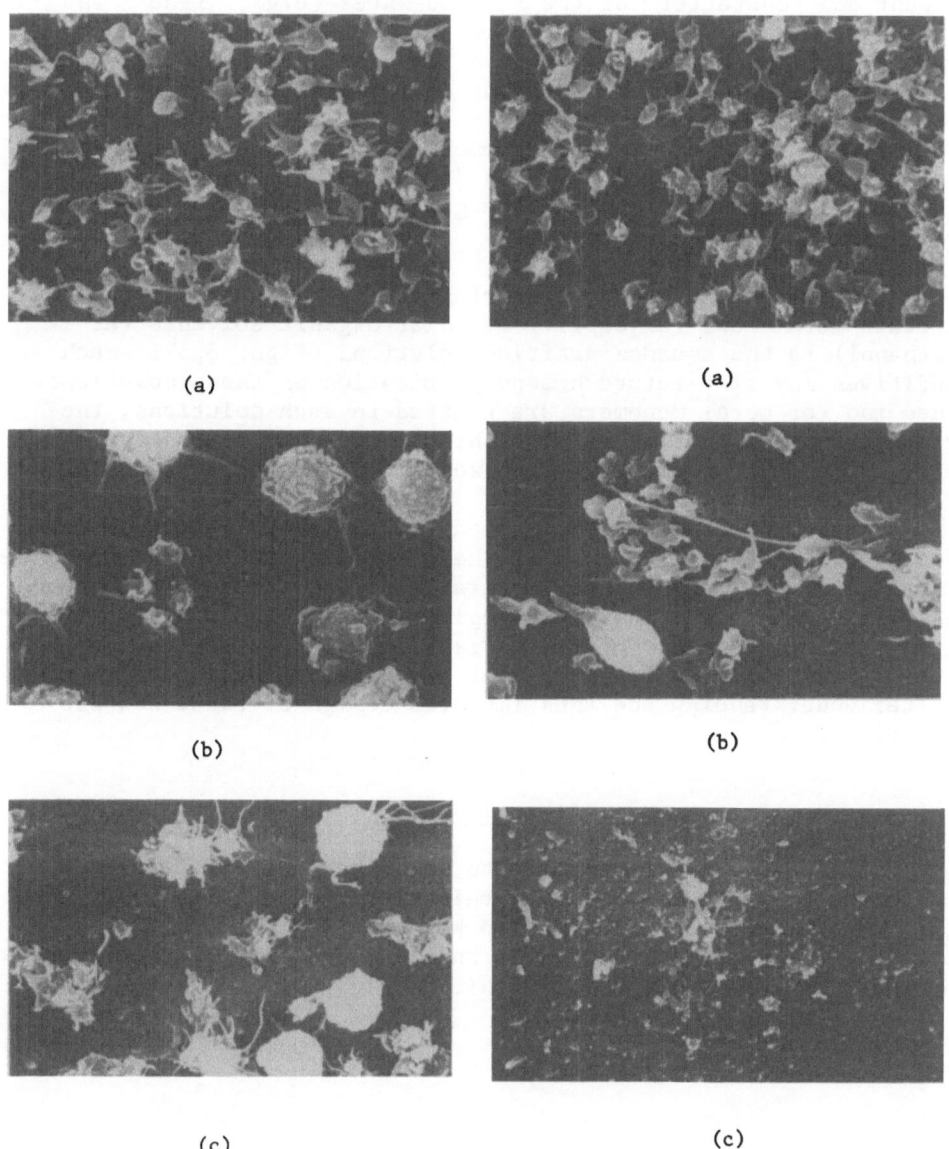

Fig. 11. Scanning microscope photos @2100 –2200X of two different
          surfaces exposed to blood in an _ex vivo_, carotid-jugular
          shunt on a dog. Series (a)-(c) at left is the inside
          surface of a Silastic shunt and series (a)-(c) at right
          is the inside surface of a grafted HEMA/N-VP Silastic
          shunt. Letters refer to time of exposure: (a) = 2 min.,
          (b) = 60 min. and (c) = 105 min. Each time is unin-
          terrupted, using a fresh shunt.

amount and "character" of the imbibed water (e.g., "free " vs.
"associated" vs. "structured," etc.) may be critical to their
ultimate biocompatibility. Fig. 3 shows how compositional variations can affect the water content of the grafted polymer.

Proteins are more readily immobilized on the higher water
content grafts, especially via an "arm" molecule (Fig. 4). The
enzyme streptokinase retained the greatest fraction of its
activity when so bonded (6).

The amount and penetration of graft may be varied by adding
certain metal ions (as cupric) or polar organic solvents (as
methanol) to the aqueous grafting solutions (Figs. 5,6). Such
additives may also retard homopolymerization in these solutions.
When two (or more) monomers are grafted in such solutions, the
composition-penetration relationship of the grafted copolymer may
also be strongly affected by the solvent composition (12) (Fig. 7).

Plasma proteins have been found to adsorb (Figs. 8, 9) and
desorb (Fig. 10) differently at the different grafted and ungrafted
interfaces. The presence of ionizable groups in the graft may be
especially important to its biological interactions. Cellular
interactions at these interfaces is also under investigation
(e.g., Fig. 11). It is hoped that such studies will lead to a
better understanding and thus improved design of these new biomaterials.

Acknowledgement

It is particularly fitting to note that Prof. Vivian Stannett
has given us valuable advice on this work. In addition, one of us
(ASH) has had the good fortune to know and to interact with this
fine man for over fifteen years, and is especially pleased to be a
part of this symposium honoring him.

* Support for these studies has been mainly by the U.S.A.E.C.,
  Contract No. AT(45-1)2225 and in part also by N.I.H.G.M.S.,
  Grant No. 16436-3 to 5.

References

1. Levowitz, B.S. et al., Trans. Amer. Soc. Artif. Internal Organs, 14, 82 (1968).
2. Bruck, S., J. Biomed. Mater. Res., 7, 337 (1973).
3. Tollar, M., Stol, M. and Kliment, K., J. Biomed. Mater. Res., 3, 305 (1969).
4. a) Chapiro, A. "Radiation Chemistry of Polymeric Systems," Interscience, N. Y. (1962).
   b) Stannett, V. and Hoffman, A.S., Amer. Dyestuff Reporter, 57, 25, 91 (1968).
5. Yasuda, H. and Refojo, M.F., J. Polymer Sci., 2, 5093 (1964).
6. Hoffman, A.S., Schmer, G., Harris C., and Kraft, W.G., Trans. Amer. Soc. Artif. Internal Organs, 18, 10 (1972).
7. Hoffman, A.S. and Kraft, W.G., Polymer Preprints (ACS), 13, 2, 723 (1972).
8. Hoffman, A.S. and Harris, C., Polymer Preprints (ACS), 13, 2, 740 (1972).
9. Lee, H.B., Shim, H.S. and Andrade, J.D., Polymer Preprints (ACS), 13, 2, 729 (1972).
10. Khaw, B.N.L., M.S. Thesis in Chem. Eng., University of Washington, June 1972.
11. Ling, T.T., M.S. Thesis in Chem. Eng., University of Washington, June 1972.
12. Ratner, B.D. and Hoffman, A.S., Org. Coatings and Plastics Chem. Preprints (ACS), 33, 2, 286 (1973).
13. Laizier, J. and Wajs, G. in "Large Radiation Sources for Industrial Processes", I.A.E.A., Vienna (1969), pp. 205
14. Kearney, J.J., Amara, I. and McDevitt, M.E., Org. Coatings and Plastics Chem. Preprints (ACS), 33, 3, 346 (1973).

IMPROVEMENT OF BLOOD COMPATIBILITY

OF MEMBRANES BY DISCHARGE POLYMERIZATION

H. Yasuda and M. O. Bumgarner   L. G. Mason

Camille Dreyfus Laboratory   Department of Pathology
Research Triangle Institute   Univ. of North Carolina
Res. Triangle Park, N.C. 27709   Chapel Hill, N.C. 27514

I. INTRODUCTION

The use of polymers in membrane moderated biomedical devices in which the polymer surface is in contact with blood requires the extra membrane property of blood compatibility. Some molecular factors that improve the mechanical and/or transport properties of polymer membranes do not necessarily influence the blood compatibility in a favorable manner. Therefore, the most logical and feasible approach to the problem is the modification of the properties of the surface by some means that does not affect the bulk properties of the membrane, both mechanical and transport, that are required for a specific application.

It is therefore not surprising to find that the grafting of hydrogels, which have good blood compatibilities but rather poor mechanical properties, onto hydrophobic polymers, which have good mechanical properties but relatively poor blood compatibilities, comprises one of the most important achievements in the development of blood compatible polymer surfaces in the last decade.

Glow discharge polymers can be effectively utilized in the surface modification of substrate polymers by depositing a small amount of glow discharge polymer onto various kinds of substrate surfaces. The main advantages (besides the blood compatibility of the polymers) of glow discharge polymerization are: 1) The thickness of the modified layer is extremely small (e.g., 500 Å to 5000 Å); consequently, the surface modification can be attained without affecting the bulk properties of the substrate. 2) Unlike most grafting reactions, glow discharge polymerization (which yields

essentially a grafted surface layer) is less dependent on the
types of substrate polymers; consequently, a desired surface
layer can be grafted on nearly all kinds of polymers regardless of
their chemical nature.

Furthermore, as found in this study, the blood compatibility
of many glow discharge polymers is excellent, ranking among the
best compatibility obtained by materials that do not incorporate
anticoagulants such as heparin.

II.   GLOW DISCHARGE POLYMERIZATION

Although glow discharge can be characterized by the plasma
(ionized gas) state, the reactive species, which can be utilized
by many practical reactions, are free radicals formed from the
ions or from the excited state of molecules associated with
ionization.

The concentration of radicals was generally found to be 10^5
times more than the concentration of ions in glow discharge.[1]
Consequently, most of the polymerization under glow discharge
conditions can be explained by free radical mechanisms in a somewhat similar manner to radiation polymerization, although the
initial step of ionization and the subsequent radical formation
are not clearly elucidated.

The elimination of hydrogen atoms from organic compounds
plays an important role in free radical formation. With some
organic compounds, free radical formation from the opening of a
double bond seems to occur.[2]

The polymerization of an organic compound (n = 1) under glow
discharge conditions, may be represented by using free radical
mechanisms as follows:

$$Mn \rightarrow Mn\cdot \qquad \text{(initiation step)}$$
$$Mn\cdot + Mm\cdot \rightarrow (Mn\text{--}Mm) \qquad \text{(recombination step)}$$
$$(Mn\text{--}Mm) \rightarrow (Mn\text{--}Mm)\cdot \qquad \text{(reinitiation step)}$$

Due to the very high rate of initiation (and reinitiation) in glow
discharge, the polymer can be formed from any organic compound
that does not polymerize by the ordinary free radical chain reaction mechanism. In this respect, the glow discharge polymerization may be regarded as an extreme case of radiation polymerization where an extremely high dose rate is employed. It was
estimated that the dose rate in glow discharge polymerization was
10^6 times higher than the dose rates ordinarily used in γ-ray
irradiation.[3]

Because of this unique process of polymerization, the monomers that can be used in the surface modification are not limited to conventional monomers (e.g., vinyl monomers). The rates of polymerization of saturated organic compounds are by and large the same as those for conventional vinyl type monomers.[4,5]

Because of this polymerization mechanism, the properties of glow discharge polymers are not necessarily the same as those of the corresponding conventional polymers or are not what might be expected from the structure of the monomer used in glow discharge polymerization. One reason for this is that the decomposition of a monomer molecule may occur in glow discharge polymerization; consequently, some groups and structures tend to be deficient in the polymer structure. On the other hand, due to a high concentration of trapped free radicals (detected by ESR spectroscopy) and their postreaction with oxygen in air, glow discharge polymers generally contain a certain amount of oxygen atoms in the forms of hydroxyl and carbonyl groups.

Glow discharge polymers can be formed in a wide variety of final forms from a monomer depending on the conditions of glow discharge. Polymers can be formed as powders with no adherence to substrates, low molecular weight liquid products, film-forming polymers that are soluble in solvents, and highly crosslinked films that adhere strongly to the substrate. The transport properties of glow discharge polymers can be controlled by the conditions of polymerization. For most membrane moderated biomedical devices that require membranes of high permeability, the method that provides an extremely thin grafted layer has a special value, since it is possible to improve the blood compatibility without deterioration of the transport properties of the substrate membrane.

III. EXPERIMENTAL

Glow Discharge Polymerization

Glow discharge polymerization was carried out by an apparatus that utilizes inductive coupling of 13.5 MHz radio frequency. The details of the equipment are described in reference 4.

A substrate polymer film, 4 in. × 4 in., is placed in a glass tube (40 mm O.D.) which is a part of the glow discharge apparatus. In most cases, Mylar film (0.001 in. thick) was used as the substrate. The film was degassed to a vacuum less than 1 μm Hg. Then a monomer was introduced into the system to establish a flow of 30 μm Hg pressure, and glow discharge was initiated. Glow discharge was continued for a predetermined time to obtain a deposition approximately 1000 Å thick. The time of the treatment was determined by separate measurements of deposition rates.

After the polymer deposition, the sample was kept in vacuum
for at least an hour to remove residual monomer. The sample was
tested for its Lindholm whole blood coagulation time without any
further treatment such as washing, extraction, etc.

Blood Compatibility Test

The Lindholm test, a whole blood clotting time test performed
directly on the test sheets or films, was used as a screening test
of the blood compatibility of glow discharge polymers in this
study. Details of this test are described in reference 7. Four
sections (2 in. × 2 in.) were cut from a single sample for the
tests. Lindholm tests were generally carried out on one control
and four different samples at a time using native human blood (no
anticoagulant added). The clotting time was always compared with
the clotting time of the control surface (silicone-coated glass).[7]
The entire test series was repeated on the same day using freshly
drawn blood (from the same or different donor).

Ten relative clotting times expressed as percent of the
average clotting time of the control were calculated for comparison of the blood compatibility behavior of each test surface.

IV.   RESULTS AND DISCUSSION

Results are summarized in Table I where the blood compatibility is expressed as percent of clotting time of control surface.
The clotting time of the control silicone surface is 64.6 $\pm$ 7.2
min.

Data presented in Table I are from polymers formed by glow
discharge of the monomer alone. The addition of some gas, particularly N_2, improves the blood compatibility of some but not all
polymers, but this data is not included in these results.

The evaluation of blood compatibility is a rather complex
problem, and the Lindholm test was used in this study only as a
screening process for a subsequent, more elaborate and costly
in vivo and/or *ex vivo* evaluations. Consequently, the values
themselves (e.g., 110 versus 105) may not have significant meaning, although they may serve as a rough means of distinguishing
"good" materials from "poor" materials (e.g., 105 versus 70).
In this context, the results shown in Table I clearly indicate
that many glow discharge polymers have promising blood compatibilities.

A preliminary indication from *in vivo* evaluation of some glow

TABLE I
BLOOD COMPATIBILITY OF POLYMERS

Glow Discharge Polymers		Conventional Polymers	
Monomer	% of Control Clotting Time	Polymer	% of Control Clotting Time
Acetylene	108	Polyethylene	77.4*
Acrylic acid	69.2	Silastic+	66.9*
Acrylonitrile	98.2	Cellulose acetate	88.2*
Allene	114	(glass)	70.9*
Chlorotrifluoroethylene	104	Polyethylene terephthalate	80.4
Ethylene	111	Parylene C	94.0
Ethylene oxide	106	Polysulfone	86.4
Hexafluorobenzene	94.4	Polyurethane++	96.7
Hexamethyldisiloxane	99.5	Polyoxymethylene+++	73.1**
Hydroxyethylmethacrylate	85.4	Cellulose+++	82.4
Methane	98.5		
Methyl acetylene	99.0		
Methyl chloride	103		
4-Picoline	96.7		
Tetrafluoroethylene	113		
Vinyl acetate	92.1		
Vinyl chloride	96.9		
Vinyl methyl ether	102		

+ Polydimethylsiloxane with SiO_2 filler, Dow Corning.
++ A copolyether urethane elastomer, Stanford Research Institute, 3-1000-1-X
+++ Cuprophane
* Data from reference 7.
** Data from RTI's Final Report to Contract NIH-70-2115, Report No.
AK-F-70-2115.

discharge polymers is also very encouraging and supports the data
shown in Table I.

Acknowledgement

This study was sponsored by The National Heart and Lung
Institute, NIH, U.S.D.E.W., Contract No. NIH-NHLI-73-2913.

References

1.   L. C. Brown and A. T. Bell, private communication, paper to
     be published in I&EC Fund.
2.   H. Yasuda, M. O. Bumgarner, and J. J. Hillman, J. Appl.
     Polymer Sci., in press.
3.   A. R. Westwood, Europ. Polymer J., 7, 363 (1971).
4.   H. Yasuda and C. E. Lamaze, J. Appl. Polymer Sci., 17, 1519
     (1973).
5.   H. Yasuda and C. E. Lamaze, J. Appl. Polymer Sci., 17, 1533
     (1973).
6.   H. Kobayashi, A. T. Bell, and M. Shen, J. Appl. Polymer Sci.,
     17, 885 (1973).
7.   R. G. Mason, Biomat., Med. Dev., Art. Org., 1, (1), 131 (1973).

BIOMEDICAL APPLICATIONS OF ANISOTROPIC MEMBRANES

Michael J. Lysaght and Cheryl A. Ford

Amicon Corporation

25 Hartwell Avenue, Lexington, Mass. 02173

INTRODUCTION

In 1966, Michaels and his co-workers developed a series of anisotropic, ultrafiltration-grade membranes constructed from refractory thermoplastic materials (1,2). Broadly speaking, the membranes were freely permeable to salts, sugars, and other low molecular weight species but retentive for intermediate and high molecular weight solutes. They were fully wet-dry reversible. Filtration rates were 50 to 500 GSFD at 50 psi, orders of magnitude higher than had been achieved with desalination grade membranes. As a consequence, practically useful separations could be achieved with small cells and convenient laboratory accessories. Problems of concentration polarization were soon identified and quantified, and this troublesome phenomenon was controlled through proper hardware design (3). By 1970, these membranes were available in hollow fiber form with that geometry's intrinsic advantages of packing density and fluid management (4).

The ultrafiltration membranes of our concern have been investigated for a wide range of applications in the gas separation (5), electrochemical (6), pervaporative (7), and industrial fields (8). However, their broadest appeal has always been in the biomedical area, initially as a useful tool for the laboratory investigator, subsequently in various forms of extracorporeal prosthesis, and most recently as an adjunct for routine clinical analysis.

MEMBRANE STRUCTURE AND PROPERTIES

The gross microstructure of these membranes is readily apparent from scanning electron micrographs, two of which are shown in Figures 1 and 2, for flat sheet and hollow fiber form. The tight skin, on the top of the flat sheet or on the inner lumen of the hollow fiber, is no more than a micron thick and is believed, on the basis of indirect evidence, to have a porosity of 1 to 5% with pore diameters in the 10 to 30 angstrom range. The bulk of the film (termed the sponge region or substrate) is a rigid and highly-voided domain containing straight through channels with diameters in the 1-5 micron range and an overall porosity as high as 80%. This region is not homogeneous, and the void volume and scale of porosity increase significantly in regions further distal from the skin.

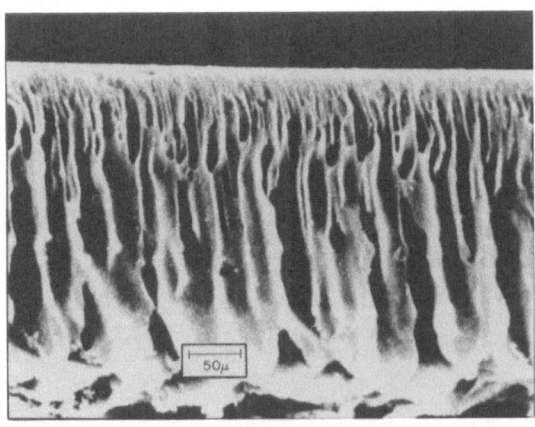

Figure 1. SEM OF FLAT SHEET MEMBRANE.

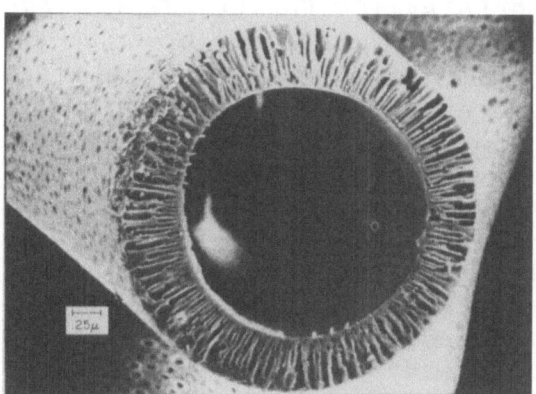

Figure 2. SEM OF HOLLOW FIBER MEMBRANE.

    As with RO membranes, the skin alone is responsible
for permselectivity and hydraulic resistance; the substrate is present merely for mechanical support.

    Flat sheet membranes have tensile strength about 500
psi but do tear easily and, for this reason, are
generally cast onto a spun-bonded polyolefin support.
Hollow fibers have burst strengths between 100 and 200
psi. All membranes are freely permeable to gases. Since
the matrix materials of the membranes do not absorb water,
they do not undergo any dimensional changes when dried.
Adsorption of solutes is sometimes observed and can cause
difficulties with aromatic permeants present in low concentration.

    Anisotropic membranes are available with nominal
molecular weight cutoffs to permeating species from 500
to 300,000, but available anisotropic membranes with cutoffs less than 10,000 are fabricated from hydrogel
materials. In practice, of course, rejection is not a
step function of molecular weight but is typified by the
broad S-shaped curve of the type shown in Figure 4.
Distilled water fluxes are in the range of 1-4 ml/
minute/cm^2 at 50 psi while fluxes achieved with actual
process fluid are usually an order of magnitude less.

PRECONCENTRATION FOR CLINICAL DIAGNOSIS

Until recently, the separative utility of the ultrafiltration membranes has been limited to laboratory
investigations or industrial applications, where it was
feasible to provide a generous amount of time and hardware (pressure sources, stirred cells or flow through
devices) for each sample to be treated. Such applications are well documented in the literature (9, 10, 11,
12). Disposable microconcentrators have now become
available which require relatively little operator
attention and no ancillary equipment. The design and
principle of operation of such devices is shown in
Figure 3. A liquid sample is introduced into a thin,
vertical, rectangular chamber which has been sealed to
the membrane surface. The driving force for ultrafiltration is provided by an absorbent pad butted against
the back side of the membrane. Water and other species
less than the molecular weight cutoff of the ultrafilter
permeate through the membrane and into the pad. The
lower portion of the membrane is rendered impermeable to
provide a "dead-stop" at the desired level of concentration. Typical rejection curves for these devices are
shown in Figure 4; it should be noted that the same
membranes would probably yield somewhat lower rejections

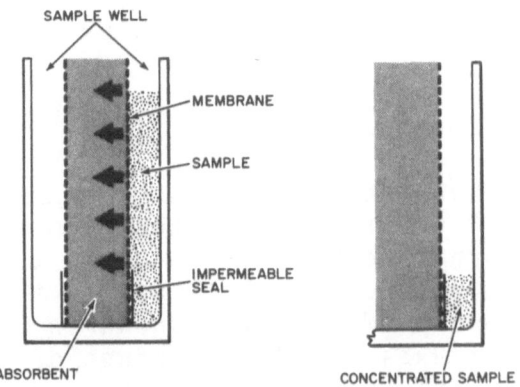

Figure 3. SCHEMATIC OF ABSORBENT
BASED MEMBRANE SYSTEM.

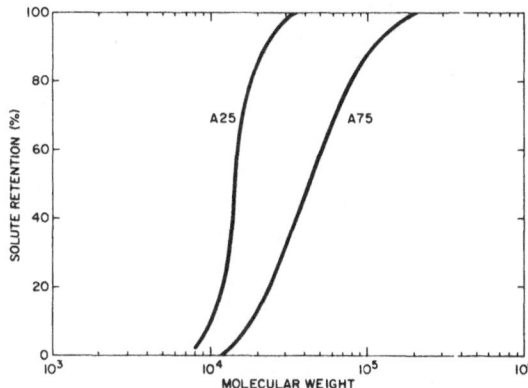

Figure 4. PLOT OF REJECTION vs
MOLECULAR WEIGHT FOR FLAT
SHEET MEMBRANES.

in a stirred cell. The concentrate is recovered
with a needle pipette. Disposable "Miricon"® microconcentrators containing multiple chambers are available
from Amicon Corporation with a variety of membrane cutoffs and preset concentration levels.

    The principal clinical application of these devices
is preconcentration of samples to allow more reliable
identification, in subsequent analysis, of proteins or
antigens. For example, the clinical laboratory is often
interested in the electrophoretic or immunophoretic
identification of proteins in urine or cerebro spinal
fluids. In many cases, the naturally occurring protein
levels are simply too low to permit reliable identification without preconcentration. This is illustrated in
Figure 5 where the electrophoretic pattern of Bence Jones
proteins in abnormal urine are shown in the naturally
occurring state and after several degrees of preconcentration. Other biological specimens requiring rapid and
convenient preconcentration, without denaturation of
proteins, are saliva, tears, amniotic fluid, urinary
gonadotrophins and, in general, any fluid in which the
macromolecule levels are too low for the analytical
method being used.

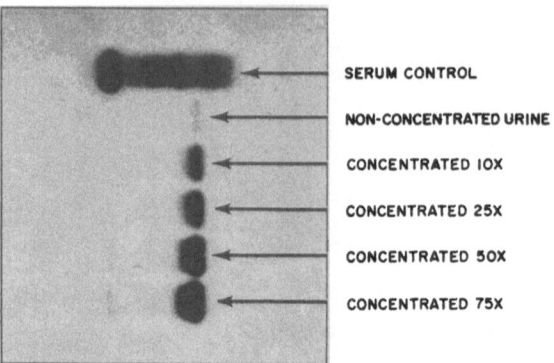

Figure 5. ELECTROPHORESIS OF ABNORMAL URINE. NOTE ENHANCEMENT OF BENCE JONES PROTEINS AFTER CONCENTRATION.

EXTRACORPOREAL PROSTHESIS

Use of these membranes in extracorporeal prosthetic devices became practical when techniques were developed preparing fibers which are nontoxic, nonpyrogenic, and nonhemolytic. (Standard anisotropic membranes produced by conventional techniques are not suited for in-vivo application.)

Perhaps the most novel approach, investigated in conjunction with Dr. Lee Henderson at the University of Pennsylvania, has been the direct use of ultrafiltration to cleanse uremic blood in an attempt to ameliorate the pathologies associated with maintenance hemodialysis (13, 14). The human kidney may be considered as an ultra-filter in series with an active readsorption region. Species smaller than proteins are all cleared by the glomerular membrane and vital solutes are reabsorbed in the tubules. Direct ultrafiltration of blood, with parental reinfusion of vital solutes, more closely replicates this natural modality than does hemodialysis, a diffusion-based process with inherently low clearance of any toxins greater than 500 M.W.

The actual circuit for blood ultrafiltration is as follows:

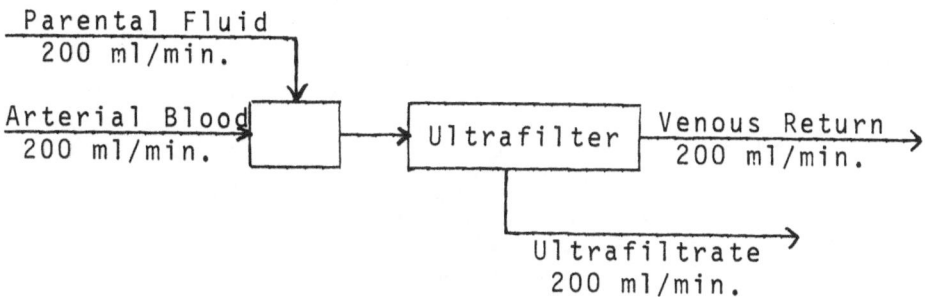

Ultrafiltrate and parental fluid are balanced exactly by an elaborate pumping mechanism. As a first approximation, the clearance of any non-rejected solute in the above circuit is 100 ml/minute. The actual clearances are favorably influenced by hematocrit, solute distribution in erythrocytes, etc., and are given in Figure 6. For comparison, the clearance achievable with standard dialysis is also shown. The region enclosed between the two curves represents the toxins removed by blood ultrafiltration but not dialysis. The process has been utilized for a full six-month clinical maintenance of one patient without intervening dialysis. Overall health and vital signs were excellent and the process studies are now being expanded to other patients.

In corollary studies, similar ultrafilters with 0.2 to 0.4 M^2 of surface areas are being utilized in clinical investigations to dewater chronically fluid-overloaded patients with cannulated blood access. This process has relevance as an adjunct to hemodialysis and in the direct treatment of ascites or pulmonary edema accompanied by renal shock.

In another area (15,16), methods have been developed for drastically reducing the ultrafiltration rate of the anisotropic membranes without affecting their diffusive and dialytic transport. The resultant fibers have a larger pore structure than conventional

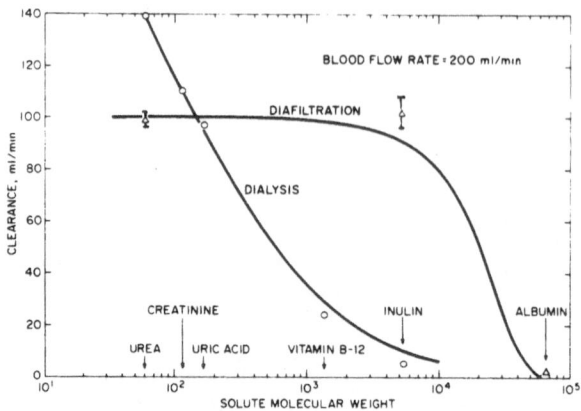

Figure 6. COMPARISON OF CLEARANCE
CURVES FOR BLOOD ULTRA-
FILTRATION AND DIALYSIS.

cellulosic dialysis membranes and, accordingly, a higher permeability to solutes in the 500 to 5,000 M.W. region. The fibers may be potted and shipped dry or wet, have a burst strength of 100 psi (versus 15 for cellulosic hollow fibers), and are fully autoclavable. Dialyzers prepared from these fibers have proven satisfactory in toxicological and acute clinical testing and are currently undergoing thorough clinical evaluation.

Anisotropic hollow fibers have also been prepared from highly hydrophobic materials and studied for blood oxygenation. Satisfactory gas transfer rates were achieved but no advantages, except possibly in cost, were found over conventional silicone or silicone-polycarbonate fibers (17).

REACTIVE SYSTEMS

Dedrick (18) and his co-workers at the National Institutes of Health have inoculated the interstices between fibers in a typical hollow fiber bundle with live cells (human choriocarcinoma JEG-7) and then nourished the cells with oxygenated media flowed through the center of the fibers. The nutrient stream also served to remove catabolic and metabolic cell products. The cells grew well in this environment and were observed to migrate into and anchor upon the fiber substrate. Dramatically higher cell densities were achieved in this configuration than are possible with conventional monolayer techniques.

Robertson (19) at Stanford University has devised a novel and highly efficient enzyme reactor by loading the sponge region of the anisotropic fibers with enzyme-active solution and flowing his substrate through the fiber lumens. Reagents diffuse through the skin and contact the enzyme in the substrate; products diffuse back into the luminal region. The procedure, which is limited to cases where the reactant and products are both of relatively low molecular weight, advantageously minimizes the diffusional resistance to reaction and also protects the enzymes from high molecular weight denaturants in the process stream.

To date, both approaches have been limited to in-vitro studies. In view of the demonstrated bio-compatibility of the fibers, extension of the techniques to treatment of enzymatic and cellular deficiencies (e.g., hepatic necrosis, diabetes) is considered a fertile area for future development.

REFERENCES

1. Michaels, A. S. Chemical Engineering Progress, 64 (12), 31, 1968.
2. Blatt, W. F., et al. Analytical Biochemistry, 18 (1), 81, 1967.
3. Porter, M. C. I&EC Product Research and Development, 11, 234, 1972.
4. Cross, R. A. AIChE Symposium Series, 68 (120), 15, 1972.
5. Bradley, D., et al. Polymer Engineering & Science, 11 (4), 284, 1971.
6. Lysaght, M. J. Electrochemical Society Symposium (Columbus), 57, 1970.
7. Desaulniers, C. W., et al. U.S. Patent 3,632,404, 1972.

8.   Porter, M. C. and Michaels, A. S.   Chemical
     Technology, Part 1, January 1971, 56; Part 2, April
     1971, 248; Part 3, July 1971, 440; Part 4, October
     1971, 633; Part 5, January 1972, 57.
9.   Strathmann, H.   Chemie-Ingenieur Technik, 42 (17),
     1095, 1970.
10. Bixler, H. J., et al.   1969 National AIChE Meeting,
     Cleveland.
11. Charm, S. E., et al.   Biotechnology & Bioengineering,
     XIII, 185, 1971.
12. Thomson, A. R.   Process Biochemistry, 35, September
     1971.
13. Livoti, L. G., Henderson, L. W., et al.   ASAIO
     Transactions XIX, 119, 1973.
14. Hamilton, R., Henderson, L. W., et al.   ASAIO
     Transactions XVII, 259, 1971.
15. Lysaght, M. J.   NIH Report AK-2-2279, 1973.
16. Lysaght, M. J., et al.   FAST Transactions, 2, 1973.
17. Cross, R. A.   NIH Report 71-2367, 1972.
18. Knazek, R. A., et al.   Science, 178, 65, 1972.
19. Robertson, C. R.   1973 National AIChE Meeting,
     Detroit.

CONTRIBUTORS

Mary A. Amini, Department of Chemical and Biochemical Engineering, Rutgers University, New Brunswick, New Jersey

R. M. Barrer, Chemistry Department, Imperial College of Science and Technology, London, England

J. A. Barrie, Department of Chemistry, Imperial College of Science and Technology, London, England

Thomas E. Brady, Owens-Illinois Technical Center, Toledo, Ohio

W. R. Brown, University of Wales Institute of Science and Technology, Cardiff, Wales

K. J. Brzozowski, Tremco Manufacturing Company, Cleveland, Ohio

M. O. Bumgarner, Camille Dreyfus Laboratories, Research Triangle Institute, Research Triangle Park, North Carolina

Daniel P. Y. Chang, Department of Civil Engineering, University of California, Davis, California

Mort Fels, Lakehead University, Thunder Bay, Ontario

Paul J. Fenelon, Borg-Warner Chemicals, Washington, West Virginia

Cheryl A. Ford, Amicon Corporation, Lexington, Massachusetts

Sheldon K. Friedlander, W. M. Keck Laboratories, California Institute of Technology, Pasadena, California

H. L. Frisch, State University of New York, Albany, New York

G. A. Gordon, Continental Can Company, Inc., Chicago, Illinois

C. Harris, Department of Chemical Engineering and Center for Bioengineering, University of Washington, Seattle, Washington

G. R. Hightower, ROGA Division, Universal Oil Products, San Diego, California

A. S. Hoffman, Department of Chemical Engineering and Center for Bioengineering, University of Washington, Seattle, Washington

F. F. Holland, Gulf-South Research Institute, New Orleans, Louisiana

H. B. Hopfenberg, Department of Chemical Engineering, North Carolina State University, Raleigh, North Carolina

T. A. Horbett, Department of Chemical Engineering and Center for Bioengineering, University of Washington, Seattle, Washington

P. R. Hsia, Continental Can Company, Inc., Chicago, Illinois

Sun-Tak Hwang, Department of Chemical and Materials Engineering, The University of Iowa, Iowa City, Iowa

J. L. Illinger, Department of the Army, Army Materials and Mechanics Research Center, Polymer and Chemistry Division, Watertown, Massachusetts

Saleh A. Jabarin, Owens-Illinois Technical Center, Toledo, Ohio

C. H. M. Jacques, Research Laboratories, General Motors Corporation, Warren, Michigan

R. N. Johnson, Union Carbide Corporation, Bound Brook, New Jersey

Karl Kammermeyer, Department of Chemical and Materials Engineering, The University of Iowa, Iowa City, Iowa

F. E. Karasz, University of Massachusetts, Amherst, Massachusetts

R. E. Kesting, Chemical Systems, Inc., Irvine, California

B. N. L. Khaw, Department of Chemical Engineering and Center for Bioengineering, University of Washington, Seattle, Washington

Elias Klein, Gulf-South Research Institute, New Orleans, Louisiana

W. G. Kraft, Department of Chemical Engineering and Center for Bioengineering, University of Washington, Seattle, Washington

C. A. Kumins, Tremco Manufacturing Company, Cleveland, Ohio

T. K. Kwei, Bell Laboratories, Murray Hill, New Jersey

Norman N. Li, Esso Research and Engineering, Linden, New Jersey

T. T. Ling, Department of Chemical Engineering and Center for Bioengineering, University of Washington, Seattle, Washington

C. R. Lyons, ROGA Division, Universal Oil Products, San Diego, California

Michael J. Lysaght, Amicon Corporation, Lexington, Massachusetts

L. G. Mason, Department of Pathology, University of North Carolina, Chapel Hill, North Carolina

T. P. Mate, Department of Chemical and Center for Bioengineering, University of Washington, Seattle, Washington

M. Matzner, Union Carbide Corporation, Bound Brook, New Jersey

J. E. McGrath, Union Carbide Corporation, Bound Brook, New Jersey

R. McGregor, School of Textiles, North Carolina State University, Raleigh, North Carolina

Alan S. Michaels, Alza Corporation, Palo Alto, California

Gerald W. Miller, Owens-Illinois Technical Center, Toledo, Ohio

A. Nunn, Department of Chemistry, Imperial College of Science and Technology, London, England

M. I. Ostler, Division of Macromolecular Sciences, Case Western Reserve University, Cleveland, Ohio

G. S. Park, University of Wales Institute of Science and Technology, Cardiff, Wales

D. R. Paul, Department of Chemical Engineering, The University of Texas, Austin, Texas

A. Peterlin, Camille Dreyfus Laboratory, Research Triangle Institute, Research Triangle Park, North Carolina

J. H. Petropoulos, Democritos Nuclear Research Center, Athens, Greece

Stephen Prager, Department of Chemistry, University of Minnesota, Minneapolis, Minnesota

W. Pusch, Max-Planck-Institut für Biophysik, Frankfurt, Germany

B. D. Ratner, Department of Chemical Engineering and Center for Bioengineering, University of Washington, Seattle, Washington

A. Rembaum, Polymer Research Section, Jet Propulsion Laboratory,
    California Institute of Technology, Pasadena, California

R. L. Riley, ROGA Division, Universal Oil Products, San Diego,
    California

L. M. Robeson, Union Carbide Corporation, Bound Brook, New Jersey

C. Robillard, Jet Propulsion Laboratory, California Institute of
    Technology, Pasadena, California

C. E. Rogers, Division of Macromolecular Sciences, Case Western
    Reserve University, Cleveland, Ohio

P. P. Roussis, Democritos Nuclear Research Center, Athens, Greece

Morris Salame, Monsanto, Bloomfield, Connecticut

G. Schmer, Department of Chemical Engineering and Center for
    Bioengineering, University of Washington, Seattle, Washington

N. S. Schneider, Department of the Army, Army Materials and
    Mechanics Research Center, Polymer and Chemistry Division,
    Watertown, Massachusetts

D. L. Schober, Union Carbide Corporation, Bound Brook, New Jersey

A. Sheer, Department of Chemistry, Imperial College of Science and
    Technology, London, England

J. K. Smith, Gulf-South Research Institute, New Orleans, Louisiana

V. Stannett, Department of Chemical Engineering, North Carolina
    State University, Raleigh, North Carolina

M. Tagami, ROGA Division, Universal Oil Products, San Diego,
    California

L. N. Teng, Department of Chemical Engineering and Center for
    Bioengineering, University of Washington, Seattle, Washington

Wolf R. Vieth, Department of Chemical and Biochemical Engineering,
    Rutgers University, New Brunswick, New Jersey

Tsuey T. Wang, Bell Laboratories, Murray Hill, New Jersey

Joel Williams, Camille Dreyfus Laboratory, Research Triangle Institute, Research Triangle Park, North Carolina

S. Yamada, Division of Macromolecular Sciences, Case Western Reserve University, Cleveland, Ohio

H. Yasuda, Camille Dreyfus Laboratory, Research Triangle Institute, Research Triangle Park, North Carolina

S. P. S. Yen, Jet Propulsion Laboratory, California Institute of Technology, Pasadena, California

INDEX